MAGNETOHYDRODYNAMICS IN BINARY STARS

ASTROPHYSICS AND SPACE SCIENCE LIBRARY

VOLUME 216

MAGNETOHYDRODYNAMICS IN BINARY STARS

by

C. G. CAMPBELL

Department of Mathematics,
University of Newcastle upon Tyne, United Kingdom

KLUWER ACADEMIC PUBLISHERS
DORDRECHT / BOSTON / LONDON

A C.I.P. Catalogue record for this book is available from the Library of Congress

ISBN 0-7923-4606-8

Published by Kluwer Academic Publishers,
P.O. Box 17, 3300 AA Dordrecht, The Netherlands.

Sold and distributed in the U.S.A. and Canada
by Kluwer Academic Publishers,
101 Philip Drive, Norwell, MA 02061, U.S.A.

In all other countries, sold and distributed
by Kluwer Academic Publishers Group,
P.O. Box 322, 3300 AH Dordrecht, The Netherlands.

Printed on acid-free paper

Printed in the Netherlands

PREFACE

Magnetic stresses were discussed as a possible means of angular momentum transport in the development of accretion disc theory, in the late sixties and early seventies. Interest in the role of magnetic fields in close binary stars steadily increased after the discovery of the nature of AM Herculis in 1976. The observed lack of an accretion disc and the synchronous rotation of the white dwarf suggested strong magnetic effects, consistent with the high degree of optical polarization. Similar systems were soon discovered. Evidence for large magnetic fields was subsequently found in the X-ray binary pulsars and the intermediate polar binaries, both believed to include systems with partially disrupted accretion discs. A magnetically channelled wind from the main sequence secondary star has been invoked to explain the higher mass transfer rates observed in binaries above the period gap, and in an explanation of the gap. Magnetically influenced winds from accretion discs have been suggested as contributing to the inflow by removing angular momentum.

Magnetism in binary stars is now an area of central importance in stellar astrophysics. Magnetic fields are believed to play a fundamental role even in apparently non-magnetic binaries. They provide the most viable means, through shear instabilities, of generating the turbulence in accretion discs necessary to drive the inflow via the resulting magnetic and viscous stresses.

The fundamental theme of this monograph is the role of magnetic fields in redistributing angular momentum in close binary stars. This involves the motions of plasmas in magnetic fields, which is the subject of magnetohydrodynamics (MHD). A basic knowledge of electromagnetism and fluid mechanics to undergraduate level is assumed. Some knowledge of stellar astrophysics is helpful, but not essential. To make the book as self-contained as possible, I have included a chapter on theoretical basics. This contains all the essentials of MHD and binary star theory, together with the relevant rotational dynamics. An appendix is also included containing useful formulae and derivations.

The four main areas of magnetism in binary stars are considered; the AM Herculis stars, systems with partial accretion discs, accretion disc magnetic fields, and magnetic winds from secondary stars and discs. A wide range of MHD problems is involved, including stellar field induction, accretion flow channelling, magneto-gravitational interactions, the effect of stellar fields on discs, field generation and angular momentum advection in discs, and magnetically-influenced stellar and disc winds. The associated stellar spin dynamics and stability are also considered.

The material is largely theoretical, but the essential observational motivations are discussed and, where possible, results are compared with the available data. It is hoped that the book will be of interest to observers, as well as theoreticians.

S.I. units are used throughout. However, magnetic field strengths are often quoted in gauss (1 tesla $= 10^4$ gauss), since astronomers usually have a more immediate feel for magnitudes in these units. Useful formulae are generally expressed in normalized forms, making any desired change of field units simple. The notations $\log = \log_{10}$ and $\ln = \log_e$ are employed.

I should like to thank Michael Beaty for invaluable help in the conversion of the original manuscript from T3 to Latex, and for help in the production of the figures. I am grateful to Klaus Beuermann and Hans Ritter for supplying me with a list of AM Herculis systems.

Table of Contents

CHAPTER 1

MAGNETISM IN BINARY STARS

1.1. Close Binaries

A close binary system is one in which the separation of the component stars is sufficiently small for them to be strongly interacting. If the size of a stellar component is a significant fraction of the orbital separation, then its outer layers will be strongly distorted by the gravitational field of its companion. The rapid stellar rotation occurring in such systems gives additional distortion due to centrifugal force.

The basic description of a binary system is given by a model proposed in 1873 by Roche. The stellar components are treated as point gravitational sources so, relative to a frame with horizontal axes rotating with the line of centres, a simple total potential results. This is the sum of the point gravitational potentials plus the centrifugal potential. For circular orbits, time-independent equipotentials can be found. Near the stellar centres these surfaces are almost spherical, but further away they become pear-shaped. Critical surfaces result when the apexes of the equipotentials due to each star touch. These surfaces contain the masses of their respective stars and are referred to as their Roche lobes. The point of contact, known as the inner Lagrangian point, L_1, is an unstable equilibrium position.

Before the present century little was known about the internal structure of stars and the Roche model was for many years only considered as a limiting case of configuration, being the extreme opposite of models with homogeneous components. By the second decade of this century it was realized that a large fraction of the mass of a main sequence star is contained in its central regions. Consequently, its gravitational potential does not deviate greatly from a point source form in its outer layers. This remains true for tidally and rotationally distorted stars, since the fractional density perturbations are greatest in their outer, most tenuous layers and these only make a small contribution to the potential. Work by Chandrasekhar (1933) and Plavec (1958) showed that the equipotentials of distorted stars in close binaries approach those of the Roche model to a very good approximation. The more detached a star is from its Roche lobe, the weaker its tidal distortion.

Wood (1950) suggested dividing close binary stars into two classes; one containing systems in which both stellar surfaces lie beneath their Roche lobes, known as detached systems, and the other containing binaries in which at least one component fills its Roche lobe. Kopal (1955) suggested dividing the latter class into two groups; the first containing systems in which only one component fills its Roche lobe, known as semi-detached systems, while in the second group both components fill their Roche lobes to form contact binaries.

Most work has been focused on semi-detached systems in which a main sequence star fills its Roche lobe and orbits with a more massive compact component, the latter usually being a white dwarf or neutron star. The main sequence component is generally referred to as the secondary star, and the compact object as the primary star. Systems with orbital periods of $\lesssim 10$ hrs have lower main sequence secondary stars.

Strong interest in close binary stars began when photometric work by Linnell (1950) on a new eclipsing variable, later designated UX Ursae Majoris, established it as the first known representative of a new class of subdwarf binaries. Subsequently, Walker (1956) discovered that Nova Herculis 1934 is an eclipsing variable (DQ Herculis) consisting of a pair of subdwarfs with an orbital period of 4 hrs and 39 mins. Further searches among stars of this type lead to the discovery of the binary nature of several other post-novae (e.g. GK Per/Nova Persei 1901, WZ Sge/Nova Sagittae 1913 and 1946). This supported the contention that nova outbursts occur only among the components of close binary systems. The stars SS Cygni (Joy, 1956) and U Geminorum (Mumford, 1962), which exhibit more frequent outbursts of light than classical or recurrent novae, were also found to be short period binaries. This, together with the foregoing discoveries, suggested that all eruptive variables, with the possible exception of supernovae, are binary objects.

Observations of close binaries led to theoretical models. Crawford and Kraft (1956) realized that the secondary star in AE Aqr fills its Roche lobe. They interpreted the observed emission lines as arising from gas passing from the secondary to orbit around the white dwarf primary. Krzeminski (1965) suggested that the cooler star in U Geminorum fills its Roche lobe, and that the hotter star is surrounded by a rotating disc. A similar model had been proposed for DQ Herculis by Kraft, Mathews and Greenstein (1962). This picture was later adopted as the standard model of short period semi-detached binaries containing white dwarf primary stars, designated the cataclysmic variables (Warner and Nather, 1971; Smak, 1971). The lobe-filling secondary transfers material, through the unstable inner Lagrangian point, to an accretion disc surrounding the compact primary.

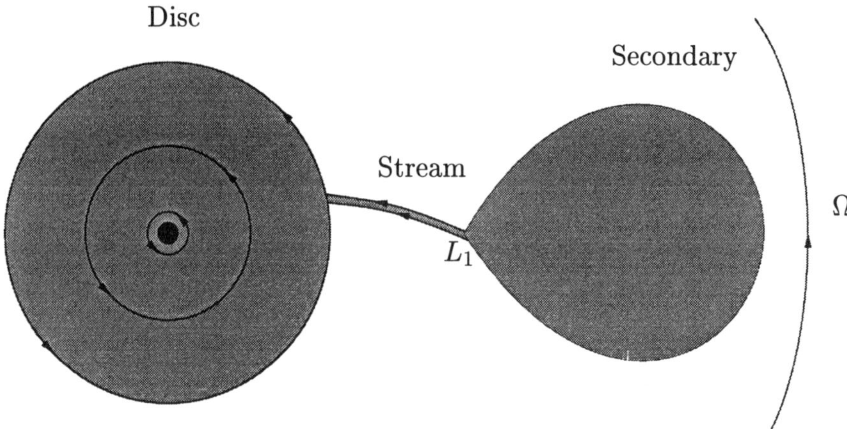

Figure 1.1. The standard model of a cataclysmic variable, viewed down the rotational pole of the system. The tidally and rotationally distorted secondary loses matter from the unstable L_1 point. The resulting stream feeds an accretion disc, centred on the white dwarf primary, through which matter slowly spirals inwards.

At the point of intersection of the stream and the disc a shock is formed, producing a bright spot on the disc. This manifests itself as a prominant hump in the optical light curve of those systems in which emission from the rest of the disc is weak (e.g. many dwarf novae in their quiescent states). Matter spiralling in through the disc liberates its binding energy and is accreted onto the surface of the primary. If the mass transfer time-scale greatly exceeds the thermal adjustment time-scale of the secondary, then the star remains in thermal equilibrium and consequently shrinks as it loses mass. However, material is kept in contact with the L_1 point since the secondary's Roche lobe also shrinks as a result of orbital angular momentum loss. This is driven by gravitational radiation losses, or by a magnetically influenced wind flowing from the secondary together with tidal coupling to the orbit. Figure 1.1 illustrates the basic structure of a cataclysmic variable. A detailed description of the observational techniques used in the study of cataclysmic variables is given in Warner (1995).

In the standard cataclysmic variable the primary star is an essentially non-magnetic white dwarf, and the accretion disc extends down to its surface. Originally the designation cataclysmic variable included dwarf novae, nova-like variables and old novae. This was subsequently extended to include systems with a strongly magnetic primary star. In such cases the accretion disc can be partially or totally disrupted due to the influence of the stellar field.

1.2. Magnetic Fields in Binaries

The presence of a strong magnetic field can significantly modify the standard structure of a close binary. A white dwarf or neutron star is capable of sustaining a far stronger magnetic field than a main sequence star. Such compact objects contain highly conducting degenerate matter and so are generally believed not to require dynamo action to sustain their magnetic fields, which have very long decay times (see Landstreet, 1994 for a discussion of the fossil field theory for white dwarfs). Hence the strongest source of magnetism in binaries is usually the compact primary. These fields are detectable via linear and circular optical polarization, resulting from cyclotron radiation emitted by accreting gas. Photospheric Zeeman split spectral lines are sometimes observed. X-rays are emitted from hot shocked gas in accretion columns, and pulsation periods in the observed intensity result due to the rotation of the primary.

The magnetic field effect is strongest in the AM Herculis systems. AM Her was identified as an X-ray source by Hearn, Richardson and Clark (1976). Similar systems were subsequently found and these objects now form an important class of close binary stars. The primary stars are white dwarfs with strong magnetic moments. The secondary stars are red dwarfs with a range of spectral types M6–M2, and the orbital periods range from 1.3 hrs to 4.6 hrs. The magnetic field completely prevents the formation of an accretion disc. Matter lost from the L_1 region of the secondary becomes channelled by the primary's magnetic field and converges to form an accretion column above its surface. Material passes through a standing shock at the top of the column and undergoes compressional heating. The intensity of the radiation emitted from the post-shock flow is observed to be modulated, due to the changing orientation of the column resulting from the primary's rotation.

One of the most striking features of the AM Her binaries is that the angular velocity of the primary appears to be the same as that of the orbit. This synchronization, unique amongst close binaries, is most likely related to the strong magnetic field. The magnetic field interacts with the accretion stream and the secondary star. It is probable that the red dwarf secondary has a dynamo-generated magnetic field, since it is convective and rapidly rotating.

The X-ray binary pulsars were discovered by Giacconi *et al* (1971). In these systems a lobe-filling secondary transfers material onto a magnetic neutron star. The magnetic moment is not as large as that of a white dwarf in an AM Her system, so a partially disrupted accretion disc forms. After passing through the disc, material is magnetically channelled to form accretion columns above the neutron star. A range of spin behaviour of the neutron star is observed. In some systems the star appears to have reached an equilibrium period, indicating that torque balance has been achieved. Magnetic interaction with the disc, as well as field channelling of material, will generate a torque on the neutron star. A similar disrupted disc model is adopted for the intermediate polars, in which the primary is a white dwarf with a magnetic field typically an order of magnitude less than in the AM Her binaries.

Accretion discs require an anomalous form of viscosity to explain their mass inflow rates. Turbulence has been employed to generate the necessary stresses for the radial advection of angular momentum. However, an instability to explain the generation of such turbulence has yet to be found in a non-magnetic disc. This dilemma is believed to have been resolved by Balbus and Hawley (1991) who showed that Keplerian discs are dynamically unstable in the presence of a weak poloidal magnetic field. Subsequent work has shown that this magnetic shear instability leads to the generation of turbulence and a self-sustaining dynamo. The resulting stresses due to the large-scale magnetic field lead to radial advection of angular momentum at least comparable to the viscous advection. Hence magnetic fields are believed to play a fundamental role in accretion discs.

It is likely that the secondary star in most close binary systems has some magnetic field. A magnetically channelled wind from a tidally synchronized secondary is invoked to generate the angular momentum loss necessary to sustain mass transfer in systems with periods $\gtrsim 3\,\mathrm{hrs}$. The field could be generated by dynamo action in the rapidly rotating star.

1.3. Magnetohydrodynamics

The theory of magnetohydrodynamics describes the fluid mechanics of plasmas and the behaviour of their magnetic fields. Maxwell's equations for the electromagnetic field are combined with the equations of hydrodynamics. For non-relativistic flow speeds, the magnetic force can be expressed as a function of the field $\mathbf{B}$ and its derivatives, and this dominates the electric force. The electric field and charge density can be eliminated from the equations, and so do not have to be explicitly considered. Faraday's law of induction shows how $\mathbf{B}$ is affected by the fluid motions, and diffused due to dissipation of the associated electric currents. The momentum equation contains the effect of the magnetic force on the motions, and the heat equation has a source due to the dissipation of currents.

Simplified forms of the equations can be used in some problems. For example, the diffusion term in the induction equation can be ignored if the diffusion time-scale is much longer than the flow time-scale. In this case a useful picture emerges in which the magnetic field lines are frozen to the plasma. Material is therefore threaded onto field lines which are advected and distorted by the flow. In the other extreme, the velocity term is small and the induction equation becomes purely diffusive, so material can move freely across field lines. Kinematic theory is appropriate when the magnetic force only has a small effect on the motions. The velocity solution of the non-magnetic momentum equation can then be used in the induction equation to calculate **B**.

Dynamo theory describes how magnetic fields can be generated and sustained by fluid motions, in the presence of dissipation. Certain conditions are necessary for dynamo action, and various types of dynamos can be defined. Dynamo generated fields are relevant to secondary stars and accretion discs.

1.4. Types of Problems

Chapter 2 contains basic magnetohydrodynamics and binary star theory, together with the spin dynamics theory subsequently required. The major role of magnetic fields in binary stars is in the redistribution of angular momentum. The AM Herculis binaries are considered in Chapters 3 to 7. Explanation of the synchronous rotation of the white dwarf leads to several magnetic problems. After an introduction to AM Herculis systems in Chapter 3, the approach to synchronism is considered in Chapter 4. This involves the interaction of the white dwarf's magnetic field with the secondary star. In the asynchronous state the secondary experiences a time-dependent **B**, resulting in induced electric currents. The magnetic force is small in the secondary, except in its very outer layers, so the kinematic problem of solving the induction equation is considered. For a synchronous secondary, the induction equation is purely diffusive. Dissipation of the induced magnetic field causes the primary to approach synchronism.

Chapter 5 addresses the nature of the accretion torque. Channelling of the accretion stream by the primary's magnetic field results because of the high conductivity of material, and the increasing field strength as the accretor is approached. The induction and momentum equations yield the magnetic force on the stream. The torque generated on the primary can be calculated in the synchronous and asynchronous cases and, in general, has components parallel and perpendicular to the orbital angular momentum vector.

Chapter 6 considers the maintenance of synchronism, which requires a non-dissipative torque to balance the accretion torque on the primary. Such a torque can be generated by the interaction of the primary with a magnetic field originating in the secondary. This field could be dynamo generated. Once a balance is found, its dynamical stability must be investigated. The case of a non-spherical primary is also considered, leading to a tidal torque on the star and a more complicated angular momentum problem in the stability analysis.

In Chapter 7 it is shown that certain conditions are necessary for the primary star to reach a synchronous state, which are independent of the torque balance condition producing such a state. A critical value for the magnetic diffusivity of the secondary arises, above which synchronism could not be attained in the presence of accretion.

Binaries with partially disrupted accretion discs are considered in Chapters 8 to 10. This is relevant to the intermediate polars and the X-ray binary pulsars. Their basic properties and observations are discussed in Chapter 8. The main MHD problem in these systems is the interaction of the magnetic primary star with the accretion disc. The aim of Chapter 9 is to explain how the disc becomes magnetically disrupted before reaching the primary star's surface. In the inner regions the magnetic field significantly perturbs the disc structure, and the full set of equations must be considered. However, the thin nature of the disc, and its high diffusivity, allow a simplification of the equations which renders the problem tractable. After disruption matter becomes field-channelled and forms accretion columns at the star's magnetic poles.

Chapter 10 considers the interaction of the stellar magnetic field with the undisrupted part of the disc. The effect of the stellar field is usually small in this region. Kinematic theory is then appropriate, so the induction equation can be solved for **B** using the velocity field of the unperturbed disc. A torque on the primary star results from the reaction to the magnetic torque exerted on the disc. A stellar torque is also generated by the inner field-channelled region. The torques can be balanced to produce an equilibrium rotation state. The nature of the magnetic diffusivity is of major significance in these problems.

Chapter 11 investigates intrinsic magnetism in accretion discs. Such fields must be generated by some form of self-sustaining dynamo process, since their decay times are much shorter than their radial advection time. A self-sustaining dynamo requires the presence of turbulence, and magnetic fields are also believed to be necessary in the generation of such turbulence in discs. Kinematic theory can be used to assess the likely mode of operation of disc dynamos. However, the presence of a magnetic field is principally of interest for its possible dynamical effects on the disc. In particular, magnetic stresses can play an important role in angular momentum advection causing inflow of material. Again, the thin nature of the disc aids the analysis, and the nature of the magnetic diffusivity is of central importance.

Chapter 12 considers the role of magnetic winds in binary stars. A magnetically influenced wind from the secondary, together with tidal coupling to the orbit, can provide a driving mechanism for mass transfer in systems with orbital periods $\gtrsim$ 3 hrs. This process may be related to the period gap in binary stars, which is the almost total absence of observable systems with periods of 2 to 3 hrs. The possibility of magnetically channelled winds from accretion discs is also investigated, as a means of angular momentum removal relating to mass inflow. The dynamical stability of self-consistent magnetic disc winds is considered.

References

Balbus, S.A. and Hawley, J.F., 1991. *Astrophys. J.*, **376**, 214.
Chandrasekhar, S., 1933. *Mon. Not. R. Astr. Soc.*, **93**, 390.
Crawford, J.A. and Kraft, R.P., 1956. *Astrophys. J.*, **123**, 44.
Giacconi, R., Gursky, H., Kellogg, E., Schreier, E. and Tananbaum, H., 1971. *Astrophys. J. Lett.*, **167**, L67.
Hearn, D.R., Richardson, J.A. and Clark, G.W., 1976. *Astrophys. J. Lett.*, **210**, L23.
Joy, A.H., 1956. *Astrophys. J.*, **124**, 317.
Kopal, Z., 1955. *Ann. d'Astrophys.*, **18**, 379.
Kraft, R.P., Mathews, J. and Greenstein, J.L., 1962. *Astrophys. J.*, **136**, 312.
Krzeminski, W., 1965. *Astrophys. J.*, **142**, 1051.
Landstreet, J.D., 1994. In *Cosmical Magnetism*, ed., Lynden-Bell, D., Kluwer Academic Publishers, Dordrecht.
Linnell, A.P., 1950. *Harvard. Obs. Circ.*, No. 455.
Mumford, G.S., 1962. *Sky and Telesc.*, **23**, 135.
Plavec, M., 1958. *Mem. Soc. R. Sci. Liege(4).*, **20**, 11.
Roche, E.N., 1873. *Ann. de l'Acad. Sci. Montpelier.*, **8**, 235.
Smak, J., 1971. *Acta. Astr.*, **21**, 15.
Walker, M., 1956. *Astrophys. J.*, **123**, 68.
Warner, B., 1995. *Cataclysmic Variable Stars*, Cambridge University Press.
Warner, B. and Nather, R.E., 1971. *Mon. Not. R. Astr. Soc.*, **152**, 219.
Wood, F.B., 1950. *Astrophys. J.*, **112**, 196.

CHAPTER 2

THEORETICAL PREREQUISITES

2.1. Introduction

This chapter contains the basic essentials of magnetohydrodynamics, binary star theory and spin dynamics. In §2.2 the necessary conditions for classical fluid dynamics to apply to plasmas are discussed (for a more detailed description of plasma kinetic theory see Mestel, 1997, and references therein). The equations describing the electromagnetic field are then combined with the equations of fluid mechanics, in the non-relativistic regime, to form the equations of MHD, and their general properties are discussed. Hydromagnetic wave motions, which redistribute energy and momentum, are derived, followed by convenient mathematical representations of the field $\mathbf{B}$. The decay of $\mathbf{B}$ and various types of diffusion processes are then considered. In §2.3 the basic dynamo problem is formulated and the main types of mean-field dynamos are defined, (for a recent review of dynamo theory see Roberts, 1994).

Section 2.4 presents the theory of close binary stars. Firstly, the Roche model is considered and the unstable nature of the inner Lagrangian point is analysed. Mass transfer and the subsequent formation of an accretion disc are then addressed. The standard steady viscous accretion disc model is presented, since this acts as a basis for understanding magnetic effects in discs. The time-dependent disc equations are formulated and various important time-scales are defined. The principles of viscous diffusion and instability are discussed.

Section 2.5 contains rigid body dynamics, which is relevant to the response of compact primary stars when they are subjected to torques. A highly magnetic primary is likely to have some distortion from spherical symmetry due to non-radial internal magnetic forces. Hence, in general, its angular velocity and angular momentum vectors will not be parallel and full spin dynamics theory is needed to analyse its motion. The relevant frame transformations are discussed. The spin evolution of such stars is of central importance and accurate observations relating to this are available in a variety of systems.

2.2. Behaviour of B Fields in Plasmas

2.2.1. PLASMA FLUIDS

A plasma is a mixture of an electron gas and an ion gas. The electrons and ions interact with each other through their Coulomb attractions and repulsions. A plasma differs from an atomic or molecular gas in that its inter-particle forces can be longer range than the short-range forces between neutral particles. A charged particle can interact with many others at any instant. The presence of a large-scale magnetic field can further complicate matters by introducing anisotropic properties. However, if certain conditions are met, the bulk motions in a plasma can be described by classical fluid mechanics, with additional effects due to a large-scale magnetic field.

If the ions result from atoms each losing Z electrons of charge magnitude e, then the total charge density is

$$\rho_{\mathrm{c}} = n_{\mathrm{i}} Z e - n_{\mathrm{e}} e, \tag{2.1}$$

where n_{i} and n_{e} are the number densities of ions and electrons, respectively. Any local charge separation cannot be large, since this would result in a very strong electric field rapidly restoring neutrality. It follows that

$$n_{\mathrm{e}} \simeq Z n_{\mathrm{i}} \tag{2.2}$$

and ρ_{c} is small. Equation (2.2) is referred to as the plasma approximation. Since $m_{\mathrm{i}} \gg m_{\mathrm{e}}$, where m_{i} and m_{e} are the ion and electron masses, electrons are relatively easily accelerated. A small local excess of electrons results in a restoring electric field leading to plasma oscillations about the neutral state with a characteristic frequency

$$\omega_{\mathrm{p}} = \left(\frac{n_{\mathrm{e}} e^2}{\epsilon_0 m_{\mathrm{e}}} \right)^{\frac{1}{2}}, \tag{2.3}$$

where n_{e} is the unperturbed electron number density and ϵ_0 the permittivity of free space.

The length-scale associated with charge oscillations is $l_{\mathrm{c}} \sim v_{\mathrm{e}}/\omega_{\mathrm{p}}$, where v_{e} is a typical electron velocity. The electric field resulting from the charge imbalance is limited in range to $\sim l_{\mathrm{c}}$ by the shielding effect of the electrons. In the absence of perturbations charge neutrality would hold, with l_{c} vanishing and the electrons totally shielding the ions. However, the formation and maintenance of the plasma state requires sufficiently high temperatures and hence charge fluctuations due to the thermal motions of the electrons are always present. A fundamental shielding length $l_{\mathrm{c}} = \lambda_{\mathrm{D}}$, known as the

Debye length, can be defined by taking v_e to be the mean thermal velocity of an electron. Hence

$$v_e \sim \omega_p \lambda_D \sim \left(\frac{kT}{m_e}\right)^{\frac{1}{2}},$$

where k is Boltzmann's constant and the electron temperature $T_e = T_i = T$ for a gas in local thermodynamic equilibrium. Use of (2.3) for the plasma frequency ω_p then yields

$$\lambda_D = \left(\frac{\epsilon_0 kT}{n_e e^2}\right)^{\frac{1}{2}}. \tag{2.4}$$

This is the length-scale of local electric fields resulting from thermally induced charge separation, and hence it gives an effective range for Coulomb collisions in a plasma. On length-scales $< \lambda_D$ thermal charge separation is significant and the plasma approximation (2.2) does not hold.

Provided the macroscopic length-scale $L \gg \lambda_D$, the electromagnetic field may be divided into a large-scale, collective field and a small-scale random field. The large-scale field has the macroscopic charge density and current density as its sources. The small-scale random field is essentially the sum of the unshielded Coulomb fields of individual particles; it acts to keep the velocity distribution function close to a Maxwellian form, and to inhibit the drift of electrons relative to ions. The plasmas occurring in binary stars generally well satisfy $L \gg \lambda_D$, so the macroscopic field equations can be employed.

In discussing plasma properties, a mean free path can be defined, using elementary kinetic theory, by

$$\lambda = \frac{1}{n\pi b^2}, \tag{2.5}$$

where n is the number density of scattering particles and b is an effective collision radius. The quantity πb^2 is a collision cross-section. For elastic scattering, such as electrons by ions, an estimate for b can be made by equating the Coulomb interaction energy to the mean thermal kinetic energy, giving

$$b \simeq \frac{Ze^2}{6\pi\epsilon_0 kT}. \tag{2.6}$$

An estimate for the average value of b including long-range scattering, employing the differential Rutherford cross-section, increases (2.6) by a factor of $\sim 9/4$.

If the plasma behaves like a nearly perfect gas, the Coulomb interaction energy at the mean inter-particle distance of $\simeq n^{-1/3}$ must be small compared to the thermal energy. This requires

$$\frac{Ze^2}{4\pi\epsilon_0 n^{-\frac{1}{3}}} \ll \frac{3}{2}kT$$

which, by use of (2.6), is equivalent to

$$bn^{\frac{1}{3}} \ll 1. \tag{2.7}$$

The ratio of the mean free path to the Debye length can be found using (2.2) and (2.4)–(2.6), noting that $n = n_{\rm i}$. The result is

$$\frac{\lambda}{\lambda_{\rm D}} \simeq \left(\frac{6}{\pi}\right)^{\frac{1}{2}} \frac{1}{(bn^{\frac{1}{3}})^{\frac{3}{2}}}. \tag{2.8}$$

In the perfect gas regime (2.7) holds for $bn^{1/3}$ and hence $\lambda \gg \lambda_{\rm D}$. So, as expected in this case, the mean free path is much greater than the Coulomb screening radius. It is simple to show that the average collision radius $b \ll \lambda_{\rm D}$, consistent with short-scale interactions in a perfect gas, and that $\lambda_{\rm D} > n^{-1/3}$. Hence the necessary ordering of length-scales is

$$\lambda \gg \lambda_{\rm D} > n^{-\frac{1}{3}} \gg b. \tag{2.9}$$

A mean collision time can be defined for scattering of electrons by ions as

$$\tau_{\rm ei} = \frac{\lambda_{\rm ei}}{(v_{\rm th})_{\rm e}}, \tag{2.10}$$

where $\lambda_{\rm ei}$ is the mean free path. This is the time-scale in which a sub-thermal electron drift velocity relative to ions is randomized. Similar time-scales can be defined for ion-ion and electron-electron encounters.

In a fully ionized gas the presence of a magnetic field can introduce an anisotropy. An electron spirals about a uniform field $\mathbf{B}$ with the gyration frequency

$$\omega_{\rm e} = \frac{eB}{m_{\rm e}}, \tag{2.11}$$

with a similar expression for ions. The gyration time-scale of $\sim 1/\omega_{\rm e}$ can be compared with the collision time-scale $\tau_{\rm ei}$, given by (2.10). If $\omega_{\rm e}\tau_{\rm ei} \gg 1$ then an electron performs a number of spiral turns before having its motion randomized by collisions with the ion gas, and hence the field introduces a microscopic anisotropy. This can result in a reduced conductivity across field lines. However, inside most stars, with plausible values of B, $\omega_{\mathrm{e}}\tau_{\mathrm{ei}} \ll 1$.

The classical fluid mechanics approach can be applied to plasmas when the macroscopic length-scale L is large compared to the particle mean free paths, and the macroscopic time-scale is long compared to the collision times. A fluid element of volume $\sim l^3$ can be defined with $\lambda \ll l \ll L$, so the gas within it contains many colliding particles representing a continuous distribution of momenta and is essentially uniform. The effect of collisions is then to randomize the particle velocities about a mean velocity $\mathbf{v}$, which is the bulk velocity of the fluid element. Viewed in a frame moving with velocity $\mathbf{v}$, the particles have a velocity distribution very close to Maxwellian, and a temperature T can be defined for the element. Small deviations, of order λ/L, from a Maxwellian distribution yield the transport quantities of resistivity and viscosity. The centre of mass of the fluid element has a local position $\mathbf{r}$, at a time t, in a global frame. Such a plasma can be regarded as a continuous fluid with velocity $\mathbf{v}(\mathbf{r}, t)$ and mass density $\rho(\mathbf{r}, t)$. Since the ion mass $m_{\mathrm{i}} \gg m_{\mathrm{e}}$, it follows that $\rho \simeq n_{\mathrm{i}} m_{\mathrm{i}}$ and the bulk velocity $\mathbf{v} \simeq \mathbf{v}_{\mathrm{i}}$ to good approximations. Because $L \gg \lambda_{\mathrm{D}}$, large-scale electric and magnetic fields can be defined, with macroscopic charge and current density sources. Magnetohydrodynamics (MHD) considers the motions, and equilibria, of such an electrically conducting fluid and the behaviour of its **B** field.

The magneto-fluid equations can be derived by applying macroscopic principles. They can also be derived from appropriately weighted integrals involving the particle distribution function $f(\mathbf{r}, \mathbf{p}, t)$. This function is a phase space, statistical description, with $f d^3\mathbf{r} d^3\mathbf{p}$ being the number of particles with position and momentum in a given six-dimensional volume element $d^3\mathbf{r} d^3\mathbf{p}$ at a time t. A separate f can be defined for each species of particle. The phase space evolution of f is described by Boltzmann's equation

$$\frac{\partial f}{\partial t} + \frac{1}{m}\mathbf{p} \cdot \nabla f + \mathbf{F} \cdot \nabla_p f = \Gamma, \tag{2.12}$$

where m is the particle mass, ∇_p the gradient operator in momentum space, and $\mathbf{F}$ is the force on a particle, excluding the short-range interactive forces associated with collisions. The term Γ gives the effect of collisions.

The number density of particles is

$$n = \int_{\mathbf{p}} f d^3\mathbf{p}, \tag{2.13}$$

where the integral is over all momentum space, and f vanishes as $p \to \pm\infty$. Since $m_{\mathrm{i}} \gg m_{\mathrm{e}}$, the velocity of a fluid element is the mean ion velocity in the real space volume element $d^3\mathbf{r}$, given by

$$\mathbf{v}(\mathbf{r},t) = \left\langle \frac{\mathbf{p}_\mathrm{i}}{m_\mathrm{i}} \right\rangle = \frac{1}{n}\int_\mathbf{p} \frac{1}{m_\mathrm{i}}\mathbf{p}_\mathrm{i} f d^3\mathbf{p}. \tag{2.14}$$

The macroscopic fluid dynamic equations can be derived by taking appropriate weighted integrals of the Boltzmann equation (2.12), for ions and electrons (e.g. Battener, 1996).

In classical fluid mechanics two types of analysis of the flow properties are possible. In the Eulerian picture fixed spatial points are considered and a fluid quantity $Q(\mathbf{r},t)$ has $\mathbf{r}$ and t as independent variables. Hence rates of change involve partial time derivatives. In the Lagrangian picture fluid elements of conserved mass are followed and hence rates of change must account for the effect of displacement due to the flow. The vector $\mathbf{r}$ now defines the instantaneous position of a fluid element, so $\mathbf{r} = \mathbf{r}(t)$ and the velocity $\mathbf{v} = d\mathbf{r}/dt$. First order Taylor expansion gives a differential change in $Q(\mathbf{r}(t),t)$ as

$$dQ = d\mathbf{r}\cdot\nabla Q + \frac{\partial Q}{\partial t}dt,$$

and hence

$$\frac{dQ}{dt} = \frac{\partial Q}{\partial t} + \mathbf{v}\cdot\nabla Q. \tag{2.15}$$

This is referred to as the material, or Lagrangian derivative of Q. The derivative also applies to vector quantities; for example, the acceleration of the fluid is

$$\mathbf{a} = \frac{d\mathbf{v}}{dt} = \frac{\partial \mathbf{v}}{\partial t} + (\mathbf{v}\cdot\nabla)\mathbf{v} \tag{2.16}$$

and this must be employed in deriving the momentum equation. The Lagrangian operator is therefore

$$\frac{d}{dt} = \frac{\partial}{\partial t} + \mathbf{v}\cdot\nabla. \tag{2.17}$$

Certain fluid properties become clearer when viewed using the Lagrangian picture.

In the remainder of this section the macroscopic electromagnetic field equations are combined with the fluid equations, in the non-relativistic regime, to obtain the equations of MHD.

2.2.2. MAXWELL'S EQUATIONS

Maxwell's equations describing the macroscopic electric and magnetic fields **E** and **B** are

$$\nabla \cdot \mathbf{E} = \frac{\rho_c}{\epsilon_0}, \tag{2.18}$$

$$\nabla \wedge \mathbf{E} = -\frac{\partial \mathbf{B}}{\partial t}, \tag{2.19}$$

$$\nabla \cdot \mathbf{B} = 0, \tag{2.20}$$

$$\nabla \wedge \mathbf{B} = \mu_0 \mathbf{J} + \frac{1}{c^2}\frac{\partial \mathbf{E}}{\partial t}, \tag{2.21}$$

where ρ_c is the charge density, **J** the current density, ϵ_0 the permittivity, μ_0 the permeability and $c = 1/(\epsilon_0\mu_0)^{1/2}$ the speed of light.

The fields **E** and **B** can be derived from scalar and vector potential functions. Equation (2.20) enables **B** to be expressed as

$$\mathbf{B} = \nabla \wedge \mathbf{A}, \tag{2.22}$$

where **A** is the magnetic vector potential. Substitution of this in (2.19) gives

$$\nabla \wedge \left(\mathbf{E} + \frac{\partial \mathbf{A}}{\partial t}\right) = \mathbf{0},$$

which is satisfied by

$$\mathbf{E} = -\nabla\psi - \frac{\partial \mathbf{A}}{\partial t}. \tag{2.23}$$

It follows from (2.22) and (2.23) that **B** and **E** are unchanged by the transformations

$$\tilde{\mathbf{A}} = \mathbf{A} + \nabla\Phi, \tag{2.24a}$$

$$\tilde{\psi} = \psi - \frac{\partial \Phi}{\partial t}, \tag{2.24b}$$

where Φ is an arbitrary scalar function. The consequent freedom of choice in the divergence of **A** enables separate equations to be derived relating **A** to **J** and ψ to ρ_c.

Substituting (2.22) and (2.23) in (2.21), and using (A7), gives

$$\nabla^2 \mathbf{A} - \frac{1}{c^2}\frac{\partial^2 \mathbf{A}}{\partial t^2} - \nabla\left(\nabla\cdot\mathbf{A} + \frac{1}{c^2}\frac{\partial\psi}{\partial t}\right) = -\mu_0 \mathbf{J}. \qquad (2.25)$$

Eliminating $\mathbf{E}$ between (2.18) and (2.23) yields

$$\nabla^2\psi + \frac{\partial}{\partial t}(\nabla\cdot\mathbf{A}) = -\frac{\rho_c}{\epsilon_0}. \qquad (2.26)$$

These equations involving $\mathbf{A}$ and ψ can be decoupled by choosing

$$\nabla\cdot\mathbf{A} = -\frac{1}{c^2}\frac{\partial\psi}{\partial t}, \qquad (2.27)$$

which is referred to as the Lorentz gauge. The separated equations are then

$$\nabla^2 \mathbf{A} - \frac{1}{c^2}\frac{\partial^2 \mathbf{A}}{\partial t^2} = -\mu_0 \mathbf{J}, \qquad (2.28)$$

$$\nabla^2 \psi - \frac{1}{c^2}\frac{\partial^2 \psi}{\partial t^2} = -\frac{\rho_c}{\epsilon_0}. \qquad (2.29)$$

If the sources $\mathbf{J}$ and ρ_c are known then, in principle, these inhomogeneous wave equations can be solved for $\mathbf{A}$ and ψ. The fields $\mathbf{B}$ and $\mathbf{E}$ then follow from (2.22) and (2.23).

The electric and magnetic fields, $\mathbf{E}'$ and $\mathbf{B}'$, measured in a frame moving with velocity $\mathbf{v}$ relative to the frame in which $\mathbf{E}$ and $\mathbf{B}$ are measured, are given by the Lorentz transformations

$$\mathbf{E}' = (1-\gamma)\frac{(\mathbf{v}\cdot\mathbf{E})}{v^2}\mathbf{v} + \gamma(\mathbf{E} + \mathbf{v}\wedge\mathbf{B}), \qquad (2.30)$$

$$\mathbf{B}' = (1-\gamma)\frac{(\mathbf{v}\cdot\mathbf{B})}{v^2}\mathbf{v} + \gamma\left(\mathbf{B} - \frac{1}{c^2}\mathbf{v}\wedge\mathbf{E}\right), \qquad (2.31)$$

where

$$\gamma = \left(1 - \frac{v^2}{c^2}\right)^{-\frac{1}{2}}. \qquad (2.32)$$

The Lorentz force on a charge q moving with velocity $\mathbf{v}_c$ in the presence of $\mathbf{E}$ and $\mathbf{B}$ fields is

$$\mathbf{F}_L = q(\mathbf{E} + \mathbf{v}_c\wedge\mathbf{B}). \qquad (2.33)$$

2.2.3. THE INDUCTION EQUATION

The magnetic fields present in binary systems interact with conducting plasma. This plasma generally consists of the secondary star, the accretion stream and disc, wind material, and sometimes a magnetosphere. The largest fluid velocities occur in the stream flow and the circular motions in the disc. These have characteristic values of $v \sim 10^6\,\mathrm{m\,s^{-1}}$ and hence fluid motions in binary stars satisfy $v/c \lesssim 3 \times 10^{-3} \ll 1$. The theory of non-relativistic magnetohydrodynamics is therefore appropriate. The induction equation is fundamental in this theory.

The equations of non-relativistic MHD are derived using the condition $v \ll c$ and working to first order in v/c. Since $\mathbf{v}$, $\mathbf{E}$ and $\mathbf{B}$ are causally inter-related, a typical velocity is $v \sim \ell/\tau$ where ℓ and τ are the length and time-scales for $\mathbf{E}$ and $\mathbf{B}$. The induction equation (2.19) then gives

$$\frac{E}{B} \sim \frac{\ell}{\tau} \sim v. \tag{2.34}$$

It therefore follows in (2.21) that

$$\frac{|\partial \mathbf{E}/\partial t|}{c^2|\nabla \wedge \mathbf{B}|} \sim \frac{\ell E}{c^2 \tau B} \sim \left(\frac{v}{c}\right)^2 \ll 1,$$

so the displacement current can be neglected and

$$\nabla \wedge \mathbf{B} = \mu_0 \mathbf{J}. \tag{2.35}$$

Taking the divergence of (2.21), noting that $\nabla \cdot (\nabla \wedge \mathbf{B}) = 0$ and using (2.18), gives

$$\nabla \cdot \mathbf{J} = -\frac{\partial \rho_{\mathrm{c}}}{\partial t}. \tag{2.36}$$

This expresses the conservation of charge. Equation (2.35) yields $\nabla \cdot \mathbf{J} = 0$ which implies that in non-relativistic MHD the electric currents flow in closed loops. Equation (2.36) then shows that the charge density is time-independent, to first order in v/c.

The current density can be expressed as the sum of the ion and electron contributions by

$$\mathbf{J} = n_{\mathrm{i}} Ze\mathbf{v}_{\mathrm{i}} - n_{\mathrm{e}} e\mathbf{v}_{\mathrm{e}}, \tag{2.37}$$

where n_{i} and n_{e} are the ion and electron number densities, $\mathbf{v}_{\mathrm{i}}$ and $\mathbf{v}_{\mathrm{e}}$ the mean drift velocities of ions and electrons, and $e > 0$. Using (2.1) for the total charge density ρ_{c}, it follows that (2.37) can be written

$$\mathbf{J} = \rho_{\mathrm{c}} \mathbf{v}_{\mathrm{i}} - n_{\mathrm{e}} e(\mathbf{v}_{\mathrm{e}} - \mathbf{v}_{\mathrm{i}}).$$

Noting that, since $m_{\text{i}} \gg m_{\text{e}}$, the bulk velocity $\mathbf{v} \simeq \mathbf{v}_{\text{i}}$ to a good approximation, the current density becomes

$$\mathbf{J} = \rho_{\text{c}}\mathbf{v} + \rho_{\text{e}}(\mathbf{v}_{\text{e}} - \mathbf{v}), \tag{2.38}$$

with $\rho_{\text{e}} = -n_{\text{e}}e$. It is noted that $(\mathbf{v}_{\text{e}} - \mathbf{v})$ is the mean drift velocity of electrons relative to the centre of mass of a fluid element, and hence

$$\mathbf{J} = \mathbf{J}' + \rho_{\text{c}}\mathbf{v}, \tag{2.39}$$

where $\mathbf{J}' = \rho_{\text{e}}(\mathbf{v}_{\text{e}} - \mathbf{v})$ is the electron current density in the fluid element frame.

Expanding (2.32) for γ, and using (2.34), it follows that the Lorentz transformations (2.30) and (2.31) become

$$\mathbf{E}' = \mathbf{E} + \mathbf{v} \wedge \mathbf{B}, \tag{2.40}$$

$$\mathbf{B}' = \mathbf{B}, \tag{2.41}$$

if terms of order $(v/c)^2$ are neglected. The current density $\mathbf{J}$ is given by

$$\mathbf{J} = \sigma\mathbf{E}' + \rho_{\text{c}}\mathbf{v}, \tag{2.42}$$

where σ is the plasma conductivity and $\mathbf{E}'$ is measured in a frame moving with the local fluid velocity $\mathbf{v}$. This expression for $\mathbf{J}$ follows from (2.39) with $\mathbf{J}' = \sigma\mathbf{E}'$ being Ohm's law in a fluid element frame, while $\rho_{\text{c}}\mathbf{v}$ is the contribution to $\mathbf{J}$ due to charge advection by the flow. It follows from (2.18) and (2.35) that

$$\frac{\rho_{\text{c}}|\mathbf{v}|}{|\mathbf{J}|} \sim \mu_0\epsilon_0\frac{|\mathbf{v}||\mathbf{E}|}{|\mathbf{B}|} \sim \left(\frac{v}{c}\right)^2,$$

where the last relation derives from (2.34). Hence, to high accuracy, charge advection is ignorable in (2.42) so $\mathbf{J} = \mathbf{J}' = \sigma\mathbf{E}'$ and use of (2.40) then gives

$$\mathbf{J} = \sigma(\mathbf{E} + \mathbf{v} \wedge \mathbf{B}). \tag{2.43}$$

Combining (2.19) with the curls of (2.35) and (2.43) gives the MHD induction equation

$$\nabla \wedge (\mathbf{v} \wedge \mathbf{B}) - \nabla \wedge (\eta\nabla \wedge \mathbf{B}) = \frac{\partial\mathbf{B}}{\partial t}, \tag{2.44}$$

where $\eta = 1/\mu_0\sigma$ is the magnetic diffusivity. Using the vector identity (A6), the induction equation can be written as

$$\frac{d\mathbf{B}}{dt} = -\mathbf{B}\ \nabla\cdot\mathbf{v} + (\mathbf{B}\cdot\nabla)\mathbf{v} - \nabla\wedge(\eta\nabla\wedge\mathbf{B}), \tag{2.45}$$

where d/dt is the Lagrangian time derivative, measuring the rate of change moving with the fluid velocity. The first and second terms on the right side of (2.45) represent the rates of change of $\mathbf{B}$ in a fluid element due to compression and stretching of the field lines by the motion, respectively. The last term gives the diffusion of $\mathbf{B}$ due to dissipation of electric currents.

The relative importance of the convective and diffusive terms in (2.44) can be gauged using the typical length and time-scales ℓ and τ. The magnetic Reynolds number is defined as

$$R_{\rm m} = \frac{\ell v}{\eta} \sim \frac{|\nabla\wedge(\mathbf{v}\wedge\mathbf{B})|}{|\nabla\wedge(\eta\nabla\wedge\mathbf{B})|}, \tag{2.46}$$

where $v \sim \ell/\tau$ is a typical fluid velocity.

In the case $R_{\rm m} \gg 1$ the induction equation becomes

$$\nabla\wedge(\mathbf{v}\wedge\mathbf{B}) = \frac{\partial\mathbf{B}}{\partial t}. \tag{2.47}$$

There are several ways of showing that in this limit the magnetic field is 'frozen' to the plasma, so fluid elements are attached to field lines. One method is to represent $\mathbf{B}$ as

$$\mathbf{B} = \nabla\alpha\wedge\nabla\beta = \nabla\wedge(\alpha\nabla\beta), \tag{2.48}$$

where α and β are referred to as Euler potentials. This satisfies $\nabla\cdot\mathbf{B} = 0$ and, since $\mathbf{B}\cdot\nabla\alpha = \mathbf{B}\cdot\nabla\beta = 0$, α and β are constant on field lines and so can be used to label these lines. Substitution of (2.48) in (2.47) gives

$$\begin{aligned}\nabla\alpha\wedge\nabla\frac{\partial\beta}{\partial t} - \nabla\beta\wedge\nabla\frac{\partial\alpha}{\partial t} &= \nabla\wedge[\mathbf{v}\wedge(\nabla\alpha\wedge\nabla\beta)]\\ &= \nabla\wedge[(\mathbf{v}\cdot\nabla\beta)\nabla\alpha - (\mathbf{v}\cdot\nabla\alpha)\nabla\beta]\\ &= \nabla(\mathbf{v}\cdot\nabla\beta)\wedge\nabla\alpha - \nabla(\mathbf{v}\cdot\nabla\alpha)\wedge\nabla\beta,\end{aligned}$$

making use of (A5). Hence

$$\nabla\wedge\left(\alpha\nabla\frac{\partial\beta}{\partial t}\right) - \nabla\wedge\left(\beta\nabla\frac{\partial\alpha}{\partial t}\right) = -\nabla\wedge[\alpha\nabla(\mathbf{v}\cdot\nabla\beta)] + \nabla\wedge[\beta\nabla(\mathbf{v}\cdot\nabla\alpha)],$$

and so

$$\nabla\wedge\left(\alpha\nabla\frac{d\beta}{dt} - \beta\nabla\frac{d\alpha}{dt}\right) = \mathbf{0}, \tag{2.49}$$

where d/dt is the Lagrangian time derivative. Equation (2.49) has a family of solutions, but the simplest is

$$\frac{d\alpha}{dt} = \frac{d\beta}{dt} = 0, \tag{2.50}$$

corresponding to magnetic field lines moving with the fluid.

Another way of demonstrating the frozen field property of (2.47) is to consider the magnetic flux through a contour c. This is given by

$$\Phi = \int_S \mathbf{B} \cdot d\mathbf{S}, \tag{2.51}$$

where S is an open surface ending on c. If the contour is taken to be always composed of the same fluid elements, then conservation of Φ is consistent with the field being frozen to the plasma. Hence the material derivative $d\Phi/dt$ should vanish if (2.47) is satisfied.

Denote the material contour at time t by c, and at time $t + \delta t$ by c'. In the time interval δt an arc length $d\mathbf{l}$ of c sweeps out an area $d\mathbf{l} \wedge \mathbf{v}\delta t$. A closed surface can be considered consisting of a surface S ending on c, a surface S' ending on c', and the band joining c and c'. The total flux through this closed surface at time $t + \delta t$ is

$$-\int_S \mathbf{B}(\mathbf{r}, t + \delta t) \cdot d\mathbf{S} + \int_{S'} \mathbf{B}(\mathbf{r}, t + \delta t) \cdot d\mathbf{S}' + \oint_c \mathbf{B}(\mathbf{r}, t + \delta t) \cdot d\mathbf{l} \wedge \mathbf{v}\delta t = 0.$$

The negative sign arises since the outward normal is required on S, and the total flux of $\mathbf{B}$ through any closed surface is zero since $\nabla \cdot \mathbf{B} = 0$. The change in flux through the moving contour is therefore

$$\begin{aligned}
\delta\Phi &= \int_{S'} \mathbf{B}(\mathbf{r}, t + \delta t) \cdot d\mathbf{S}' - \int_S \mathbf{B}(\mathbf{r}, t) \cdot d\mathbf{S} \\
&= \int_S [\mathbf{B}(\mathbf{r}, t + \delta t) - \mathbf{B}(\mathbf{r}, t)] \cdot d\mathbf{S} + \delta t \oint_c \mathbf{B}(\mathbf{r}, t) \cdot \mathbf{v} \wedge d\mathbf{l} \\
&= \delta t \left\{ \int_S \frac{\partial \mathbf{B}}{\partial t} \cdot d\mathbf{S} - \oint_c \mathbf{v} \wedge \mathbf{B} \cdot d\mathbf{l} \right\}.
\end{aligned}$$

Dividing by δt, letting $\delta t \to 0$ and using Stokes' integral theorem, gives

$$\frac{d\Phi}{dt} = \int_S \left(\frac{\partial \mathbf{B}}{\partial t} - \nabla \wedge (\mathbf{v} \wedge \mathbf{B}) \right) \cdot d\mathbf{S}. \tag{2.52}$$

Equation (2.47) then shows $d\Phi/dt = 0$, which is consistent with frozen $\mathbf{B}$.

In the general case, with diffusion present, (2.44) and (2.52) yield

$$\frac{d\Phi}{dt} = -\int_S \nabla \wedge (\eta \nabla \wedge \mathbf{B}) \cdot d\mathbf{S}, \tag{2.53}$$

showing that the flux through a material contour changes due to diffusion of **B**.

When $R_m \ll 1$ the induction equation becomes

$$\nabla \wedge (\eta \nabla \wedge \mathbf{B}) = -\frac{\partial \mathbf{B}}{\partial t}. \tag{2.54}$$

For constant η this yields

$$\eta \nabla^2 \mathbf{B} = \frac{\partial \mathbf{B}}{\partial t}, \tag{2.55}$$

which is a standard diffusion equation for **B**. The rate of dissipation of the electric current in a fluid element can be found by considering the rate at which the electric field does work on the electrons. This rate of work, per unit volume, is

$$Q_m = (\mathbf{v}_e - \mathbf{v}_i) \cdot (\rho_e \mathbf{E}'),$$

where $(\mathbf{v}_e - \mathbf{v}_i)$ and $\mathbf{E}'$ are the electron mean drift velocity and the electric field both measured in the centre of mass frame of the element, remembering $\mathbf{v}_i \simeq \mathbf{v}$. Since $\mathbf{E}' = \mu_0 \eta \mathbf{J}'$ and $\mathbf{J}' = \rho_e(\mathbf{v}_e - \mathbf{v}_i)$ while, to first order in v/c, $\mathbf{J}' = \mathbf{J}$, it follows that

$$Q_m = \mu_0 \eta \mathbf{J}^2. \tag{2.56}$$

The kinetic energy imparted to the electrons is dissipated due to collisions with the ions, resulting in heating of the plasma. Hence (2.56) represents the rate of dissipation of electric current per unit volume. In the absence of significant fluid motions and external sources, (2.55) shows that a magnetic field with length-scale ℓ decays on a characteristic diffusion time-scale

$$\tau_d = \frac{\ell^2}{\eta} \sim \frac{B^2/2\mu_0}{Q_m}, \tag{2.57}$$

where the last relation derives from (2.35) and (2.56).

It follows from (2.46) and (2.57) that the magnetic Reynolds number is

$$R_m = \frac{\ell^2/\eta}{\ell/v} = \frac{\tau_d}{\tau_{dyn}}, \tag{2.58}$$

where τ_{dyn} is the flow time-scale. When $R_m \gg 1$ then $\tau_{dyn} \ll \tau_d$ and the flow distorts the magnetic field faster than it can diffuse. Conversely, when $R_m \ll 1$ rapid diffusion prevents the flow from significantly affecting the field.

An expression for the rate of change of the magnetic energy density can be derived from the induction equation (2.19). Taking the scalar product of this with $\mathbf{B}$ and using identity (A4), together with (2.35), gives

$$\frac{\partial}{\partial t}\left(\frac{B^2}{2\mu_0}\right) = -\nabla\cdot(\mathbf{E}\wedge\mathbf{H}) - \mathbf{E}\cdot\mathbf{J},$$

where $\mathbf{H} = \mathbf{B}/\mu_0$. Use of (2.43) yields

$$\mathbf{E}\cdot\mathbf{J} = \mu_0\eta\mathbf{J}^2 - (\mathbf{v}\wedge\mathbf{B})\cdot\mathbf{J} = \mu_0\eta\mathbf{J}^2 + \mathbf{v}\cdot(\mathbf{J}\wedge\mathbf{B}).$$

It is shown below that $\mathbf{J}\wedge\mathbf{B}$ is the magnetic force density $\mathbf{F}_\mathrm{m}$, so

$$\frac{\partial}{\partial t}\left(\frac{B^2}{2\mu_0}\right) = -\nabla\cdot(\mathbf{E}\wedge\mathbf{H}) - \mu_0\eta\mathbf{J}^2 - \mathbf{v}\cdot\mathbf{F}_\mathrm{m}. \qquad (2.59)$$

The first term on the right hand side of this equation represents the divergence of the Poynting energy flux $\mathbf{E}\wedge\mathbf{H}$. A positive divergence corresponds to magnetic energy flowing away from a point, leading to a decrease in the energy density. The second term is the dissipation of currents, always causing a decrease in field energy. The term $\mathbf{v}\cdot\mathbf{F}_\mathrm{m}$ is the rate at which the magnetic force does work on the fluid motions; the reaction to this leads to stretching and compression of $\mathbf{B}$ and hence to a change in its energy density.

The concept of magnetic flux tubes can be useful. A flux tube is the volume enclosed by the set of field lines which intercept a simple closed curve. Since no field lines cross the surface of the tube, it follows from $\nabla\cdot\mathbf{B} = 0$ that for any volume of it contained between two cross-sections as much flux leaves as enters. Hence the flux Φ through the tube, give by (2.51) with $d\mathbf{S}$ having the same sense as $\mathbf{B}$, is constant. When $R_\mathrm{m} \gg 1$ matter can flow along flux tubes but not across them.

It is sometimes convenient to express the induction equation, or part of it, in terms of the vector potential $\mathbf{A}$. In general electromagnetism, $\mathbf{A}$ is related to $\mathbf{J}$ by (2.28), subject to the Lorentz gauge (2.27). In non-relativistic MHD the potentials $\mathbf{A}$ and ψ, like $\mathbf{E}$ and $\mathbf{B}$, vary with length and time-scales ℓ and τ and $v \sim \ell/\tau$. It then follows in (2.28) that

$$\frac{|\partial^2\mathbf{A}/\partial t^2|}{c^2|\nabla^2\mathbf{A}|} \sim \frac{\ell^2/\tau^2}{c^2} \sim \frac{v^2}{c^2},$$

and hence, to first order in v/c, the magnetic vector potential is related to the current density by

$$\nabla^2\mathbf{A} = -\mu_0\mathbf{J}. \qquad (2.60)$$

The Lorentz gauge (2.27) has a ratio of terms given by

$$\frac{|\partial\psi/\partial t|}{c^2|\nabla\cdot\mathbf{A}|}\sim\frac{(\ell/\tau)\psi}{c^2A}\sim\frac{(\ell/\tau)\ell E}{c^2\ell B}\sim\frac{(\ell/\tau)v}{c^2}\sim\frac{v^2}{c^2},$$

using (2.22), (2.23) and (2.34) to estimate the fields. The gauge condition therefore becomes

$$\nabla\cdot\mathbf{A}=0. \tag{2.61}$$

It will be seen that in axisymmetric problems it is particularly useful to express the poloidal magnetic field in terms of a vector potential.

2.2.4. MAGNETIC FORCE

The Lorentz force density on ions and electrons is

$$\mathbf{F}=n_{\mathrm{i}}Ze(\mathbf{E}+\mathbf{v}_{\mathrm{i}}\wedge\mathbf{B})-n_{\mathrm{e}}e(\mathbf{E}+\mathbf{v}_{\mathrm{e}}\wedge\mathbf{B}). \tag{2.62}$$

Use of (2.1) and (2.37) for ρ_{c} and $\mathbf{J}$ then gives

$$\mathbf{F}=\rho_{\mathrm{c}}\mathbf{E}+\mathbf{J}\wedge\mathbf{B}. \tag{2.63}$$

The ratio of the electric to magnetic force is

$$\frac{\rho_{\mathrm{c}}|\mathbf{E}|}{|\mathbf{J}\wedge\mathbf{B}|}\sim\epsilon_0\mu_0\left(\frac{E}{B}\right)^2\sim\left(\frac{v}{c}\right)^2,$$

where the first relation follows from (2.18) and (2.35) and the second from (2.34). The electric force density is therefore negligible so

$$\mathbf{F}=\mathbf{F}_{\mathrm{m}}=\mathbf{J}\wedge\mathbf{B}, \tag{2.64}$$

and use of (2.35) gives

$$\mathbf{F}_{\mathrm{m}}=\frac{1}{\mu_0}(\nabla\wedge\mathbf{B})\wedge\mathbf{B}. \tag{2.65}$$

Employing the vector identity (A2), this can be written as

$$\mathbf{F}_{\mathrm{m}}=\frac{1}{\mu_0}(\mathbf{B}\cdot\nabla)\mathbf{B}-\nabla\left(\frac{B^2}{2\mu_0}\right). \tag{2.66}$$

Hence $\mathbf{F}_{\mathrm{m}}$ can be expressed in the Cartesian tensor form

$$F_{\mathrm{m}i}=\frac{\partial M_{ij}}{\partial x_j}, \tag{2.67a}$$

where the summation convention is used and

$$M_{ij} = \frac{1}{\mu_0} B_i B_j - \frac{B^2}{2\mu_0} \delta_{ij} \tag{2.67b}$$

is the Maxwell stress tensor. The $B_i B_j / \mu_0$ term represents a tension along the magnetic field, while $B^2/2\mu_0$ is a pressure.

Writing $\mathbf{B} = B\hat{\mathbf{s}}$ and $\hat{\mathbf{s}} \cdot \nabla = d/ds$, then

$$(\mathbf{B} \cdot \nabla)\mathbf{B} = B^2 \frac{d}{ds}(\hat{\mathbf{s}}) + \hat{\mathbf{s}} \frac{d}{ds}\left(\frac{B^2}{2}\right) = \frac{B^2}{R_c}\hat{\mathbf{n}} + \nabla_{\parallel}\left(\frac{B^2}{2}\right),$$

where R_c is the local radius of curvature of a field line, $\hat{\mathbf{n}}$ is a unit vector directed at the centre of curvature and $\nabla_{\parallel}$ measures the spatial rate of change along $\mathbf{B}$. Substitution in (2.66) gives

$$\mathbf{F}_{\mathrm{m}} = \frac{B^2}{\mu_0 R_c}\hat{\mathbf{n}} - \nabla_{\perp}\left(\frac{B^2}{2\mu_0}\right). \tag{2.68}$$

The magnetic force therefore results from curvature of the field lines, which are stressed along their length, and from the variation of magnetic pressure across field lines.

There are two non-trivial cases in which the magnetic force density vanishes. A current-free region has $\mathbf{J} = \mathbf{0}$ and hence zero $\mathbf{F}_{\mathrm{m}}$. Equation (2.35) then shows that $\mathbf{B}$ can be expressed as

$$\mathbf{B} = -\nabla\psi_{\mathrm{m}}, \tag{2.69}$$

where, by virtue of $\nabla \cdot \mathbf{B} = 0$, ψ_{m} satisfies Laplace's equation

$$\nabla^2\psi_{\mathrm{m}} = 0. \tag{2.70}$$

The second case of vanishing $\mathbf{F}_{\mathrm{m}}$ arises when $\mathbf{J}$ is finite but parallel to $\mathbf{B}$. Such a field is referred to as force-free and satisfies

$$\nabla \wedge \mathbf{B} = f(\mathbf{r})\mathbf{B}. \tag{2.71}$$

Taking the divergence of this equation, and using (A3), shows that

$$\mathbf{B} \cdot \nabla f = 0, \tag{2.72}$$

so $f(\mathbf{r})$ is constant along lines of $\mathbf{B}$. The curl of (2.71), together with (A5) and (A7), yields

$$\nabla^2\mathbf{B} + f^2\mathbf{B} = -\nabla f \wedge \mathbf{B}. \tag{2.73}$$

Force-free fields tend to occur when the plasma has low density, $v \ll (B^2/\mu_0\rho)^{1/2}$ and $B^2/2\mu_0 \gg P$, where P is the thermal pressure. In the degenerate case, with $f = 0$, the force-free condition (2.71) reduces to the current-free condition.

2.2.5. THE MAGNETO-FLUID EQUATIONS

The plasma motion is governed by the equations of continuity, momentum and thermal energy. The equation of mass conservation is

$$\nabla \cdot (\rho \mathbf{v}) = -\frac{\partial \rho}{\partial t}. \tag{2.74}$$

It expresses the fact that the density at a point increases if mass flows into the surrounding volume element, while it decreases if there is a positive divergence of mass flux. Expanding the divergence gives

$$\rho \nabla \cdot \mathbf{v} = -\left(\frac{\partial \rho}{\partial t} + \mathbf{v} \cdot \nabla \rho\right),$$

so it follows that

$$\nabla \cdot \mathbf{v} = -\frac{1}{\rho}\frac{d\rho}{dt}, \tag{2.75}$$

where d/dt is the Lagrangian time derivative. In general, a fluid element contracts or expands as it moves with the flow, corresponding to negative and positive $\nabla \cdot \mathbf{v}$ respectively. In the special case when fluid elements conserve their volume, $d\rho/dt = 0$. Equation (2.75) shows that such incompressible flows have $\nabla \cdot \mathbf{v} = 0$.

The equation of motion equates the rate of change of momentum of a fluid element to the sum of the forces acting on it. In an inertial frame, this gives

$$\frac{\partial \mathbf{v}}{\partial t} + (\mathbf{v} \cdot \nabla)\mathbf{v} = -\frac{1}{\rho}\nabla P - \nabla \psi + \frac{1}{\mu_0 \rho}(\nabla \wedge \mathbf{B}) \wedge \mathbf{B} + \frac{1}{\rho}\mathbf{F}_{\mathrm{v}}, \tag{2.76}$$

where P is the pressure, ψ the gravitational potential, and the acceleration of the element is the Lagrangian derivative of $\mathbf{v}$ given by (2.16). The first two forces per unit mass result from the pressure gradient and gravitational field, while the third and fourth are the specific magnetic and viscous forces, respectively.

In a frame rotating with instantaneous angular velocity $\mathbf{\Omega}$ the equation of motion has apparent force terms which are part of the inertial acceleration. A particle has a position in a Cartesian frame given by

$$\mathbf{r} = x_i \hat{\mathbf{x}}_i,$$

where the tensor summation convention is employed. If the rotating frame is used, then in inertial space the unit vector $\hat{\mathbf{x}}_i$ traces a circle about $\boldsymbol{\Omega}$ at a rate

$$\frac{d}{dt}(\hat{\mathbf{x}}_i) = \boldsymbol{\Omega} \wedge \hat{\mathbf{x}}_i.$$

The inertial space particle velocity is therefore

$$\frac{d\mathbf{r}}{dt} = \frac{d}{dt}(x_i\hat{\mathbf{x}}_i) = v_i\hat{\mathbf{x}}_i + x_i\boldsymbol{\Omega} \wedge \hat{\mathbf{x}}_i,$$

where $v_i = \dot{x}_i$ are the velocity components relative to the rotating axes. The inertial acceleration is then

$$\frac{d^2\mathbf{r}}{dt^2} = \dot{v}_i\hat{\mathbf{x}}_i + 2v_i\boldsymbol{\Omega} \wedge \hat{\mathbf{x}}_i + x_i\left[\frac{d\boldsymbol{\Omega}}{dt} \wedge \hat{\mathbf{x}}_i + \boldsymbol{\Omega} \wedge (\boldsymbol{\Omega} \wedge \hat{\mathbf{x}}_i)\right].$$

Hence the accelerations in the inertial and rotating frames are related by

$$\frac{d^2\mathbf{r}}{dt^2} = \ddot{\mathbf{r}} + 2\boldsymbol{\Omega} \wedge \mathbf{v} + \boldsymbol{\Omega} \wedge (\boldsymbol{\Omega} \wedge \mathbf{r}) + \frac{d\boldsymbol{\Omega}}{dt} \wedge \mathbf{r}. \tag{2.77}$$

In a binary star orbital frame the time-scale for changes in $\boldsymbol{\Omega}$ far exceeds any flow time-scales and so the term $(d\boldsymbol{\Omega}/dt) \wedge \mathbf{r}$ is ignorable. This leaves the Coriolis and centrifugal forces. The latter can be expressed as a gradient by noting that

$$(\boldsymbol{\Omega} \wedge \mathbf{r}) \cdot (\boldsymbol{\Omega} \wedge \mathbf{r}) = [(\boldsymbol{\Omega} \wedge \mathbf{r}) \wedge \boldsymbol{\Omega}] \cdot \mathbf{r} = \left[\Omega^2 r^2 - (\boldsymbol{\Omega} \cdot \mathbf{r})^2\right],$$

and hence

$$\begin{aligned}\nabla(|\boldsymbol{\Omega} \wedge \mathbf{r}|^2) &= 2\Omega^2 r \nabla r - 2(\boldsymbol{\Omega} \cdot \mathbf{r})\nabla(\boldsymbol{\Omega} \cdot \mathbf{r}) \\ &= 2\Omega^2 \mathbf{r} - 2(\boldsymbol{\Omega} \cdot \mathbf{r})\boldsymbol{\Omega} \\ &= -2\boldsymbol{\Omega} \wedge (\boldsymbol{\Omega} \wedge \mathbf{r}).\end{aligned}$$

The inertial acceleration of a fluid element can therefore be expressed as

$$\frac{d^2\mathbf{r}}{dt^2} = \ddot{\mathbf{r}} + 2\boldsymbol{\Omega} \wedge \mathbf{v} - \nabla\left(\frac{1}{2}|\boldsymbol{\Omega} \wedge \mathbf{r}|^2\right). \tag{2.78}$$

Substitution of this in the inertial frame equation of motion (2.76) gives the equation of motion in the rotating frame as

$$\begin{aligned}\frac{\partial \mathbf{v}}{\partial t} + (\mathbf{v} \cdot \nabla)\mathbf{v} = &-\frac{1}{\rho}\nabla P - \nabla\left(\psi - \frac{1}{2}|\boldsymbol{\Omega} \wedge \mathbf{r}|^2\right) - 2\boldsymbol{\Omega} \wedge \mathbf{v} \\ &+ \frac{1}{\mu_0\rho}(\nabla \wedge \mathbf{B}) \wedge \mathbf{B} + \frac{1}{\rho}\mathbf{F}_{\mathrm{V}}.\end{aligned} \tag{2.79}$$

The force due to ordinary molecular viscosity is usually too small to be significant. However, the presence of turbulence can lead to a much larger effective viscous force. Turbulence is a field of random macroscopic velocities superposed on a mean laminar flow. The standard picture of hydrodynamic turbulence has an inertial range in which the non-linear inertial terms cause a dissipation-free cascade of energy from the larger to the smaller eddies, until scales are reached at which the micro-viscosity operates. The micro-viscosity fixes the minimum scale at which turbulent motions persist. The time-scale of the dissipative process is that of the cascade, being typically the turn-over time of the larger eddies. The micro-viscosity is replaced by a macro eddy-viscosity acting on the mean flow, with a macro length-scale and a turbulent velocity replacing the mean free path and the thermal velocity, respectively. Hence the turbulent viscous force per unit volume is expressed in the Cartesian tensor form

$$F_{\mathrm{v}i} = \frac{\partial s_{ij}}{\partial x_j}, \tag{2.80a}$$

where

$$s_{ij} = \rho\nu\left(2e_{ij} - \frac{2}{3}\nabla\cdot\mathbf{v}\,\delta_{ij}\right) \tag{2.80b}$$

is the stress tensor, with

$$e_{ij} = \frac{1}{2}\left(\frac{\partial v_i}{\partial x_j} + \frac{\partial v_j}{\partial x_i}\right) \tag{2.80c}$$

the rate of strain tensor. The viscosity coefficient ν is usually expressed as

$$\nu = \frac{1}{3}v_{\mathrm{T}}\lambda_{\mathrm{T}}, \tag{2.81}$$

where v_{T} and λ_{T} are the rms turbulent speed and the mixing length, respectively. This turbulent form of ν has values which exceed those of the molecular form by several orders of magnitude, since λ_{T} greatly exceeds a typical molecular mean free path. The use of (2.80) for $\mathbf{F}_{\mathrm{v}}$ in (2.76) assumes subsonic turbulence, so the associated density fluctuations are modest.

The viscous force can be expressed in vector operator terms by expanding (2.80) and using (A6). This gives, (see Appendix),

$$\mathbf{F}_{\mathrm{v}} = \rho\nu\nabla^2\mathbf{v} - [\nabla^2(\rho\nu)]\mathbf{v} + \nabla\left(\nabla\cdot(\rho\nu\mathbf{v}) - \frac{2}{3}\rho\nu\nabla\cdot\mathbf{v}\right) + \nabla\wedge[\mathbf{v}\wedge\nabla(\rho\nu)]. \tag{2.82}$$

It is often the case in stars and discs that the shears due to differential rotation dominate those due to poloidal flows, so $\mathbf{F}_{\mathrm{v}}$ is principally azimuthal. For an axisymmetric flow, expressed in cylindrical coordinates (ϖ, ϕ, z), with angular velocity $\Omega(\varpi, z)$, (2.82) yields

$$F_{\mathrm{v}\phi} = \frac{1}{\varpi^2}\frac{\partial}{\partial \varpi}\left(\rho\nu\varpi^3\frac{\partial\Omega}{\partial\varpi}\right) + \frac{\partial}{\partial z}\left(\rho\nu\varpi\frac{\partial\Omega}{\partial z}\right). \tag{2.83}$$

The rate of work done, per unit volume, by the viscous force on the fluid at a fixed point is

$$\mathbf{v}\cdot\mathbf{F}_{\mathrm{v}} = v_i\frac{\partial s_{ij}}{\partial x_j} = \frac{\partial}{\partial x_j}(v_i s_{ij}) - s_{ij}\frac{\partial v_i}{\partial x_j}.$$

The first term on the right hand side is the divergence of the viscous stress flow vector $v_i s_{ij}$. This represents an advection of viscous stress. By Gauss' divergence theorem, the volume integral of this term will vanish if the normal component of $v_i s_{ij}$ vanishes on the surface of the region containing the fluid. This is usually the case for astrophysical bodies. The second term represents a dissipation of energy due to the work done by shears in deforming fluid elements, and hence is a source of heat. The magnitude of this dissipation can be written

$$Q_{\mathrm{v}} = s_{ij}\frac{\partial v_i}{\partial x_j} = \frac{1}{2}s_{ij}\left(\frac{\partial v_i}{\partial x_j} + \frac{\partial v_j}{\partial x_i}\right) = s_{ij}e_{ij},$$

where the second equality employs the symmetry of s_{ij}. It then follows from (2.80b,c) that

$$Q_{\mathrm{v}} = 2\rho\nu\left[e_{ij}e_{ij} - \frac{1}{3}(\nabla\cdot\mathbf{v})^2\right]. \tag{2.84}$$

This is the viscous dissipation per unit volume.

The heat equation represents the conservation of thermal energy and is obtained by equating the rate of change of heat for a fluid element to its net energy losses and gains. The first law of thermodynamics gives the rate of change of heat per unit mass in a fluid element as

$$\frac{dQ}{dt} = \frac{dE}{dt} - \frac{P}{\rho^2}\frac{d\rho}{dt}, \tag{2.85}$$

where E is the internal energy per unit mass. In general, E will be a function of P and ρ, so (2.85) gives

$$dQ = P\left(\frac{\partial E}{\partial P}\right)_\rho d\ln P + \rho\left[\left(\frac{\partial E}{\partial \rho}\right)_P - \frac{P}{\rho^2}\right] d\ln\rho. \tag{2.86}$$

Adiabatic exponents Γ_1 and Γ_3 are defined as

$$\Gamma_1 = \left(\frac{d\ln P}{d\ln\rho}\right)_{\rm ad} = \frac{\gamma\rho}{P}\left(\frac{\partial P}{\partial\rho}\right)_T, \tag{2.87a,b}$$

$$\Gamma_3 - 1 = \left(\frac{d\ln T}{d\ln\rho}\right)_{\rm ad} = \frac{1}{\rho c_V}\left(\frac{\partial P}{\partial T}\right)_\rho, \tag{2.88a,b}$$

where $\gamma = c_P/c_V$ with c_P and c_V the specific heat capacities at constant pressure and volume, given by

$$c_V = \left(\frac{\partial E}{\partial T}\right)_\rho, \qquad c_P = \left(\frac{\partial E}{\partial T}\right)_P - \frac{P}{\rho^2}\left(\frac{\partial\rho}{\partial T}\right)_P. \tag{2.89a,b}$$

Setting $dQ = 0$ in (2.86) and using (2.87a) shows that dQ can be expressed as

$$dQ = P\left(\frac{\partial E}{\partial P}\right)_\rho (d\ln P - \Gamma_1 d\ln\rho). \tag{2.90}$$

The derivative $(\partial E/\partial P)_\rho$ can be related to $(\Gamma_3 - 1)$. From an equation of state $P = P(\rho, T)$ it follows that E can be expressed as a function of ρ and T, then setting $dQ = 0$ in (2.85) and using (2.88a) yields

$$\Gamma_3 - 1 = \frac{\rho\left[P/\rho^2 - (\partial E/\partial\rho)_T\right]}{T\,(\partial E/\partial T)_\rho}. \tag{2.91}$$

Since

$$\left(\frac{\partial E}{\partial T}\right)_\rho = \left(\frac{\partial E}{\partial P}\right)_\rho \left(\frac{\partial P}{\partial T}\right)_\rho,$$

(2.91) gives

$$\left(\frac{\partial E}{\partial P}\right)_\rho = \frac{\left[P - \rho^2(\partial E/\partial\rho)_T\right]}{(\Gamma_3 - 1)\rho T\,(\partial P/\partial T)_\rho}. \tag{2.92}$$

For reversible changes $dQ = TdS$, where S is the entropy per unit mass. Hence it follows from (2.85) that

$$dE = TdS + \frac{P}{\rho^2}d\rho,$$

which yields

$$T = \left(\frac{\partial E}{\partial S}\right)_\rho, \quad P = \rho^2\left(\frac{\partial E}{\partial\rho}\right)_S.$$

Then using $E(\rho, T) = E(S(\rho, T), \rho)$ and the Maxwell relation

$$\left(\frac{\partial P}{\partial T}\right)_\rho = -\rho^2 \left(\frac{\partial S}{\partial \rho}\right)_T ,$$

gives

$$\left(\frac{\partial E}{\partial \rho}\right)_T = \frac{P}{\rho^2} - \frac{T}{\rho^2}\left(\frac{\partial P}{\partial T}\right)_\rho .$$

It therefore follows from (2.92) that

$$\left(\frac{\partial E}{\partial P}\right)_\rho = \frac{1}{\rho(\Gamma_3 - 1)}, \tag{2.93}$$

which proves (2.88b), since $c_V = (\partial E/\partial T)_\rho$. Hence (2.90) gives

$$\frac{dQ}{dt} = \frac{P}{(\Gamma_3 - 1)\rho}\left(\frac{d}{dt}\ln P - \Gamma_1 \frac{d}{dt}\ln \rho\right) . \tag{2.94}$$

The heat equation is

$$\rho \frac{dQ}{dt} = \mathfrak{L} , \tag{2.95}$$

where $\mathfrak{L}$ represents heat sources and sinks. Hence (2.17) and (2.94) yield

$$\frac{\partial P}{\partial t} - \frac{\Gamma_1 P}{\rho}\frac{\partial \rho}{\partial t} + \mathbf{v}\cdot\left(\nabla P - \frac{\Gamma_1 P}{\rho}\nabla\rho\right) = (\Gamma_3 - 1)\mathfrak{L}, \tag{2.96a}$$

where

$$\mathfrak{L} = \rho\epsilon + \mu_0 \eta \mathbf{J}^2 + Q_{\mathrm{v}} - \nabla \cdot \mathbf{F}. \tag{2.96b}$$

The first three terms on the right hand side of (2.96b) are due to nuclear sources, dissipation of electric currents and viscous dissipation, the latter being given by (2.84). The last term is the heat flow per unit volume.

The heat flux $\mathbf{F}$ will depend on the energy transport mechanism. In an optically thick region the radiative flux is given by

$$\mathbf{F}_{\mathrm{R}} = -\frac{16\sigma_{\mathrm{B}} T^3}{3\kappa\rho}\nabla T, \tag{2.97}$$

where σ_{B} is the Stefan-Boltzmann constant and κ is the Rosseland mean opacity. When the radiative gradient is unstable convection occurs, leading to a flux

$$\mathbf{F}_{\mathrm{c}} = \rho c_P v_{\mathrm{T}} \lambda_{\mathrm{T}} \Delta\nabla T\, \mathbf{e}, \tag{2.98}$$

where v_T is the rms turbulent velocity, λ_T the mixing length, and $\Delta\nabla T$ is the temperature gradient excess over the adiabatic value (see Cox and Giuli, 1968, for an account of mixing length theory). The unit vector $\mathbf{e}$ is antiparallel to the vertical gravity.

It is noted that for a perfect gas $\Gamma_3 = \gamma$, so (2.93) then integrates to give the thermal energy per unit mass as

$$E = \frac{P}{\rho(\gamma - 1)}. \tag{2.99}$$

Since the thermal energy per unit volume is $\rho E \sim P$, the net heat flow time-scale for a fluid element is $\tau_{\rm th} \sim P/|\mathfrak{L}|$. The dynamical time-scale associated with the fluid motion is $\tau_{\rm dyn} \sim \ell/v$, where ℓ is its length-scale. The ratio of a term on the left of (2.96a) to $|\mathfrak{L}|$ is therefore typically

$$\frac{|\mathbf{v}\cdot\nabla P|}{|\mathfrak{L}|} \sim \frac{vP}{\ell|\mathfrak{L}|} \sim \frac{\tau_{\rm th}}{\tau_{\rm dyn}}. \tag{2.100}$$

If $\tau_{\rm th} \gg \tau_{\rm dyn}$, then net heat flows can be ignored on the dynamical time-scale and the motion is essentially adiabatic. The ratio (2.100) then gives $|\mathbf{v}\cdot\nabla P| \gg |\mathfrak{L}|$ and (2.96a) is satisfied by the vanishing of both its sides. Hence $dQ/dt = 0$ and (2.94) yields

$$\frac{dP}{dt} - \frac{\Gamma_1 P}{\rho}\frac{d\rho}{dt} = 0. \tag{2.101}$$

For constant Γ_1, this integrates to give the adiabatic relation

$$P = K\rho^{\Gamma_1}, \tag{2.102}$$

where K is a constant following the motion.

For a non-degenerate gas the equation of state is

$$P = \frac{\Re}{\mu}\rho T + \frac{4\sigma_{\rm B}}{3c}T^4, \tag{2.103}$$

where $\Re$ and μ are the gas constant and mean molecular weight. The second term represents radiation pressure, which can become important in upper main sequence stars and in the inner parts of some accretion discs. When radiation pressure is ignorable, (2.87b) and (2.103) give $\Gamma_1 = \gamma$. For a fully ionized plasma, μ is given by

$$\frac{1}{\mu} = \frac{3}{2}X + \frac{1}{4}Y + \frac{1}{2}, \tag{2.104}$$

where X and Y are the fractions by mass of hydrogen and helium, respectively (e.g. Cox and Giuli, 1968).

2.2.6. HYDROMAGNETIC WAVES

Disturbances introduced into a conducting fluid containing a magnetic field are propagated away by the stresses in the field and fluid. If a stable equilibrium state exists, perturbations propagate in the form of hydromagnetic waves. When gravitational and viscous forces are ignorable, and adiabatic motions are considered, (2.76) and (2.96) become

$$\frac{\partial \mathbf{v}}{\partial t} + (\mathbf{v}\cdot\nabla)\mathbf{v} = -\frac{1}{\rho}\nabla P + \frac{1}{\mu_0\rho}(\nabla\wedge\mathbf{B})\wedge\mathbf{B}, \tag{2.105}$$

$$\frac{\partial P}{\partial t} - \frac{\gamma P}{\rho}\frac{\partial \rho}{\partial t} + \mathbf{v}\cdot\left(\nabla P - \frac{\gamma P}{\rho}\nabla\rho\right) = 0, \tag{2.106}$$

where $\Gamma_1 = \gamma$ for a perfect gas. The simplest wave solutions arise in the ideal conductivity case, in which (2.47) applies. If the properties of the undisturbed medium vary with a length-scale ℓ, then these can be considered essentially uniform over a length-scale $\lambda_0 \ll \ell$. Wave solutions with wavelengths $\lambda \lesssim \lambda_0$ will be considered here.

The equilibrium state has $\mathbf{v} = \mathbf{0}$ and a uniform magnetic field $\mathbf{B}_0$. It has uniform density $\rho = \rho_0$ and sound speed $c_s = (dP/d\rho)^{1/2}$. The magnetic field in the perturbed state can be written

$$\mathbf{B} = \mathbf{B}_0 + \mathbf{B}_1(\mathbf{r}, t), \tag{2.107}$$

so $\mathbf{B}_1$ is the field perturbation. Similar expressions hold for other quantities. The Alfvén velocity for the medium is defined as

$$\mathbf{v}_A = \frac{\mathbf{B}_0}{(\mu_0\rho)^{\frac{1}{2}}}. \tag{2.108}$$

Consider, first, a medium in which the sound speed c_s is much greater than v_A and v. The motions are then essentially incompressible, so that $\nabla\cdot\mathbf{v} = 0$. Defining

$$\mathbf{w} = \frac{\mathbf{B}_1}{(\mu_0\rho)^{\frac{1}{2}}}, \tag{2.109}$$

(2.66), (2.105), (2.47) and (2.107) give

$$\frac{\partial \mathbf{v}}{\partial t} + (\mathbf{v}\cdot\nabla)\mathbf{v} - [(\mathbf{v}_A + \mathbf{w})\cdot\nabla]\mathbf{w} = -\nabla\left(\frac{P}{\rho} + \frac{B^2}{2\mu_0\rho}\right), \tag{2.110}$$

$$\frac{\partial \mathbf{w}}{\partial t} + (\mathbf{v}\cdot\nabla)\mathbf{w} - [(\mathbf{v}_A + \mathbf{w})\cdot\nabla]\mathbf{v} = \mathbf{0}, \tag{2.111}$$

where ρ is uniform. These equations become identical when

$$P + \frac{1}{2}\rho(\mathbf{v}_{\mathrm{A}} + \mathbf{w})^2 = C, \tag{2.112}$$

with C a constant, so the thermal and magnetic pressure gradients cancel, and

$$\mathbf{v} = \pm\mathbf{w}. \tag{2.113}$$

Then $\mathbf{v}$ satisfies the equations

$$\frac{\partial \mathbf{v}}{\partial t} \mp (\mathbf{v}_{\mathrm{A}} \cdot \nabla)\mathbf{v} = \mathbf{0}, \tag{2.114}$$

which have the general solutions

$$\mathbf{v} = \mathbf{v}(\mathbf{r} \pm \mathbf{v}_{\mathrm{A}} t). \tag{2.115}$$

These represent arbitrary waveforms moving with the Alfvén velocity $\mathbf{v}_{\mathrm{A}}$, defined by (2.108). Such Alfvén waves are non-linear solutions, since $|\mathbf{B}_1|$ does not necessarily have to be small relative to $|\mathbf{B}_0|$. It can be shown that the wave has a net energy transport given by the flux

$$\left(\frac{1}{2}\rho v^2 + \frac{B_1^2}{2\mu_0}\right)\mathbf{v}_{\mathrm{A}} = \rho v^2 \mathbf{v}_{\mathrm{A}}. \tag{2.116}$$

Kinetic and magnetic energies are transported along $\mathbf{B}_0$ at the Alfvén speed v_{A}, with equal contributions.

Consider, now, the more general case in which c_{s} is not necessarily large compared to v_{A} and v, so compressible motions can occur. Take small perturbations about the equilibrium state, so the equations can be linearized. Equations (2.105), (2.47), (2.74) and (2.106) then respectively yield

$$\frac{\partial \mathbf{v}}{\partial t} = -\frac{1}{\rho_0}\nabla P_1 + \frac{1}{\mu_0 \rho_0}(\nabla \wedge \mathbf{B}_1) \wedge \mathbf{B}_0, \tag{2.117}$$

$$\frac{\partial \mathbf{B}_1}{\partial t} = \nabla \wedge (\mathbf{v} \wedge \mathbf{B}_0), \tag{2.118}$$

$$\frac{\partial \rho_1}{\partial t} = -\rho_0 \nabla \cdot \mathbf{v}, \tag{2.119}$$

$$\frac{\partial P_1}{\partial t} = c_{\mathrm{s}}^2 \frac{\partial \rho_1}{\partial t}, \tag{2.120}$$

where

$$c_s^2 = \frac{\gamma P_0}{\rho_0}, \tag{2.121}$$

and the equilibrium state is uniform. Differentiating (2.117) with respect to time, and using (2.118)–(2.120), gives the linear wave equation

$$\frac{\partial^2 \mathbf{v}}{\partial t^2} = c_s^2 \nabla(\nabla \cdot \mathbf{v}) + \frac{1}{\mu_0 \rho_0} (\nabla \wedge [\nabla \wedge (\mathbf{v} \wedge \mathbf{B}_0)]) \wedge \mathbf{B}_0. \tag{2.122}$$

For an unbounded medium, plane wave solutions can be sought with the form

$$\mathbf{v} = \mathbf{u} \exp(i\mathbf{k} \cdot \mathbf{r} - i\omega t), \tag{2.123}$$

where $\mathbf{u}$ is a constant vector. The operator ∇ can then be replaced by $\mathbf{ik}$, so (2.122) and (2.123) yield the time-independent equation

$$\omega^2 \mathbf{u} = c_s^2 (\mathbf{k} \cdot \mathbf{u}) \mathbf{k} + \frac{1}{\mu_0 \rho_0} (\mathbf{k} \wedge [\mathbf{k} \wedge (\mathbf{u} \wedge \mathbf{B}_0)]) \wedge \mathbf{B}_0,$$

which, using (A1), can be written

$$\begin{aligned} \omega^2 \mathbf{u} = v_A^2 \Big[& (\mathbf{k} \cdot \hat{\mathbf{B}}_0)^2 \mathbf{u} - (\mathbf{k} \cdot \mathbf{u})(\mathbf{k} \cdot \hat{\mathbf{B}}_0) \hat{\mathbf{B}}_0 \\ & + \{(1 + c_s^2/v_A^2)(\mathbf{k} \cdot \mathbf{u}) - (\mathbf{k} \cdot \hat{\mathbf{B}}_0)(\hat{\mathbf{B}}_0 \cdot \mathbf{u})\} \mathbf{k} \Big], \end{aligned} \tag{2.124}$$

where $\hat{\mathbf{B}}_0 = \mathbf{B}_0 / |\mathbf{B}_0|$ and v_A is given by (2.108). Taking the scalar product of (2.124) with $\mathbf{k}$ and then with $\hat{\mathbf{B}}_0$ gives

$$(-\omega^2 + k^2 c_s^2 + k^2 v_A^2)(\mathbf{k} \cdot \mathbf{u}) = k^3 v_A^2 \cos\theta_B (\hat{\mathbf{B}}_0 \cdot \mathbf{u}), \tag{2.125}$$

$$k c_s^2 \cos\theta_B (\mathbf{k} \cdot \mathbf{u}) = \omega^2 (\hat{\mathbf{B}}_0 \cdot \mathbf{u}), \tag{2.126}$$

where

$$\cos\theta_B = \hat{\mathbf{k}} \cdot \hat{\mathbf{B}}_0. \tag{2.127}$$

An incompressible solution exists with $\nabla \cdot \mathbf{v} = 0$ and hence $\mathbf{k} \cdot \mathbf{u} = 0$. Equations (2.125) and (2.126) show that for this mode

$$\hat{\mathbf{B}}_0 \cdot \mathbf{u} = 0, \tag{2.128}$$

so the fluid velocity is perpendicular to the equilibrium magnetic field. Equation (2.124) then yields the dispersion relation

$$\omega = \pm v_{\rm A}\hat{\mathbf{B}}_0 \cdot \mathbf{k} = \pm k v_{\rm A} \cos\theta_B, \tag{2.129}$$

where the different signs of ω correspond to oppositely directed waves, when substituted in (2.123). This is the linear form of the foregoing Alfvén wave solution. The phase speed is

$$v_p = \frac{\omega}{k} = \pm v_{\rm A} \cos\theta_B, \tag{2.130}$$

so propagation is not possible in directions perpendicular to $\mathbf{B}_0$. The group velocity follows from (2.129) as

$$\mathbf{v}_{\rm g} = \frac{\partial\omega}{\partial\mathbf{k}} = \pm v_{\rm A}\hat{\mathbf{B}}_0, \tag{2.131}$$

showing that the energy is carried along $\mathbf{B}_0$ at the Alfvén velocity $\mathbf{v}_{\rm A}$, consistent with the non-linear result (2.116). Equations (2.118) and (2.123) give

$$-\omega\mathbf{B}_1 = \mathbf{k} \wedge (\mathbf{v} \wedge \mathbf{B}_0) = (\mathbf{k} \cdot \mathbf{B}_0)\mathbf{v},$$

since $\mathbf{k} \cdot \mathbf{v} = 0$. Use of (2.129) for ω then shows

$$\mathbf{v} = \mp\frac{\mathbf{B}_1}{(\mu_0\rho_0)^{\frac{1}{2}}}, \tag{2.132}$$

in agreement with (2.113). It follows from (2.128) and (2.132) that

$$\mathbf{B}_0 \cdot \mathbf{B}_1 = 0. \tag{2.133}$$

The first order magnetic force for this mode is therefore

$$\mathbf{J}_1 \wedge \mathbf{B}_0 = \frac{1}{\mu_0}(\mathbf{B}_0 \cdot \nabla)\mathbf{B}_1 = \frac{i}{\mu_0}(\mathbf{B}_0 \cdot \mathbf{k})\mathbf{B}_1, \tag{2.134}$$

illustrating that the restoring force is purely due to magnetic tension in this incompressible transverse Alfvén mode.

Equations (2.125) and (2.126) yield two compressible modes, for which $\mathbf{k} \cdot \mathbf{u} \neq 0$. Elimination of $(\mathbf{k} \cdot \mathbf{u})/(\hat{\mathbf{B}}_0 \cdot \mathbf{u})$ between these equations leads to the dispersion relation

$$\omega^4 - (c_{\rm s}^2 + v_{\rm A}^2)k^2\omega^2 + c_{\rm s}^2 v_{\rm A}^2 k^4 \cos^2\theta_B = 0. \tag{2.135}$$

The two solutions for ω^2 have phase speeds

$$v_f = \left[\frac{1}{2}(c_{\rm S}^2 + v_{\rm A}^2) + \frac{1}{2}(c_{\rm S}^4 + v_{\rm A}^4 - 2c_{\rm S}^2 v_{\rm A}^2 \cos 2\theta_B)^{\frac{1}{2}}\right]^{\frac{1}{2}}, \tag{2.136}$$

$$v_{\rm S} = \left[\frac{1}{2}(c_{\rm S}^2 + v_{\rm A}^2) - \frac{1}{2}(c_{\rm S}^4 + v_{\rm A}^4 - 2c_{\rm S}^2 v_{\rm A}^2 \cos 2\theta_B)^{\frac{1}{2}}\right]^{\frac{1}{2}}, \tag{2.137}$$

and are referred to as the fast and slow magnetosonic modes, respectively. The Alfvén mode phase speed lies between v_f and $v_{\rm S}$, and hence this mode is sometimes referred to as the intermediate mode. It can be shown that the fast mode propagates most energy in directions near to that of the wave vector $\mathbf{k}$, while the slow mode transports most energy in directions nearer to $\hat{\mathbf{B}}_0$.

The importance of the pressure gradient and magnetic force terms in the momentum equation (2.76) can be gauged by defining the dimensionless numbers

$$\mathfrak{M}_{\rm S} = \frac{v_0}{c_{\rm S}}, \qquad \mathfrak{M}_{\rm A} = \frac{v_0}{v_{\rm A}}, \tag{2.138a,b}$$

where v_0 is a characteristic fluid speed. If the characteristic length-scale for variations in the medium is ℓ, then the ratio of the inertial term to the thermal pressure term in the momentum equation is

$$\frac{|(\mathbf{v}\cdot\nabla)\mathbf{v}|}{|\nabla P|/\rho} \sim \frac{v_0^2}{P/\rho} \sim \left(\frac{v_0}{c_{\rm S}}\right)^2 = \mathfrak{M}_{\rm S}^2. \tag{2.139}$$

For highly supersonic flow $\mathfrak{M}_{\rm S} \gg 1$, and hence the pressure term is ignorable. The physical reason for this is that the sound speed determines how fast pressure disturbances propagate through the gas. The magnitude of $c_{\rm S}$ therefore limits the rapidity with which the flowing gas can respond to pressure changes. For $\mathfrak{M}_{\rm S} \gg 1$, the flow time-scale, ℓ/v_0, is much shorter than the sound travel time, $\ell/c_{\rm S}$, so the gas does not have time to respond to pressure changes. In energy terms, the kinetic energy density, $\rho v^2/2$, far exceeds the thermal energy density, $\sim P$, when $\mathfrak{M}_{\rm S} \gg 1$, so the flow is not significantly affected by the pressure.

A similar argument applies for the effect of the magnetic force term in (2.76). Then

$$\frac{|(\mathbf{v}\cdot\nabla)\mathbf{v}|}{|(\nabla\wedge\mathbf{B})\wedge\mathbf{B}|/\mu_0\rho} \sim \frac{v_0^2}{B^2/\mu_0\rho} = \left(\frac{v_0}{v_{\rm A}}\right)^2 = \mathfrak{M}_{\rm A}^2, \tag{2.140}$$

using (2.108) for the Alfvén speed $v_{\rm A}$. When $\mathfrak{M}_{\rm A} \gg 1$, the flow is highly super-Alfvénic and the gas does not have time to respond to spatial changes in the magnetic field (i.e. to the $\mathbf{J} \wedge \mathbf{B}$ force), since the Alfvén travel time, ℓ/v_A, is much longer than the flow time, ℓ/v_0. The effect of the magnetic force is therefore ignorable in such flows. The velocity is then determined by the non-magnetic equations and this $\mathbf{v}$ can be used in the induction equation to determine $\mathbf{B}$. The hydrodynamic and hydromagnetic problems are essentially decoupled in such kinematic situations.

2.2.7. THE FREE-FALL ALFVÉN RADIUS

An estimate of where the magnetic field becomes dynamically significant (i.e. $\mathfrak{M}_A \sim 1$) for accretion onto a magnetized star can be made for the case of spherically symmetric, radial inflow. The simplest case is that without significant thermal pressure support, so the resulting supersonic radial velocity has a free-fall variation with distance r from the centre of the accretor.

Taking the star to have a dipole magnetic field, and ignoring angular variations, gives

$$B \simeq \frac{B_0 R_p^3}{r^3},$$

where R_p is the stellar radius and B_0 the surface field. The free-fall radial velocity has magnitude

$$|v_r| = \left(\frac{2GM_p}{r}\right)^{\frac{1}{2}},$$

where M_p is the stellar mass. Ignoring field distortion, an Alfvén radius r_A can be defined by equating the kinetic and magnetic energy densities, so

$$\rho v_r^2 = \frac{B^2}{\mu_0},$$

where $|v_r| = v_A$. The above equations, together with

$$\rho|v_r| = \frac{\dot{M}_p}{4\pi r^2},$$

where $\dot{M}_p$ is the accretion rate, yield

$$r_A = \left(\frac{8\pi^2}{\mu_0^2 G}\right)^{\frac{1}{7}} \frac{B_0^{\frac{4}{7}} R_p^{\frac{12}{7}}}{M_p^{\frac{1}{7}} \dot{M}_p^{\frac{2}{7}}}. \tag{2.141}$$

Normalizing with respect to parameters typical of accreting magnetic neutron stars, gives

$$r_{\mathrm{A}} = 3.7 \times 10^{6} \frac{\left(\frac{B_0}{10^{12}\,\mathrm{G}}\right)^{\frac{4}{7}} \left(\frac{R_{\mathrm{p}}}{10^{4}\,\mathrm{m}}\right)^{\frac{12}{7}}}{\left(\frac{M_{\mathrm{p}}}{1.4\,M_{\odot}}\right)^{\frac{1}{7}} \left(\frac{\dot{M}_{\mathrm{p}}}{10^{-9}\,M_{\odot}\,\mathrm{yr}^{-1}}\right)^{\frac{2}{7}}}\ \mathrm{m}. \tag{2.142}$$

For highly conducting material, this gives an estimate of the radius at which the stellar magnetic field starts to affect the accretion flow when this has velocities close to free-fall values.

2.2.8. POLOIDAL AND TOROIDAL REPRESENTATION OF B

The magnetic field, being divergenceless, can be separated into poloidal and toroidal parts as

$$\mathbf{B} = \mathbf{B}_p + \mathbf{B}_T = \nabla \wedge [\nabla \wedge (\Phi_p \hat{\mathbf{r}})] + \nabla \wedge (\Phi_T \hat{\mathbf{r}}), \tag{2.143}$$

where Φ_p and Φ_T are the poloidal and toroidal scalars.

In spherical polar coordinates (r, θ, ϕ) the components of the poloidal and toroidal fields are

$$B_{pr} = \frac{1}{r^2} L^2 \Phi_p, \tag{2.144a}$$

$$B_{p\theta} = \frac{1}{r} \frac{\partial^2 \Phi_p}{\partial r \partial \theta}, \tag{2.144b}$$

$$B_{p\phi} = \frac{1}{r \sin\theta} \frac{\partial^2 \Phi_p}{\partial r \partial \phi}, \tag{2.144c}$$

and

$$B_{Tr} = 0, \tag{2.145a}$$

$$B_{T\theta} = \frac{1}{r \sin\theta} \frac{\partial \Phi_T}{\partial \phi}, \tag{2.145b}$$

$$B_{T\phi} = -\frac{1}{r} \frac{\partial \Phi_T}{\partial \theta}, \tag{2.145c}$$

where

$$L^2 = -\frac{1}{\sin\theta} \frac{\partial}{\partial \theta} \left(\sin\theta \frac{\partial}{\partial \theta}\right) - \frac{1}{\sin^2\theta} \frac{\partial^2}{\partial \phi^2}. \tag{2.146}$$

It is noted that these fields are generalizations of the canonical axisymmetric poloidal and toroidal fields. Equations (2.144) and (2.145) show that only in an axisymmetric case will $\mathbf{B}_p$ lie in a meridional plane and $\mathbf{B}_T$ be purely azimuthal. Nevertheless, the generalized fields have similar useful analytic properties to the axisymmetric fields.

It is seen from (2.143) that the curl of a toroidal vector is a poloidal vector, the converse also being true since

$$\nabla \wedge \mathbf{B}_p = \nabla \wedge \left[\left\{ -r\nabla^2 \left(\frac{\Phi_p}{r} \right) \right\} \hat{\mathbf{r}} \right].$$

The current density can therefore be written as

$$\mathbf{J} = \mathbf{J}_p + \mathbf{J}_T = \frac{1}{\mu_0} \nabla \wedge [\nabla \wedge (\Phi_T \hat{\mathbf{r}})] + \frac{1}{\mu_0} \nabla \wedge \left[\left\{ -r\nabla^2 \left(\frac{\Phi_p}{r} \right) \right\} \hat{\mathbf{r}} \right]. \quad (2.147)$$

In an insulator this shows, in general, that

$$\nabla^2 \left(\frac{\Phi_p}{r} \right) = 0, \qquad \Phi_T = 0. \quad (2.148a,b)$$

Since

$$\nabla^2 = \frac{1}{r^2} \frac{\partial}{\partial r} \left(r^2 \frac{\partial}{\partial r} \right) - \frac{1}{r^2} L^2, \quad (2.149)$$

it follows from (2.69), (2.144) and (2.148) that in an insulator **B** can be expressed as

$$\mathbf{B} = \nabla \left(\frac{\partial \Phi_p}{\partial r} \right). \quad (2.150)$$

It is often convenient to express Φ_p and Φ_T as expansions in radial functions and spherical harmonics, so

$$\Phi_p = \sum_{l=1}^{\infty} \sum_{m=-l}^{l} U_{lm}(r) Y_l^m(\theta, \phi) f_{lm}(t), \quad (2.151)$$

$$\Phi_T = \sum_{l=1}^{\infty} \sum_{m=-l}^{l} V_{lm}(r) Y_l^m(\theta, \phi) g_{lm}(t), \quad (2.152)$$

where $f_{lm}(t)$ and $g_{lm}(t)$ are time dependencies appropriate to the specific problem.

Spherical harmonics are eigenfunctions of L^2 since

$$L^2 Y_l^m(\theta,\phi) = l(l+1)Y_l^m(\theta,\phi), \tag{2.153}$$

(see Appendix). They are given by

$$Y_l^m(\theta,\phi) = P_l^{|m|}(\theta)e^{im\phi}, \tag{2.154}$$

where

$$P_l^{|m|}(\mu) = \frac{(1-\mu^2)^{\frac{|m|}{2}}}{2^l l!}\left(\frac{d}{d\mu}\right)^{l+|m|}(\mu^2-1)^l, \tag{2.155}$$

with $\mu = \cos\theta$, are associated Legendre functions obeying (A25) and having the orthogonality relation (A28). If the problem has an appropriate small dimensionless parameter, finite expansions involving the first few terms in (2.151) and (2.152) can be used.

2.2.9. DECAY MODES OF **B** IN A CONDUCTOR

In a plasma with no significant motions and a constant magnetic diffusivity, **B** obeys the diffusion equation (2.55)

$$\eta\nabla^2\mathbf{B} = \frac{\partial\mathbf{B}}{\partial t}.$$

In the absence of an externally applied field, **B** decays on the time-scale τ_d, given by (2.57). Exact solutions of (2.55) can be found for decay modes of **B** in a spherical conductor. The magnetic field can be split into poloidal and toroidal parts, as in (2.143). If the external medium is taken as a vacuum, only the poloidal field extends beyond the sphere, so the decay of this component will be considered.

In spherical coordinates (r,θ,ϕ) the poloidal scalar can be expressed in harmonics of the form

$$\Phi_p = U_l(r)Y_l^m(\theta,\phi)\exp(-\lambda^2 t), \tag{2.156}$$

where λ^2 is the decay rate to be determined. The poloidal magnetic field is

$$\mathbf{B}_p = \nabla\wedge[\nabla\wedge(\Phi_p\hat{\mathbf{r}})]. \tag{2.157}$$

Substituting (2.156) and (2.157) in (2.55) shows that $U_l(r)$ must satisfy

$$\frac{d^2U_l}{dr^2} + \left[\frac{\lambda^2}{\eta} - \frac{l(l+1)}{r^2}\right]U_l = 0. \tag{2.158}$$

The solution of this equation, free of singularity at $r = 0$, is

$$U_l(r) = r^{\frac{1}{2}} J_{l+\frac{1}{2}}\left(\frac{\lambda}{\eta^{\frac{1}{2}}} r\right), \tag{2.159}$$

where $J_{l+1/2}$ is a Bessel function of the first kind of order $l + 1/2$.

In the surrounding vacuum $\eta \to \infty$ and the solution of (2.158) which vanishes as $r \to \infty$ is

$$U_l(r) = \frac{A}{r^l}, \tag{2.160}$$

where A is a constant. It follows from (2.144) that the continuity of $\mathbf{B}$ at the surface requires U_l and dU_l/dr to be continuous at $r = R_s$. Applying these conditions, using (2.159) and (2.160) and eliminating A, gives

$$R_s J'_{l+\frac{1}{2}}\left(\frac{\lambda}{\eta^{\frac{1}{2}}} R_s\right) + \left(l + \tfrac{1}{2}\right) J_{l+\frac{1}{2}}\left(\frac{\lambda}{\eta^{\frac{1}{2}}} R_s\right) = 0, \tag{2.161}$$

where the prime denotes differentiation with respect to r. Using the recurrence relation

$$x\frac{dJ_\nu}{dx} + \nu J_\nu(x) = x J_{\nu-1}(x),$$

with $x = (\lambda/\eta^{1/2})r$, (2.161) written as

$$J_{l-\frac{1}{2}}\left(\frac{\lambda}{\eta^{\frac{1}{2}}} R_s\right) = 0.$$

Denoting the jth zero of $J_{l-\frac{1}{2}}$ by α_{lj}, (2.159) becomes

$$U_{lj}(r) = r^{\frac{1}{2}} J_{l+\frac{1}{2}}\left(\alpha_{lj}\frac{r}{R_s}\right). \tag{2.162}$$

The corresponding decay times of these modes are λ^{-2} so (2.159) and (2.162) give

$$\tau_{lj} = \frac{R_s^2}{\alpha_{lj}^2 \eta}. \tag{2.163}$$

The length-scales of the corresponding magnetic fields are R_s/α_{lj}. The higher modes have shorter length-scales and hence higher field derivatives, resulting in larger values of $|\mathbf{J}|$. They therefore dissipate more rapidly, in accordance with (2.56), and consequently have shorter decay times.

2.2.10. MAGNETIC DIFFUSIVITY

The electrical conductivity of a fully-ionized, collision-dominated plasma is given by Spitzer (1962) as

$$\sigma = 1.5 \times 10^{-2} \frac{T^{\frac{3}{2}}}{\ln \Lambda} \text{ (ohm-metre)}^{-1}. \tag{2.164}$$

The Coulomb logarithm is typically $5 < \ln \Lambda < 20$, having a weak dependence on density and temperature. The magnetic diffusivity $\eta = 1/\mu_0 \sigma$ and hence

$$\eta_{\text{ohm}} = 5.2 \times 10^7 \ln \Lambda \, T^{-\frac{3}{2}} \, \text{m}^2\text{s}^{-1}. \tag{2.165}$$

As discussed in §2.2.1, the conductivity becomes anisotropic if $\omega_\text{e} \tau_\text{ei} \gg 1$, where ω_e is the electron gyro-frequency and τ_ei the collision time for scattering of electrons by ions.

When the hydrogen plasma is partially ionized (2.165) for η_ohm should be multiplied by $(1 + \tau_\text{ei}/\tau_\text{en})$, where the ratio of effective electron collision times with ions and neutral particles is

$$\frac{\tau_\text{ei}}{\tau_\text{en}} = 5.2 \times 10^{-11} \frac{n_\text{n}}{n_\text{e}} \frac{T^2}{\ln \Lambda}, \tag{2.166}$$

with n_n and n_e being the number densities of neutral particles and electrons, respectively. The diffusivity η_ohm is used for non-turbulent plasmas.

The secondary stars in close binaries have masses $\lesssim 0.7 M_\odot$ and hence possess deep convective envelopes. Equation (2.165) is then inappropriate for η, since the presence of turbulence is believed to greatly enhance the diffusion of magnetic fields in plasmas. Due to the absence of a rigorous theory of stellar turbulence, the mixing length approach is usually adopted to estimate η. If the turbulent eddies have an rms speed v_T and a length-scale λ_T, an eddy magnetic diffusivity can be defined as

$$\eta_\text{T} = \frac{1}{3} v_\text{T} \lambda_\text{T}. \tag{2.167}$$

This is analogous to the eddy viscosity coefficient given by (2.81). The form (2.167) for η_T arises when the cascade picture of hydrodynamic turbulence is combined with field freezing, provided the field does not react back on the dynamics of the turbulence. For a perfect gas equation of state, the convective heat flux is given by

$$F_\text{c} = 10 \rho v_\text{T}^3 \frac{\lambda_P}{\lambda_\text{T}}, \tag{2.168}$$

(Cox and Giuli, 1968) where λ_P is the pressure scale height. It follows that

$$v_{\mathrm{T}} = \left(\frac{\lambda_{\mathrm{T}}}{\lambda_P} \frac{L}{40\pi r^2 \rho} \right)^{\frac{1}{3}}, \tag{2.169}$$

where L is the luminosity and the star is taken as non-rotating. The mixing length is usually taken to be λ_P in the main body of the star. In a fully convective star a different form for λ_{T} must be used near the stellar centre, since λ_P diverges there.

The secondary stars in close binary systems are expected to be synchronized with the orbital motion, due to tidal effects (e.g. Campbell and Papaloizou, 1983). Their rotation periods are therefore $P \lesssim 7\,\mathrm{hrs}$, while a typical convective time-scale in the stellar interior is $\tau_{\mathrm{T}} \sim \lambda_P/v_{\mathrm{T}} \sim 4$ months. Since $P \ll \tau_{\mathrm{T}}$ in the inner regions some modification should be made in the calculation of η_{T}, to estimate the effect of rotation on convection. A Rossby number can be defined, as a measure of the effect of rotation on the turbulence, as

$$R_{\mathrm{T}} = \frac{v_{\mathrm{T}}}{\ell \Omega} \sim \frac{|(\mathbf{v}_{\mathrm{T}} \cdot \nabla)\mathbf{v}_{\mathrm{T}}|}{|2\mathbf{\Omega} \wedge \mathbf{v}_{\mathrm{T}}|} \sim \frac{P}{\tau_{\mathrm{T}}}, \tag{2.170}$$

where $\mathbf{v}_{\mathrm{T}}$ is measured in the orbital frame. Goldreich and Keeley (1977) adopt the local formula

$$\eta_{\mathrm{T}} = \eta_0 G_{\mathrm{R}}, \tag{2.171}$$

where

$$G_{\mathrm{R}} = \begin{cases} 1, & R_{\mathrm{T}} > 1, \\ R_{\mathrm{T}}^2, & R_{\mathrm{T}} < 1. \end{cases}$$

Both v_{T} and λ_{T} are modified and (2.169) gives

$$\eta_0 = \frac{1}{3} \lambda_P \left(\frac{\lambda_{\mathrm{T}}}{\lambda_P} \right)^{\frac{4}{3}} \left(\frac{L}{40\pi r^2 \rho} \right)^{\frac{1}{3}}. \tag{2.172}$$

In turbulent accretion discs a simple prescription similar to (2.167) is often used for η_{T}, with v_{T} being a fraction of the sound speed and λ_{T} a fraction of the disc height h. Hence

$$\eta_{\mathrm{T}} = \epsilon c_{\mathrm{s}} h, \tag{2.173}$$

with $\epsilon < 1$. A modification similar to (2.171) can be made for the effect of rotation.

Magnetic buoyancy is believed to play a role in the diffusion of the field in stars and discs. The basic mechanism can be understood by considering a horizontal flux tube with uniform field $\mathbf{B}$ surrounded by a gas with a locally weak magnetic field. The vertical gravitational field $\mathbf{g}$ is taken to be uniform, and the medium is assumed to be isothermal with temperature T. If $P_{\rm in}$ and $P_{\rm ex}$ are the thermal pressures in the tube and in the exterior medium, respectively, horizontal force balance across $\mathbf{B}$ yields

$$P_{\rm ex} = P_{\rm in} + \frac{B^2}{2\mu_0}.$$

The gas equation and isothermality then give

$$\frac{\rho_{\rm ex} - \rho_{\rm in}}{\rho_{\rm ex}} = \frac{B^2}{2\mu_0 P_{\rm ex}}. \tag{2.174}$$

This implies $\rho_{\rm in} < \rho_{\rm ex}$ so the tube experiences a buoyancy force per unit volume of

$$F_{\rm B} = (\rho_{\rm ex} - \rho_{\rm in})g = \frac{g\rho_{\rm ex}B^2}{2\mu_0 P_{\rm ex}}. \tag{2.175}$$

As the tube rises it will become curved and feel a force due to magnetic tension of $\sim B^2/\mu_0 R_c$, where R_c is its characteristic radius of curvature. This will be small compared to $F_{\rm B}$ if

$$R_c > \frac{B^2}{\mu_0 g(\rho_{\rm ex} - \rho_{\rm in})} = \frac{2P_{\rm ex}}{g\rho_{\rm ex}} = \frac{2P_{\rm ex}}{|dP_{\rm ex}/dz|},$$

where the last equality uses vertical equilibrium in the surroundings. Hence magnetic tension is small if $R_c > 2\lambda_P$, where λ_P is the pressure scale height. In the presence of turbulence the rising tube will experience a frictional force opposing $F_{\rm B}$. Writing the resultant force density as $\zeta F_{\rm B}$, where $\zeta < 1$, the tube velocity $v_{\rm B}$ attained in moving a distance ℓ can be found by equating the kinetic energy gained to the work done, so

$$\frac{1}{2}\rho_{\rm in} v_{\rm B}^2 \simeq \zeta F_{\rm B}\ell = \frac{\zeta g \ell \rho_{\rm ex} B^2}{2\mu_0 P_{\rm ex}} = \frac{\ell}{\lambda_P}\frac{\zeta B^2}{2\mu_0},$$

giving

$$v_{\rm B} \simeq \left(\frac{\ell}{\lambda_P}\zeta\right)^{\frac{1}{2}} \left(\frac{B^2}{\mu_0 \rho_{\rm in}}\right)^{\frac{1}{2}} = \left(\frac{\ell}{\lambda_P}\zeta\right)^{\frac{1}{2}} v_{\rm A}, \tag{2.176}$$

where the Alfvén speed $v_{\mathrm{A}} = (B^2/\mu_0\rho_{\mathrm{in}})^{1/2}$. Taking the magnetic field to be dissipated over the length-scale $\ell \sim \lambda_P$, which has an associated tube velocity $v_{\mathrm{B}} \simeq \sqrt{\zeta} v_{\mathrm{A}}$, leads to a magnetic buoyancy transport coefficient

$$\eta_{\mathrm{B}} = v_{\mathrm{B}}\lambda_P = \frac{\xi|B_\perp|}{(\mu_0\rho)^{\frac{1}{2}}}\lambda_P, \tag{2.177}$$

where $\xi = \sqrt{\zeta}$, the subscript has been dropped from ρ_{in}, and $B_\perp$ is the horizontal field. The field is dissipated via small-scale turbulent reconnection causing friction between rising tubes. The detailed magnetic buoyancy instability was analysed by Parker (1979).

Usually η_{T} and η_{B} are several orders of magnitude larger than η_{ohm}.

2.3. Basic Dynamo Theory

2.3.1. FIELD GENERATION

The induction equation, given by (2.44), describes the evolution of a magnetic field in a plasma with finite conductivity and fluid motions. In the absence of significant motions (i.e. $R_{\mathrm{m}} \ll 1$), the diffusive limit (2.54) holds. Without an externally imposed field, this equation has decaying solutions with $|\mathbf{B}|$ decreasing on the diffusion time-scale τ_{d}, given by (2.57). It is often the case with astrophysical bodies that τ_{d} is shorter than their age, so internal motions are necessary to explain observed magnetic fields which would otherwise have decayed to negligible values. This is especially true in turbulent bodies in which the diffusion of $\mathbf{B}$ is believed to be greatly enhanced compared to that in ohmic conductors, as discussed above.

This situation results in the need for a dynamo mechanism to generate $\mathbf{B}$ and counter its dissipation. It was seen that the $\mathbf{v}\wedge\mathbf{B}$ term in (2.44) by itself (i.e. $R_{\mathrm{m}} \gg 1$) corresponds to the plasma being threaded on the field lines. With significant diffusion (i.e. $R_{\mathrm{m}} \sim 1$) a fluid element can move across field lines, but there is still some freezing effect with motions affecting the field structure. The induction equation shows how this process operates. In general, the magnetic force $\mathbf{J} \wedge \mathbf{B}$ affects the motions and a full non-linear MHD problem results. However, the basic dynamo mechanism can be understood by just considering the induction equation, which illustrates what types of motion are necessary for sustained field generation.

Consider, first, an axisymmetric situation so quantities are independent of ϕ in the cylindrical coordinate system (ϖ, ϕ, z). The magnetic field can be split into poloidal and toroidal components, as

$$\mathbf{B} = \mathbf{B}_p + B_\phi\hat{\boldsymbol{\phi}}. \tag{2.178}$$

The velocity $\mathbf{v}$ can be resolved in a similar way. The poloidal magnetic field may be expressed in terms of a toroidal vector potential by

$$\mathbf{B}_p = \nabla \wedge (A\hat{\boldsymbol{\phi}}), \tag{2.179}$$

so $\nabla \cdot \mathbf{B} = 0$ is satisfied. The induction equation then splits into two components, one for the time derivative of B_ϕ and the other for that of A.

The toroidal equation follows from taking the scalar product of (2.44) with $\hat{\boldsymbol{\phi}}/\varpi$ and employing the result

$$\nabla \wedge \left(\frac{1}{\varpi}\hat{\boldsymbol{\phi}}\right) = \mathbf{0},$$

together with standard vector indentities. This gives

$$\frac{\partial B_\phi}{\partial t} + \varpi \mathbf{v}_p \cdot \nabla \left(\frac{B_\phi}{\varpi}\right) = \varpi \mathbf{B}_p \cdot \nabla \Omega - B_\phi \nabla \cdot \mathbf{v}_p + \eta \left(\nabla^2 B_\phi - \frac{B_\phi}{\varpi^2}\right) \tag{2.180}$$

where $\Omega = v_\phi/\varpi$ and η is taken as constant. It is noted that, due to axisymmetry, $\mathbf{v}_p$ can be replaced by $\mathbf{v}$ in this equation. After division by ϖ, the left hand side represents the rate of change of B_ϕ/ϖ following the motion of a fluid element (i.e. the Langrangian derivative given by (2.17)). The first term on the right hand side gives the rate of change of the angular velocity along the poloidal field. When Ω varies along $\mathbf{B}_p$ the shearing motion draws out $\mathbf{B}_p$ in the azimuthal direction. Hence this term represents the creation of B_ϕ from $\mathbf{B}_p$. The second term, involving $\nabla \cdot \mathbf{v}$, corresponds to a rate of change of B_ϕ due to compressibility in the flow. Compression of a fluid element causes an increase in $|B_\phi|$, due to azimuthal field lines being squeezed together. The final term is the diffusion of B_ϕ, this being a decay term resulting from the dissipation of poloidal electric currents.

The equation describing the evolution of A, the vector potential corresponding to $\mathbf{B}_p$, follows from substituting (2.179) in the induction equation, giving

$$\frac{\partial A}{\partial t} + \frac{1}{\varpi}\mathbf{v}_p \cdot \nabla(\varpi A) = \eta \left(\nabla^2 A - \frac{A}{\varpi^2}\right). \tag{2.181}$$

After multiplying by ϖ, the left hand side of this equation is the Lagrangian time derivative of ϖA, while the right hand side represents the diffusion of A. The vector potential A, and hence $\mathbf{B}_p$, therefore decays in a fluid element due to diffusion, since (2.181) lacks a field creation term. Although (2.180) contains a term corresponding to the creation of B_ϕ from $\mathbf{B}_p$, this term will decay with $\mathbf{B}_p$ and hence B_ϕ will also decay. This situation is summarized in the theorem due to Cowling (1934) which states that 'a steady axisymmetric magnetic field cannot be maintained by dynamo action'.

Since many astrophysical bodies, such as stars and discs, are essentially axisymmetric and magnetic fields of this symmetry were observed, Cowling's theorem cast serious doubts on dynamo theory for over twenty years. A suggestion for resolving the problem was made by Parker (1955). He noted that if small-scale convective motions are present in a rotating body then rising plasma elements tend to rotate due to the Coriolis force. With significant advection of the magnetic field, such twisting motions convert toroidal field to a poloidal component. Falling elements rotate the field in the opposite direction and there must be some asymmetry between the up and down motions for a net effect. Stratification, with rising elements expanding and falling ones contracting, was used to provide the asymmetry. Parker modelled the net effect of many convective cells as a toroidal electric field

$$E_\phi = \alpha B_\phi. \tag{2.182}$$

This is the so-called α-effect. The quantity α is a measure of the mean helicity of the turbulence. Since the anticyclonic motions have odd symmetry about the equatiorial plane, α is an odd function of the vertical coordinate.

In astrophysical bodies turbulence is usually assumed to generate the α-effect. Mean-field electrodynamics can be used to formulate the effect. The velocity and magnetic fields are expressed as sums of mean and turbulent fluctuating parts, so

$$\mathbf{v} = \mathbf{v}_0 + \mathbf{v}_\mathrm{T}, \quad \mathbf{v}_0 = \langle \mathbf{v} \rangle, \tag{2.183}$$

$$\mathbf{B} = \mathbf{B}_0 + \mathbf{B}_\mathrm{T}, \quad \mathbf{B}_0 = \langle \mathbf{B} \rangle, \tag{2.184}$$

where the spatial average is over a volume large enough to contain many turbulent eddies, but with a length-scale small compared to that of the mean fields. A local time average, or an ensemble average, can also be adopted. Substituting (2.183) and (2.184) in the induction equation and taking averages, noting that $\langle \mathbf{v}_0 \rangle = \mathbf{v}_0$ and $\langle \mathbf{B}_0 \rangle = \mathbf{B}_0$ while $\langle \mathbf{v}_\mathrm{T} \rangle = \langle \mathbf{B}_\mathrm{T} \rangle = \mathbf{0}$, leads to the mean-field equation

$$\frac{\partial \mathbf{B}_0}{\partial t} = \nabla \wedge (\mathbf{v}_0 \wedge \mathbf{B}_0) + \nabla \wedge \boldsymbol{\mathcal{E}} - \nabla \wedge (\eta \nabla \wedge \mathbf{B}_0), \tag{2.185a}$$

where

$$\boldsymbol{\mathcal{E}} = \langle \mathbf{v}_\mathrm{T} \wedge \mathbf{B}_\mathrm{T} \rangle. \tag{2.185b}$$

The fluctuating part of the induction equation is obtained by subtracting (2.185a) from the total equation for $\mathbf{B}$, giving

$$\frac{\partial \mathbf{B}_\mathrm{T}}{\partial t} = \nabla \wedge (\mathbf{v}_0 \wedge \mathbf{B}_\mathrm{T} + \mathbf{v}_\mathrm{T} \wedge \mathbf{B}_0) + \nabla \wedge \mathbf{G} - \nabla \wedge (\eta \nabla \wedge \mathbf{B}_\mathrm{T}), \tag{2.186a}$$

where

$$\mathbf{G} = \mathbf{v}_{\mathrm{T}} \wedge \mathbf{B}_{\mathrm{T}} - \langle \mathbf{v}_{\mathrm{T}} \wedge \mathbf{B}_{\mathrm{T}} \rangle. \tag{2.186b}$$

The electric field $\boldsymbol{\mathcal{E}}$ is of crucial importance since it leads to the generation of toroidal electric currents which are the source of the poloidal magnetic field. Equation (2.186a) is a linear relationship between $\mathbf{B}_{\mathrm{T}}$ and $\mathbf{B}_0$, while (2.185b) shows that $\boldsymbol{\mathcal{E}}$ is linear in $\mathbf{B}_{\mathrm{T}}$ and hence in $\mathbf{B}_0$. Since the turbulent eddies are assumed to have length-scales much smaller than that of $\mathbf{B}_0$, $\boldsymbol{\mathcal{E}}$ should be expressible as a functional involving the local value of $\mathbf{B}_0$. This takes the form

$$\mathcal{E}_i = \alpha_{ij} B_{0j} + \beta_{ijk} \frac{\partial B_{0j}}{\partial x_k}, \tag{2.187}$$

where the summation convention is used. If the first order smoothing approximation is valid then the tensors α_{ij} and β_{ijk} can be calculated, if the statistical properties of the turbulence are prescribed. This approximation involves ignoring $\mathbf{G}$ in (2.186a), and can be justified only if the turbulence does not significantly distort the magnetic field, so that $|\mathbf{B}_{\mathrm{T}}| \ll |\mathbf{B}_0|$. This is the case if the diffusion time is small compared to the turnover time, τ_0, for a turbulent eddy, or if the correlation time, τ_{cor}, is much less than τ_0.

For locally isotropic turbulence

$$\alpha_{ij} = \alpha \delta_{ij}, \quad \beta_{ijk} = \beta \epsilon_{ijk}. \tag{2.188a,b}$$

When $\tau_{\mathrm{cor}} \ll \tau_0$ mean-field theory yields

$$\alpha = -\frac{1}{3}\tau_{\mathrm{cor}} \langle \mathbf{v}_{\mathrm{T}} \cdot \nabla \wedge \mathbf{v}_{\mathrm{T}} \rangle, \quad \beta = \frac{1}{3}\tau_{\mathrm{cor}} \langle \mathbf{v}_{\mathrm{T}} \cdot \mathbf{v}_{\mathrm{T}} \rangle. \tag{2.189a,b}$$

The quantity $\langle \mathbf{v}_{\mathrm{T}} \cdot \nabla \wedge \mathbf{v}_{\mathrm{T}} \rangle$ is referred to as the mean helicity of the turbulence. Mirror-asymmetric turbulence is required to produce finite α. Turbulence that is statistically isotropic, but non-mirror symmetric, is referred to as pseudo-isotropic. Gravity and rotation are believed to be important in the generation of such turbulence.

The β coefficient in (2.189b) can be found, in principle, if the spectrum of the turbulence is known. This quantity is equivalent to a turbulent diffusivity η_{T}. In the absence of a detailed knowledge of the turbulence, η_{T} is usually estimated from the simple form (2.167). It should be noted that if the Lorentz force $\mathbf{J} \wedge \mathbf{B}$ is significant in the momentum equation for $\mathbf{v}$, then α_{ij} and β_{ijk} will depend on the local value of $\mathbf{B}_0$.

Use of (2.188) in (2.187) gives

$$\boldsymbol{\mathcal{E}} = \alpha \mathbf{B}_0 - \eta_{\mathrm{T}} \nabla \wedge \mathbf{B}_0. \tag{2.190}$$

Dropping the subscript zero from the mean fields and substituting $\boldsymbol{\mathcal{E}}$ in (2.185a) yields

$$\frac{\partial \mathbf{B}}{\partial t} = \nabla \wedge (\mathbf{v} \wedge \mathbf{B}) - \nabla \wedge (\eta_{\mathrm{T}} \nabla \wedge \mathbf{B}) + \nabla \wedge (\alpha \mathbf{B}), \tag{2.191}$$

writing $\eta + \eta_{\mathrm{T}}$ as η_{T} for a dominant turbulent diffusivity. Steady, axisymmetric solutions are now possible for the mean magnetic field.

2.3.2. TYPES OF MEAN-FIELD DYNAMOS

Two extreme types of dynamo can be identified, by considering the field creation terms in (2.191). The equations for B_ϕ and A can now be written in the forms

$$\begin{aligned}\frac{\partial B_\phi}{\partial t} + \varpi \mathbf{v}_p \cdot \nabla \left(\frac{B_\phi}{\varpi}\right) =& \varpi \mathbf{B}_p \cdot \nabla \Omega - \alpha \Delta A - \frac{1}{\varpi} \nabla \alpha \cdot \nabla (\varpi A) \\ & - B_\phi \nabla \cdot \mathbf{v}_p + \eta_{\mathrm{T}} \Delta B_\phi, \end{aligned} \tag{2.192}$$

$$\frac{\partial A}{\partial t} + \frac{1}{\varpi} \mathbf{v}_p \cdot \nabla (\varpi A) = \alpha B_\phi + \eta_{\mathrm{T}} \Delta A, \tag{2.193}$$

where $\Delta = \nabla^2 - \varpi^{-2}$. The first term on the right hand side of (2.192) is the rate of creation of B_ϕ by the shearing of $\mathbf{B}_p$, This is referred to as the ω-effect. The α-effect terms in the above equations generate toroidal and poloidal field.

The magnetic Reynolds number, defined by (2.46), measures the ratio of the magnitudes of the velocity and diffusion terms in the induction equation. Reynolds numbers for the ω and α-effects can be defined as

$$R_\omega = \frac{\Omega_0 \ell^2}{\eta_{\mathrm{T}}}, \qquad R_\alpha = \frac{\alpha_0 \ell}{\eta_{\mathrm{T}}}, \tag{2.194}$$

where subscripts zero denote characteristic values and ℓ is the length-scale of field variation.

The axisymmetric dynamo equations (2.192) and (2.193) are usually solved in a finite region enclosed by a surface S. If the external region is a vacuum then continuity of $\mathbf{B}$ requires B_ϕ to vanish on S, together with continuity of A and its normal derivative.

The relative importance of the ω and α-effects depends on the ratio R_ω / R_α. There are two extreme types of dynamo possible. If $R_\omega \ll R_\alpha$ then the ω-effect is negligible and (2.192) and (2.193) show that the α-effect acts not only to produce A from B_ϕ, but also B_ϕ from A. This is referred to as an α^2-dynamo. The dynamo becomes self-sustaining when R_α attains a critical value. The other extreme is when $R_\omega \gg R_\alpha$ so the α-effect can be ignored in the creation of B_ϕ from A, described by (2.192). However, the α-term in (2.193) remains crucial in creating A from B_ϕ. In this $\alpha\omega$-dynamo the product

$$D = R_\alpha R_\omega, \tag{2.195}$$

referred to as the dynamo number, must attain a critical value for field maintenance. If the poloidal flow is insignificant then (2.193) and (2.179) show that $|B_p| \sim R_\alpha |B_\phi|$, while (2.192) gives $|B_\phi| \sim R_\omega |B_p|$. It follows that the critical dynamo number is $D_c \sim 1$ and

$$\frac{|B_\phi|}{|B_p|} \sim \left(\frac{R_\omega}{R_\alpha}\right)^{\frac{1}{2}} \gg 1. \tag{2.196}$$

If v_ϖ and Ω are symmetric with v_z and α antisymmetric under reflection in the equatorial plane (i.e. $z \to -z$), then the eigenfunctions of (2.192) and (2.193) fall into one of the following groups;
Dipole mode:

$$A(\varpi, -z, t) = A(\varpi, z, t),$$

$$B_\phi(\varpi, -z, t) = -B_\phi(\varpi, z, t). \tag{2.197}$$

Quadrupole mode:

$$A(\varpi, -z, t) = -A(\varpi, z, t),$$

$$B_\phi(\varpi, -z, t) = B_\phi(\varpi, z, t). \tag{2.198}$$

2.4. Close Binary Stars

2.4.1. THE ROCHE MODEL

The basic model of a close binary system, due to Roche (1873), corresponds to the restricted three-body problem. This consists of a system of three particles in which one particle has a mass negligible compared to those of the other two. The two massive particles constitute a two-body problem having orbital solutions which are conic sections focused on the centre of mass. Hence the bound solutions are circles or ellipses. The motion of the third particle is then determined by its initial position and velocity, together with the gravitational force exerted on it by the orbiting bodies.

In the Roche model of a close binary the stellar components are the massive bodies, while transferred material can be considered as low mass particles. The stars are treated as gravitational point sources. This is essentially exact for a spherical white dwarf or neutron star, and is a reasonable first approximation for main sequence stars since even a star of mass $0.2M_\odot$ has a central density $\rho_c \sim 6\bar{\rho}$, where $\bar{\rho}$ is the mean density. Circular orbits are considered in the standard model, this being consistent with observations which show that the eccentricities are usually small. Tidal interactions circularize orbits on time-scales shorter than the binary lifetime.

Since the mean free path of ions in the accretion stream is many orders of magnitude smaller than the orbital separation, a hydrodynamical analysis of the flow, including the pressure gradient, is appropriate. However, Lubow and Shu (1975) showed that, provided the pressure gradient along the stream is small, the locus of its centre can be described accurately by purely ballistic considerations. In thin accretion discs, in which self-gravity is small, the radial pressure gradient is ignorable and consequently the azimuthal motion of material corresponds to that of individual particle orbits. However, pressure gradients normal to the orbital plane are important in streams and discs, opposing gravitational collapse due to the vertical component of stellar gravity. Although the ballistic approach allows the following basic model of a binary system to be formulated, ultimately hydrodynamics is required to fully describe the structure of accretion streams and discs.

The orbital frame $Oxyz$ is defined to have its origin O at the centre of mass, its y-axis along the line joining the stellar centres and its z-axis along the orbital angular velocity $\boldsymbol{\Omega}$. The total potential, Ψ, in this rotating frame is

$$\Psi = -\frac{GM_{\rm s}}{|\mathbf{r}-\mathbf{r}_{\rm s}|} - \frac{GM_{\rm p}}{|\mathbf{r}-\mathbf{r}_{\rm p}|} - \frac{1}{2}|\boldsymbol{\Omega}\wedge\mathbf{r}|^2, \tag{2.199}$$

where $M_{\rm s}$ is the mass of the main sequence secondary star and $M_{\rm p}$ the mass of the compact primary, while $\mathbf{r}_{\rm s}$ and $\mathbf{r}_{\rm p}$ are the positions of the centres of mass of the respective stars. The last term is the centrifugal potential. The equipotential surfaces of Ψ are therefore given by the equation

$$\frac{GM_{\rm s}}{[x^2+(y-y_{\rm s})^2+z^2]^{\frac{1}{2}}} + \frac{GM_{\rm p}}{[x^2+(y-y_{\rm p})^2+z^2]^{\frac{1}{2}}} + \frac{1}{2}\Omega^2(x^2+y^2) = K, \tag{2.200}$$

where K is a constant. Close to the stellar centres these surfaces are nearly spherical, but at larger distances they become pear-shaped. A critical surface occurs when the stellar lobes meet on the line of centres. The point of contact is referred to as the inner Lagrangian point, denoted L_1, and

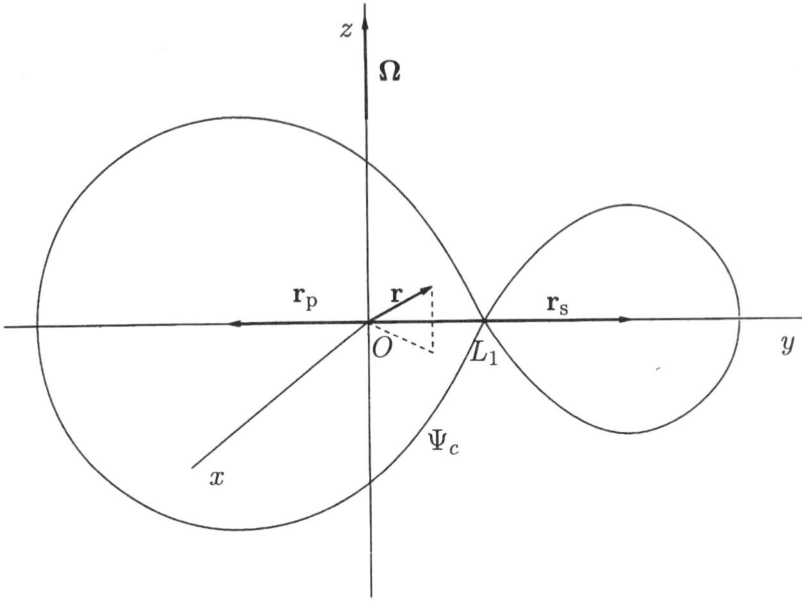

Figure 2.1. The orbital frame $Oxyz$, showing the critical equipotential Ψ_c and the inner Lagrangian point L_1. The system centre of mass is at O, and the secondary and primary centres of mass are at $\mathbf{r}_S$ and $\mathbf{r}_P$ respectively.

the critical surfaces are known as the Roche lobes. A particle placed at L_1 experiences zero force since $(\nabla\Psi)_{L_1} = \mathbf{0}$. Figure 2.1 shows the critical equipotentials and coordinates used.

The stability of a particle at the L_1 point can be investigated by considering a small displacement $\Delta\mathbf{r}$. The equation of motion of a particle relative to the orbital frame is

$$\ddot{\mathbf{r}} = -\nabla\Psi - 2\mathbf{\Omega} \wedge \dot{\mathbf{r}}. \tag{2.201}$$

A small displacement from L_1 then satisfies the equations

$$\Delta\ddot{x} + \omega_x^2 \Delta x = 2\Omega\Delta\dot{y}, \tag{2.202}$$

$$\Delta\ddot{y} - \omega_y^2 \Delta y = -2\Omega\Delta\dot{x}, \tag{2.203}$$

$$\Delta\ddot{z} + \omega_z^2 \Delta z = 0, \tag{2.204}$$

where

$$\omega_x^2 = \left(\frac{\partial^2 \Psi}{\partial x^2}\right)_{L_1} = -\Omega^2 + \frac{GM_s}{r_{1s}^3} + \frac{GM_p}{r_{1p}^3}, \tag{2.205}$$

$$\omega_y^2 = -\left(\frac{\partial^2 \Psi}{\partial y^2}\right)_{L_1} = \Omega^2 + \frac{2GM_s}{r_{1s}^3} + \frac{2GM_p}{r_{1p}^3}, \tag{2.206}$$

$$\omega_z^2 = \left(\frac{\partial^2 \Psi}{\partial z^2}\right)_{L_1} = \frac{GM_s}{r_{1s}^3} + \frac{GM_p}{r_{1p}^3}, \tag{2.207}$$

and r_{1s} and r_{1p} are the distances of the L_1 point from the primary and secondary centres.

Equations (2.202)–(2.204) show that small vertical motions are independent of those in a horizontal plane. Since $\omega_z^2 > 0$, it follows that the vertical motions are stable. The stability of the horizontal motions can be investigated by taking

$$\Delta x = Ae^{i\omega t}, \qquad \Delta y = Be^{i\omega t}, \tag{2.208}$$

where A and B are constants, and finding the sign of ω^2. Substitution in (2.202)–(2.204) gives

$$\omega^4 - (\omega_x^2 - \omega_y^2 + 4\Omega^2)\omega^2 - \omega_x^2\omega_y^2 = 0,$$

which has roots

$$\omega^2 = \frac{1}{2}(\omega_x^2 - \omega_y^2 + 4\Omega^2) \pm \frac{1}{2}\left[(\omega_x^2 - \omega_y^2 + 4\Omega^2)^2 + 4\omega_x^2\omega_y^2\right]^{\frac{1}{2}}. \tag{2.209}$$

Since

$$\Omega^2 = \frac{G(M_s + M_p)}{D^3}, \tag{2.210}$$

where $D = r_{1s} + r_{1p}$, it follows from (2.205) that

$$\omega_x^2 = GM_p\left(\frac{1}{r_{1p}^3} - \frac{1}{D^3}\right) + GM_s\left(\frac{1}{r_{1s}^3} - \frac{1}{D^3}\right) > 0,$$

and (2.206) shows $\omega_y^2 > 0$. Hence $\omega_x^2\omega_y^2 > 0$ and (2.209) yields positive and negative eigenvalues, the unstable negative case being

$$\omega^2 = \frac{1}{2}(\omega_x^2 - \omega_y^2 + 4\Omega^2) - \frac{1}{2}\left[(\omega_x^2 - \omega_y^2 + 4\Omega^2)^2 + 4\omega_x^2\omega_y^2\right]^{\frac{1}{2}}. \tag{2.211}$$

This can be used to find the direction of the instability.

Equations (2.202), (2.203) and (2.208) relate A and B, giving the horizontal displacement as

$$\Delta \mathbf{r}_h = B\left[-i\frac{(\omega^2+\omega_y^2)}{2\omega\Omega}\hat{\mathbf{x}}+\hat{\mathbf{y}}\right]e^{i\omega t}. \tag{2.212}$$

The unstable eigenvalue then yields

$$\Delta \mathbf{r}_h = B\left[\frac{(\omega^2+\omega_y^2)}{2\Omega\sqrt{Q}}\hat{\mathbf{x}}+\hat{\mathbf{y}}\right]e^{i\omega t}, \tag{2.213a}$$

where

$$Q=\frac{1}{2}\left[(\omega_x^2-\omega_y^2+4\Omega^2)^2+4\omega_x^2\omega_y^2\right]^{\frac{1}{2}}-\frac{1}{2}(\omega_x^2-\omega_y^2+4\Omega^2). \tag{2.213b}$$

The direction of motion therefore follows from the components of (2.213b) as

$$\frac{\Delta x}{\Delta y}=\frac{\omega^2+\omega_y^2}{2\Omega\sqrt{Q}}. \tag{2.214}$$

It can now be shown that this direction lies well inside the critical lobe. The change in Ψ due to a small displacement from L_1 in the $z=0$ plane is

$$\Delta\Psi=\frac{1}{2}\omega_x^2(\Delta x)^2-\frac{1}{2}\omega_y^2(\Delta y)^2.$$

Along the critical lobe, $\Delta\Psi=0$ and hence

$$\left(\frac{\Delta x}{\Delta y}\right)_c=\pm\left(\frac{\omega_y^2}{\omega_x^2}\right)^{\frac{1}{2}}. \tag{2.215}$$

Use of (2.205), (2.206), (2.211) and (2.213b) shows that the ratios (2.214) and (2.215) can be expressed as functions of

$$\delta=\frac{\epsilon_{\rm s}\epsilon_{\rm p}}{\epsilon_{\rm s}+\epsilon_{\rm p}}, \tag{2.216}$$

where

$$\epsilon_{\rm s}=\frac{M}{M_{\rm s}}\left(\frac{r_{1\rm s}}{D}\right)^3,\qquad \epsilon_{\rm p}=\frac{M}{M_{\rm p}}\left(\frac{r_{1\rm p}}{D}\right)^3, \tag{2.217a, b}$$

with $M=M_{\rm s}+M_{\rm p}$. Typically $\epsilon_{\rm s}\sim\epsilon_{\rm p}\sim 0.2$ and hence $\delta\sim 0.1$. Expansion to first order in δ then yields

$$\frac{\Delta x}{\Delta y}=\frac{4}{3}\left(\frac{\delta}{2}\right)^{\frac{1}{2}}\left(1+\frac{17}{36}\delta\right), \tag{2.218}$$

and

$$\left|\left(\frac{\Delta x}{\Delta y}\right)_c\right| = \sqrt{2}\left(1 + \frac{3}{4}\delta\right). \qquad (2.219)$$

It follows that

$$\frac{\Delta x/\Delta y}{|(\Delta x/\Delta y)_c|} = \frac{2}{3}\delta^{\frac{1}{2}}\left(1 - \frac{5}{18}\delta\right). \qquad (2.220)$$

This ratio is always less than unity, being typically 0.2, so the unstable mode lies well inside the critical lobe. For $\delta = 0.1$ the critical lobe makes an angle of $\theta_c = \tan^{-1}(\Delta x/\Delta y)_c = 56.7^\circ$ with the line of stellar centres at L_1 in the orbital plane, and the unstable eigenvector makes an angle of $\theta = 17.3^\circ$ leading the orbital motion of this line.

The distance of the L_1 point from the centre of the compact primary, denoted $r_{1\mathrm{p}}$, can be found from the condition $(\partial\Psi/\partial y)_{L_1} = 0$ together with the y-coordinates shown in Figure 2.1, noting that $y_\mathrm{s} = (M_\mathrm{p}/M)D$. This yields

$$\frac{M_\mathrm{s}}{M} - \frac{r_{1\mathrm{p}}}{D} + \frac{M_\mathrm{p}}{M}\left(\frac{D}{r_{1\mathrm{p}}}\right)^2 - \frac{M_\mathrm{s}}{M}\left(1 - \frac{r_{1\mathrm{p}}}{D}\right)^{-2} = 0, \qquad (2.221)$$

from which $r_{1\mathrm{p}}$ can be found numerically. It follows from (2.221) that $r_{1\mathrm{p}}/D$ is a pure function of the mass ratio $M_\mathrm{s}/M_\mathrm{p}$ and that, since $r_{1\mathrm{s}} = D - r_{1\mathrm{p}}$, $r_{1\mathrm{s}}$ is also a pure function of this ratio. The shapes of the Roche lobes depend only on ϵ_s and ϵ_p and hence, from (2.217a,b), are pure functions of $M_\mathrm{s}/M_\mathrm{p}$ (e.g. the critical angle $\theta_c = \tan^{-1}(\Delta x/\Delta y)_c$, with $(\Delta x/\Delta y)_c$ given by (2.219)). A good accuracy fitted formula for $r_{1\mathrm{p}}$ is given by Plavec and Kratochvil (1964) as

$$\frac{r_{1\mathrm{p}}}{D} = 0.500 - 0.227\log\left(\frac{M_\mathrm{s}}{M_\mathrm{p}}\right). \qquad (2.222)$$

A mean radius, R_L, can be defined for the secondary star's critical lobe as the radius of a sphere of equal volume. Eggleton (1983) gives a formula for this radius as

$$\frac{R_\mathrm{L}}{D} = \frac{0.49q^{\frac{2}{3}}}{0.6q^{\frac{2}{3}} + \ln(1 + q^{\frac{1}{3}})}, \qquad (2.223)$$

where $q = M_\mathrm{s}/M_\mathrm{p}$, accurate to $\sim 99\%$ for all q values. A simpler formula for R_L, valid for $0.1 \lesssim M_\mathrm{s}/M_\mathrm{p} \lesssim 0.8$ to $\sim 98\%$ accuracy, is

$$\frac{M}{M_\mathrm{s}}\left(\frac{R_\mathrm{L}}{D}\right)^3 = 0.1, \qquad (2.224)$$

due to Paczyński (1967). If the mean radius of the secondary star equals R_{L} then it fills its Roche lobe and material is in contact with L_1.

Lower main sequence stars obey the approximate mass-radius relation

$$\frac{R_{\mathrm{s}}}{R_{\odot}} \simeq q \frac{M_{\mathrm{s}}}{M_{\odot}}, \tag{2.225}$$

with $q \sim 1.1$ (Kippenhahn and Weigert, 1990). Using this, together with (2.224) and $\Omega^2 = GM/D^3$, gives a relation between the orbital period and the mass of a lobe-filling secondary as

$$\left(\frac{P}{\mathrm{hr}}\right) \simeq 9q^{\frac{3}{2}} \left(\frac{M_{\mathrm{s}}}{M_{\odot}}\right). \tag{2.226}$$

The equipotentials of Ψ correspond to the zero-velocity surfaces of the restricted three-body problem. These surfaces arise from the energy integral which is derived from the equation of motion of a particle relative to the orbital frame, given by (2.201). Taking the scalar product of this with $\dot{\mathbf{r}}$ removes the Coriolis force and gives

$$\frac{d}{dt}\left(\frac{1}{2}v^2 + \Psi\right) = 0,$$

in which the conserved quantity is the energy per unit mass, E, so

$$\frac{1}{2}v^2 + \Psi = E. \tag{2.227}$$

The particle speed is therefore

$$v = [2(E - \Psi)]^{\frac{1}{2}}, \tag{2.228}$$

requiring $|\Psi| \geq |E|$. The velocity vanishes on the surface $\Psi = E$, with E being fixed by the particle's initial position and velocity.

2.4.2. MASS TRANSFER

If the secondary star fills its Roche lobe material will be in contact with the L_1 point. Thermal motions allow particles to pass over the potential maximum at L_1 and leave the star. The initial velocity of material is $\sim c_{\mathrm{S}}$, the local sound speed, but matter rapidly gains speeds of $\sim \Omega D$, characteristic of orbital speeds. Since $c_{\mathrm{S}} \ll \Omega D$, the speed of material at L_1 can be taken as effectively zero in (2.228) which therefore yields

$$v = [2(\Psi_c - \Psi)]^{\frac{1}{2}}, \tag{2.229}$$

for the subsequent speed, where Ψ_c is the critical potential value on the Roche lobes. Matter is then trapped within the Roche lobe of the primary, since this is its zero-velocity surface with $\Psi_c = E$.

The mass loss rate from the secondary can be expressed as

$$\dot{M}_{\rm s} = -\rho_{L_1} c_{\rm s} \Delta S, \tag{2.230}$$

where ΔS is an effective cross-sectional area perpendicular to the line of stellar centres, at a small distance Δy into the secondary from the L_1 point. This effective area is determined by considering the density scale height normal to the line of centres at $y = y_{L_1} + \Delta y$.

For a synchronous star, hydrostatic equilibrium is

$$-\frac{1}{\rho}\nabla P = \nabla \Psi, \tag{2.231}$$

where Ψ is the sum of the gravitational and rotational potentials. In the tenuous outer layers of the secondary it is a good approximation to take Ψ as the Roche potential, since the main gravitational sources are the compact primary and the central regions of the secondary. The scale height in the orbital plane, H_x, slightly exceeds the vertical scale height, H_z, due to the effect of centrifugal force. However, since the difference is only a few per cent, the area ΔS will be essentially circular with radius $H \simeq H_x \simeq H_z$.

To calculate H, consider the z-direction at $x = 0$, $y = y_{L_1} + \Delta y$. It is reasonable to take the gas as isothermal in this direction. The gas equation of state and (2.231) then give

$$\frac{c_{\rm s}^2}{\rho}\frac{\partial \rho}{\partial z} = -\frac{\partial \Psi}{\partial z}, \tag{2.232a}$$

where

$$c_{\rm s}^2 = \frac{\Re}{\mu} T. \tag{2.232b}$$

Using the fact that $|z|$ is small for points in the star near L_1 and that $\Delta y \ll y$, (2.199) for the Roche potential gives

$$\frac{\partial \Psi}{\partial z} = \left(\frac{GM_{\rm s}}{r_{1\rm s}^3} + \frac{GM_{\rm p}}{r_{1\rm p}^3} \right) z,$$

along the required line, with $r_{1\rm s} = y_{\rm s} - y_{L_1}$ and $r_{1\rm p} = y_{L_1} - y_{\rm p}$ (see Figure 2.1 for coordinates). Equation (2.232a) then has the solution

$$\rho = \rho_c e^{-z^2/H^2}, \tag{2.233}$$

where ρ_c is the density in the central plane and the scale height is

$$H = \sqrt{2} c_{\rm s} \left(\frac{GM_{\rm s}}{r_{1\rm s}^3} + \frac{GM_{\rm p}}{r_{1\rm p}^3} \right)^{-\frac{1}{2}} . \tag{2.234}$$

This can be expressed as

$$H = f \frac{c_{\rm s}}{\Omega D} R_{\rm s}, \tag{2.235a}$$

where

$$f = \left[\left(\frac{2\epsilon_{\rm s}\epsilon_{\rm p}}{\epsilon_{\rm s} + \epsilon_{\rm p}} \right) \left(\frac{D}{R_{\rm s}} \right)^2 \right]^{\frac{1}{2}}, \tag{2.235b}$$

with $\epsilon_{\rm s}$ and $\epsilon_{\rm p}$ given by (2.217a,b). The dimensionless factor f is a slowly varying function of the orbital parameters, with $f \sim 1$. The mass loss rate is

$$\dot{M}_{\rm s} = -\pi \rho_{L_1} c_{\rm s} H^2. \tag{2.236}$$

The mass transfer process can only continue if the secondary remains in contact with its Roche lobe. However, the radius of a main-sequence star in thermal equilibrium decreases as its mass decreases so, unless the Roche lobe shrinks, material will lose contact with L_1. To investigate the change in the Roche lobe size it is necessary to consider the angular momentum of the system. The orbital angular momentum can be written as

$$\mathbf{L}_{\rm orb} = \left(\frac{GD}{M} \right)^{\frac{1}{2}} M_{\rm s} M_{\rm p} \hat{\mathbf{z}}. \tag{2.237}$$

The angular momentum of the secondary is $L_{\rm s} \sim k_{\rm s}^2 M_{\rm s} R_{\rm s}^2 \Omega$, where $k_{\rm s}^2 \sim 0.1$. Using (2.210) for Ω, it follows that

$$\frac{L_{\rm s}}{L_{\rm orb}} \sim k_{\rm s}^2 \frac{M}{M_{\rm p}} \left(\frac{R_{\rm s}}{D} \right)^2, \tag{2.238}$$

which is typically $\sim 10^{-2}$. The primary's spin angular momentum is typically $\lesssim 10^{-2} L_{\rm s}$, so the stellar rotations only make a small contribution to the total angular momentum.

Consider continuous mass transfer from the secondary to the primary with $L_{\rm orb}$ and M conserved. Differentiating (2.237) with respect to time then gives

$$\frac{\dot{D}}{D} = -2 \left(1 - \frac{M_{\rm s}}{M_{\rm p}} \right) \frac{\dot{M}_{\rm s}}{M_{\rm s}}. \tag{2.239}$$

Since $\dot{M}_s$ is negative it follows that, for $M_p \geq M_s$, the orbital separation increases. The mean radius of the secondary's Roche lobe is given by (2.224) which, together with (2.239), yields

$$\frac{\dot{R}_L}{R_L} = -\frac{5}{3}\left(1 - \frac{6}{5}\frac{M_s}{M_p}\right)\frac{\dot{M}_s}{M_s}. \tag{2.240}$$

Hence, for $M_p > (6/5)M_s$, the secondary's lobe expands as the star losses mass. However, a main-sequence secondary in thermal equlibrium contracts as its mass decreases, obeying the approximate lower main-sequence mass-radius relation (2.225). The secondary therefore becomes detached from its Roche lobe and mass transfer ceases, if angular momentum is conserved. Continuous mass transfer requires the orbital angular momentum to decrease with time, and there are two main mechanisms for achieving this.

Gravitational Radiation: This mechanism was proposed by Kraft, Mathews and Greenstein (1962) and Paczyński (1967). Gravitational waves, a general relativistic effect, remove orbital angular momentum at a rate

$$\frac{dL_{orb}}{dt} = -\frac{32G}{5c^5}\left(\frac{M_s M_p}{M}\right)^2 D^4 \Omega^5, \tag{2.241}$$

(Landau and Lifshitz, 1951). Faulkner (1971) calculated the resulting mass transfer rate from a secondary in thermal equilibrium. Using (2.224), (2.225), (2.237) and (2.241), with $R_s = R_L$, gives

$$\dot{M}_s = -1.9 \times 10^{-10} \frac{(1-\mu)^2}{(4-7\mu)\mu^{\frac{2}{3}}} M_\odot \,\mathrm{yr}^{-1}, \tag{2.242}$$

where $\mu = M_s/M$.

Magnetic Braking: If the secondary has a magnetic field and a stellar wind then rotational angular momentum is removed. Material flows along magnetic field lines and the associated stresses lead to a braking torque on the star (see Chapter 12). Since the secondary is kept close to orbital corotation by tidal torques, the magnetic wind drains angular momentum from the orbit.

A simple expression can be obtained for the braking of the secondary by using angular velocity observations of single stars and assuming them to be magnetically braked. This law, due to Skumanich (1972), is

$$R_s \Omega_s(t) = 10^{12} \bar{f} t^{-\frac{1}{2}} \,\mathrm{m\,s^{-1}}, \tag{2.243}$$

where $0.73 < \bar{f} < 1.78$. Verbunt and Zwaan (1981) applied this to find $\dot{M}_s$. Taking the star's spin angular momentum as

$$L_s = k_s^2 M_s R_s^2 \Omega_s,$$

and using $\dot{L}_{\rm s} = \dot{L}_{\rm orb}$, yields

$$\dot{M}_{\rm s} = -1.9 \times 10^{-7} k_{\rm s}^2 \bar{f}^{-2} \frac{M}{M_\odot} \frac{\mu^{\frac{5}{3}}}{(4-7\mu)} M_\odot \, {\rm yr}^{-1}. \tag{2.244}$$

Alternative expressions for $\dot{M}_{\rm s}$ are derived in Chapter 12, using fast rotator magnetic braking theory.

2.4.3. ACCRETION DISCS

Disc Formation and Luminosity

In the process of Roche lobe overflow material is lost from the secondary through the L_1 region, at a distance $r_{\rm 1p}$ from the primary. Matter passing through L_1 has a small velocity component ($\sim c_{\rm s}$) along the line of stellar centres. Relative to the orbital frame, the stream will experience a Coriolis force causing some initial deflection from the line of centres. As material approaches the compact primary its central force becomes dominant and the stream orbits around it, subsequently intersecting itself. An estimate of the size of this initial orbit can be made by considering the specific angular momentum of material.

In an inertial frame centred on the primary, the specific angular momentum of material at L_1 is $r_{\rm 1p}^2 \Omega$. Matter in the stream experiences the gravitational fields of the primary and secondary stars. The angular momentum of material about the primary is not exactly conserved, since the stream deviates from the line of stellar centres so generating a gravitational torque on the secondary, and angular momentum can be exchanged with the orbit. However, for a free stream, the angular momentum exchange is small ($\sim 6\%$) because the stream does not make large angles to the line of centres in the region of L_1, in which its gravitational interaction with the secondary is greatest. The angular momentum of material can therefore be taken to be approximately conserved. The radius, R_c, of a circular orbit about the primary then follows from

$$R_c \left(\frac{GM_{\rm p}}{R_c} \right)^{\frac{1}{2}} = r_{\rm 1p}^2 \Omega.$$

Since $\Omega^2 = GM/D^3$, this yields

$$\frac{R_c}{D} = \left(1 + \frac{M_{\rm s}}{M_{\rm p}} \right) \left(\frac{r_{\rm 1p}}{D} \right)^4. \tag{2.245}$$

This gives an estimate of the characteristic size of the initial orbit of material about the primary. Taking the mean radius of the primary's Roche lobe, R_1, to be given by an expression similar to (2.224), yields

$$\frac{R_1}{D} = 0.46 \left(\frac{M_\mathrm{p}}{M}\right)^{\frac{1}{3}},$$

and hence

$$\frac{R_c}{R_1} = 2.16 \left(\frac{M}{M_\mathrm{p}}\right)^{\frac{4}{3}} \left(\frac{r_{1\mathrm{p}}}{D}\right)^4. \qquad (2.246)$$

The radius R_c is always smaller than R_1, typically $R_c = 0.4R_1$, so material orbits the primary well inside its Roche lobe.

As matter is fed into the initial ring it spreads in the orbital plane, due to dissipation and angular momentum transfer, to form a differentially rotating disc. Internal friction causes material to spiral in through the disc, losing energy and angular momentum. Dissipation heats the disc and energy is radiated from its surfaces. It is shown below that the azimuthal velocity of disc material is close to a Keplerian distribution, with v_K given by (2.261). Since the self-gravity is ignorable, and $|v_\varpi| \ll v_\mathrm{K}$, the energy per unit mass in a thin disc is

$$E = \frac{1}{2}v_\mathrm{K}^2 - \frac{GM_\mathrm{p}}{\varpi} = -\frac{GM_\mathrm{p}}{2\varpi}, \qquad (2.247)$$

where ϖ is the distance from the centre of the accreting primary star. A ring of mass dM therefore has energy

$$dE = -\frac{GM_\mathrm{p}}{2\varpi}dM.$$

In a time interval dt this mass moves inwards through a distance $d\varpi$ releasing energy at a rate

$$d\dot{E} = \frac{GM_\mathrm{p}}{2\varpi^2}\frac{d\varpi}{dt}dM = \frac{GM_\mathrm{p}\dot{M}}{2\varpi^2}d\varpi,$$

where the constant mass transfer rate $\dot{M}$ is positive and $d\varpi < 0$. Since $\dot{M} = \dot{M}_\mathrm{p}$, the total rate of energy release is

$$\dot{E} = \frac{1}{2}GM_\mathrm{p}\dot{M}_\mathrm{p}\int_{R_\mathrm{D}}^{R_\mathrm{p}} \frac{d\varpi}{\varpi^2} = -\frac{GM_\mathrm{p}\dot{M}_\mathrm{p}}{2R_\mathrm{p}},$$

using $R_\mathrm{D} \gg R_\mathrm{p}$, where R_D is the disc radius. The disc luminosity is therefore

$$L_\mathrm{D} = \frac{GM_\mathrm{p}\dot{M}_\mathrm{p}}{2R_\mathrm{p}}. \qquad (2.248)$$

This shows that matter releases one half of its gravitational accretion energy per unit mass, $GM_\mathrm{p}/R_\mathrm{p}$, in spiralling in through the disc. The other half is retained as azimuthal kinetic energy, available for release through an inner boundary layer.

The Steady Viscous Disc

The basic model of an accretion disc was formulated by Shakura and Sunyaev (1973). The internal friction is generated by viscosity, allowing angular momentum advection through the disc. Values of the viscosity, ν, characteristic of a turbulent origin are necessary to transport matter through the disc at the externally imposed rate. Steady, thin discs are considered first.

In a cylindrical coordinate system (ϖ, ϕ, z) centred on the accreting primary, the components of the steady, axisymmetric momentum equation are

$$\frac{\partial}{\partial \varpi}\left(\frac{v_\varpi^2}{2}\right) + v_z \frac{\partial v_\varpi}{\partial z} - \frac{v_\phi^2}{\varpi} = -\frac{1}{\rho}\frac{\partial P}{\partial \varpi} - \frac{\partial \psi}{\partial \varpi}, \tag{2.249}$$

$$\frac{v_\varpi}{\varpi}\frac{\partial}{\partial \varpi}(\varpi^2 \Omega) + \frac{v_z}{\varpi}\frac{\partial}{\partial z}(\varpi^2\Omega) = \frac{1}{\varpi^2 \rho}\frac{\partial}{\partial \varpi}\left(\rho \nu \varpi^3 \frac{\partial \Omega}{\partial \varpi}\right), \tag{2.250}$$

$$\frac{\partial}{\partial z}\left(\frac{v_z^2}{2}\right) + v_\varpi \frac{\partial v_z}{\partial \varpi} = -\frac{1}{\rho}\frac{\partial P}{\partial z} - \frac{\partial \psi}{\partial z}, \tag{2.251}$$

where $\Omega = v_\phi/\varpi$, ν is the viscosity coefficient, ψ is the gravitational potential, and other symbols have their usual meanings. It is assumed that the viscous force due to turbulence has the standard form given by (2.82). Since the dominant disc shear is that due to differential rotation, the poloidal components of $\mathbf{F}_\mathrm{v}$ are ignorable compared to $F_{\mathrm{v}\phi}$, where the latter is given by (2.83). The vertical derivative term in this expression is dropped here since, for vanishing ρ at the disc surfaces, it gives zero net torque on rings of material. For thin discs in binary systems self-gravity is negligible, so ψ is the potential of the compact primary star given by

$$\psi = -\frac{GM_\mathrm{p}}{(\varpi^2 + z^2)^{\frac{1}{2}}}. \tag{2.252}$$

The main body of the disc lies well inside the primary's Roche lobe, so tidal distortion due to the secondary is small, allowing an axisymmetric structure to be considered.

The continuity equation (2.74) becomes

$$\frac{1}{\varpi}\frac{\partial}{\partial \varpi}(\varpi \rho v_\varpi) + \frac{\partial}{\partial z}(\rho v_z) = 0. \tag{2.253}$$

The equations of the disc surfaces can be written as

$$z = \pm h(\varpi). \tag{2.254}$$

A thin disc will satisfy the condition

$$\frac{dh}{d\varpi} \sim \frac{h}{\varpi} \ll 1. \tag{2.255}$$

The disc pressure and density are expected to have maximum values in the central plane $z = 0$, and fall vertically with the length-scale h. The medium surrounding the disc is taken to be essentially a vacuum, so surface conditions can be defined as

$$P(\varpi, \pm h) = \rho(\varpi, \pm h) = 0. \tag{2.256a,b}$$

Equations (2.253) and (2.255) give

$$\left| \frac{v_z}{v_\varpi} \right| \lesssim \frac{h}{\varpi} \ll 1. \tag{2.257}$$

The thin nature of the disc enables the equations to be simplified. The azimuthal component of the momentum equation, given by (2.250), yields an order of magnitude estimate for the inflow speed as

$$|v_\varpi| \sim \frac{\nu}{\varpi}. \tag{2.258}$$

Firstly, the importance of the velocity terms in the vertical momentum equation (2.251) can be examined. Equation (2.257) gives

$$\left| \frac{\partial}{\partial z} \left(\frac{v_z^2}{2} \right) \right| \sim \left| v_\varpi \frac{\partial v_z}{\partial \varpi} \right| \sim \frac{v_\varpi^2}{\varpi} \frac{h}{\varpi}. \tag{2.259}$$

Differentiation of (2.252) shows that, to first order in h/ϖ,

$$\left| \frac{\partial \psi}{\partial z} \right| \sim \frac{v_{\mathrm{K}}^2}{\varpi} \frac{h}{\varpi}, \tag{2.260}$$

where the square of the Keplerian speed is

$$v_{\mathrm{K}}^2 = \frac{GM_{\mathrm{p}}}{\varpi}. \tag{2.261}$$

It follows that

$$\frac{|v_\varpi \partial v_z / \partial \varpi|}{|\partial \psi / \partial z|} \sim \left(\frac{v_\varpi}{v_{\mathrm{K}}} \right)^2, \tag{2.262}$$

and similarly for the v_z^2 inertial term. Equations (2.258) and (2.261) give

$$\left| \frac{v_\varpi}{v_{\mathrm{K}}} \right| \sim \frac{\nu}{(GM_{\mathrm{p}}\varpi)^{\frac{1}{2}}}. \tag{2.263}$$

Consider the situation $|v_\varpi| \sim |v_{\mathrm{K}}|$, which requires $\nu \sim (GM_{\mathrm{p}}\varpi)^{1/2}$. Taking $M_{\mathrm{p}} = 1M_\odot$ and $\varpi = 10^8$m yields $\nu \sim 10^{14}\mathrm{m}^2\mathrm{s}^{-1}$. This is a viscosity far larger than that attainable via atomic transport processes, so a turbulent origin is considered. The turbulent viscosity can be expressed in a parameterized form as

$$\nu = \epsilon c_{\mathrm{S}} \lambda_{\mathrm{T}}, \tag{2.264}$$

where c_{S} is the isothermal sound speed, given by

$$c_{\mathrm{S}} = \left(\frac{P}{\rho}\right)^{\frac{1}{2}} = \left(\frac{\Re}{\mu}T\right)^{\frac{1}{2}}, \tag{2.265}$$

λ_{T} is a mixing length ($\lesssim h$) and $\epsilon < 1$ for subsonic turbulence. Hence $|v_{\varpi}| \sim |v_{\mathrm{K}}|$ requires

$$\nu = \epsilon \left(\frac{\Re}{\mu}T\right)^{\frac{1}{2}} \lambda_{\mathrm{T}} \sim 10^{14}\mathrm{m}^2\mathrm{s}^{-1}.$$

Taking $\epsilon = 0.1$, $\mu = 0.6$ and $T = 10^5$ K yields $\lambda_{\mathrm{T}} \sim 2.5 \times 10^{10}$m, implying that at $\varpi = 10^8$m, $h/\varpi \gtrsim 250$. This strongly violates the thin disc condition (2.255), and hence it follows that for a thin disc

$$\left|\frac{v_{\varpi}}{v_{\mathrm{K}}}\right| \ll 1 \tag{2.266}$$

is required. Nevertheless, the disc viscosity coefficient must still be orders of magnitude larger than a molecular form. Hence the turbulent parameterization (2.264) is still employed, but having $\lambda_{\mathrm{T}} \lesssim 3 \times 10^6$ m in a thin disc. Equations (2.262) and (2.266) show that the inertial terms are ignorable in the vertical momentum equation (2.251), which therefore becomes

$$\frac{1}{\rho}\frac{\partial P}{\partial z} + \frac{\partial \psi}{\partial z} = 0, \tag{2.267}$$

corresponding to hydrostatic equilibrium.

Consider, next, the radial momentum equation (2.249). Using (2.252) for ψ, the gravity term can be written

$$\frac{\partial \psi}{\partial \varpi} = \frac{v_{\mathrm{K}}^2}{\varpi}.$$

Then

$$\frac{|\partial(v_{\varpi}^2/2)/\partial\varpi|}{\partial\psi/\partial\varpi} \sim \frac{|v_z \partial v_{\varpi}/\partial z|}{\partial\psi/\partial\varpi} \sim \left(\frac{v_{\varpi}}{v_{\mathrm{K}}}\right)^2 \ll 1,$$

so these inertial terms are small and (2.249) becomes

$$\frac{v_\phi^2}{\varpi} = \frac{v_{\rm K}^2}{\varpi} + \frac{1}{\rho}\frac{\partial P}{\partial \varpi}.$$

It follows from (2.265) that

$$\frac{|\partial P/\partial \varpi|}{\rho v_{\rm K}^2/\varpi} \sim \left(\frac{c_{\rm S}}{v_{\rm K}}\right)^2.$$

However, the vertical equilibrium equation (2.267), together with (2.252) and (2.265), gives

$$\frac{c_{\rm S}}{v_{\rm K}} \sim \frac{h}{\varpi} \ll 1, \tag{2.268}$$

so the radial momentum for a thin disc yields

$$v_\phi = v_{\rm K} = \left(\frac{GM_{\rm p}}{\varpi}\right)^{\frac{1}{2}}, \tag{2.269}$$

with a corresponding angular velocity

$$\Omega = \Omega_{\rm K} = \left(\frac{GM_{\rm p}}{\varpi^3}\right)^{\frac{1}{2}}. \tag{2.270}$$

The continuity equation (2.253) can be integrated vertically through the disc, using the surface condition (2.256b), to give

$$\frac{d}{d\varpi}\int_{-h}^{h} \varpi \rho v_\varpi dz = 0.$$

The thin disc condition (2.255) allows the radial differential operator to be taken out of the integral, even though its limits depend on ϖ. The integral is then a constant, being related to the steady mass transfer rate through the disc by

$$\dot{M} = -2\pi \int_{-h}^{h} \varpi \rho v_\varpi dz, \tag{2.271}$$

where the factor of 2π arises from azimuthal integration.

The azimuthal momentum equation (2.250) describes the advection of specific angular momentum $\varpi^2\Omega$ through the disc. Combining this with the continuity equation (2.253) gives

$$\frac{\partial}{\partial \varpi}(\varpi \rho v_\varpi \varpi^2 \Omega) + \frac{\partial}{\partial z}(\varpi \rho v_z \varpi^2 \Omega) = \frac{\partial}{\partial \varpi}\left(\rho \nu \varpi^3 \frac{d\Omega}{d\varpi}\right). \tag{2.272}$$

Integrating vertically, taking ν as z-independent, yields

$$-\frac{d}{d\varpi}(\dot{M}\varpi^2\Omega) = \frac{d}{d\varpi}\left(2\pi\varpi^3\nu\Sigma\frac{d\Omega}{d\varpi}\right), \tag{2.273}$$

where

$$\Sigma = \int_{-h}^{h} \rho dz \tag{2.274}$$

is twice the mass per unit surface area. Radial integration of (2.273) gives

$$\dot{M}\varpi^2\Omega + 2\pi\varpi^3\nu\Sigma\frac{d\Omega}{d\varpi} = C, \tag{2.275}$$

where C is a constant equal to the total radial flux of angular momentum, being the sum of material and viscous contributions. For a non-magnetic primary, the disc extends to the stellar surface $\varpi = R_{\rm p}$. For material to accrete the star must be rotating below break-up speed at its equator, requiring $\Omega_{\rm p} < \Omega_{\rm K}(R_{\rm p})$. It follows that Ω must turn over and decrease to $\Omega_{\rm p}$ through a boundary layer. If this layer has width δ, then $d\Omega/d\varpi = 0$ at $\varpi = R_{\rm p} + \delta$ where $\delta \ll R_{\rm p}$. For a sharp turn over at the edge of the layer, Ω will be close to its Keplerian value so (2.275) yields

$$C = \dot{M}(R_{\rm p} + \delta)^2\Omega_{\rm K}(R_{\rm p} + \delta).$$

Since $\delta \ll R_{\rm p}$, this can be written as

$$C = \dot{M}R_{\rm p}^2\Omega_{\rm K}(R_{\rm p}), \tag{2.276}$$

provided it is remembered that $\varpi > R_{\rm p} + \delta$ in the subsequent disc solution. Equations (2.270) and (2.275) then give

$$\nu\Sigma = \frac{\dot{M}}{3\pi}\left[1 - \left(\frac{R_{\rm p}}{\varpi}\right)^{\frac{1}{2}}\right]. \tag{2.277}$$

For a thin disc, it follows from (A18) that the dominant elements in the rate of strain tensor are

$$e_{\varpi\phi} = e_{\phi\varpi} = \frac{1}{2}\varpi\frac{\partial\Omega}{\partial\varpi}.$$

Hence (2.84) gives the viscous dissipation rate per unit volume as

$$Q_{\rm v} = 4\rho\nu e_{\varpi\phi}e_{\varpi\phi} = \rho\nu\left(\varpi\frac{\partial\Omega}{\partial\varpi}\right)^2. \tag{2.278}$$

The viscous dissipation per unit area of disc surface is then

$$U(\varpi) = \int_0^h Q_{\rm v} dz = \frac{1}{2}\nu\Sigma(\varpi\Omega')^2, \tag{2.279}$$

so use of (2.270) and (2.277) gives

$$U(\varpi) = \frac{3GM_{\rm p}\dot{M}_{\rm p}}{8\pi\varpi^3}\left[1 - \left(\frac{R_{\rm p}}{\varpi}\right)^{\frac{1}{2}}\right]. \tag{2.280}$$

Integration of this radially across the disc yields the luminosity $L_{\rm D}$, in agreement with (2.248).

The inflow speed v_ϖ can now be found. Since $\Omega_{\rm K}$ is independent of z, the azimuthal momentum equation (2.250) gives

$$v_\varpi \frac{d}{d\varpi}(\varpi^2\Omega_{\rm K}) = \frac{1}{\varpi\rho}\frac{\partial}{\partial\varpi}\left(\rho\nu\varpi^3\frac{d\Omega_{\rm K}}{d\varpi}\right). \tag{2.281}$$

This shows that if ν varies slowly with z, then so does v_ϖ. The mass transfer rate integral (2.271) then yields

$$\dot{M} = -2\pi\varpi\Sigma v_\varpi. \tag{2.282}$$

The inflow speed follows by eliminating $\Sigma/\dot{M}$ between this and (2.277), so

$$v_\varpi = -\frac{3\nu}{2\varpi}\left[1 - \left(\frac{R_{\rm p}}{\varpi}\right)^{\frac{1}{2}}\right]^{-1}. \tag{2.283}$$

The thermal problem remains to be solved. In an optically thick disc with heat transport via radiation, (2.97) gives the radiative flux through a z =constant surface as

$$F_z = -\frac{16\sigma_{\rm B}T^3}{3\kappa\rho}\frac{\partial T}{\partial z}, \tag{2.284}$$

where $\sigma_{\rm B}$ is the Stefan-Boltzmann constant and κ the Rosseland mean opacity. For thermal equilibrium, the divergence of the total flux is equal to the rate of viscous dissipation per unit volume, $Q_{\rm v}$. Since the vertical derivative terms dominate in $\nabla \cdot \mathbf{F}$, the equilibrium is

$$\frac{\partial F_z}{\partial z} = Q_{\rm v}.$$

Then, noting that $F_z(\varpi, 0) = 0$,

$$F_z(\varpi, h) = \int_0^h Q_{\rm v} dz = U(\varpi), \tag{2.285}$$

where $U(\varpi)$ is the dissipation per unit surface area, given by (2.280). Equation (2.284) yields

$$\int_0^h \kappa\rho F_z dz = \frac{4\sigma_{\rm B}}{3}T_c^4.$$

Approximating the integral as $\kappa_c\rho_c h F_z(\varpi, h)$, and taking the optical depth as

$$\tau = \kappa_c\rho_c h, \tag{2.286}$$

gives

$$F_z(\varpi, h) = \frac{4\sigma_{\rm B}}{3\tau}T_c^4. \tag{2.287}$$

The foregoing equations describing a thin viscous disc are;

$$v_\varpi = -\frac{3\nu}{2\varpi}\left[1-\left(\frac{R_{\rm p}}{\varpi}\right)^{\frac{1}{2}}\right]^{-1}, \tag{2.288}$$

$$v_\phi = \left(\frac{GM_{\rm p}}{\varpi}\right)^{\frac{1}{2}}, \tag{2.289}$$

$$h = \sqrt{2}\frac{c_{\rm s}}{v_\phi}\varpi, \tag{2.290}$$

$$c_{\rm s} = \left(\frac{P}{\rho}\right)^{\frac{1}{2}}, \tag{2.291}$$

$$\rho_c = \Sigma/h, \tag{2.292}$$

$$P_c = \frac{\Re}{\mu}\rho_c T_c, \tag{2.293}$$

$$\frac{4\sigma_{\rm B}}{3\tau}T_c^4 = \frac{3GM_{\rm p}\dot{M}_{\rm p}}{8\pi\varpi^3}\left[1-\left(\frac{R_{\rm p}}{\varpi}\right)^{\frac{1}{2}}\right], \tag{2.294}$$

where (2.290) employs the mean value theorem using the vertical average $\langle z\rho\rangle = \rho_c h/2$ in (2.267), and (2.292) uses $\langle\rho\rangle = \rho_c/2$ in (2.274).

The standard disc model uses a simple mixing length prescription for the viscosity due to turbulence, given by

$$\nu = \epsilon c_{\mathrm{s}} h, \tag{2.295}$$

where $\epsilon < 1$ for subsonic turbulence. The opacity is specified by Kramers' law as

$$\kappa = 6.6 \times 10^{18} \rho T^{-\frac{7}{2}} \,\mathrm{m}^2\,\mathrm{Kg}^{-1}. \tag{2.296}$$

The set of disc equations is then algebraic and their solution yields,

$$h = 1.7 \times 10^6 \epsilon^{-\frac{1}{10}} \dot{M}_{13}^{\frac{3}{20}} M_1^{-\frac{3}{8}} \varpi_8^{\frac{9}{8}} f^{\frac{3}{20}} \,\mathrm{m}, \tag{2.297}$$

$$\rho_c = 3.1 \times 10^{-5} \epsilon^{-\frac{7}{10}} \dot{M}_{13}^{\frac{11}{20}} M_1^{\frac{5}{8}} \varpi_8^{-\frac{15}{8}} f^{\frac{11}{20}} \,\mathrm{Kg}\,\mathrm{m}^{-3}, \tag{2.298}$$

$$T_c = 1.4 \times 10^4 \epsilon^{-\frac{1}{5}} \dot{M}_{13}^{\frac{3}{10}} M_1^{\frac{1}{4}} \varpi_8^{-\frac{3}{4}} f^{\frac{3}{10}} \,\mathrm{K}, \tag{2.299}$$

$$\tau = 33 \epsilon^{-\frac{4}{5}} \dot{M}_{13}^{\frac{1}{5}} f^{\frac{1}{5}}, \tag{2.300}$$

$$\nu = 1.8 \times 10^{10} \epsilon^{\frac{4}{5}} \dot{M}_{13}^{\frac{3}{10}} M_1^{-\frac{1}{4}} \varpi_8^{\frac{3}{4}} f^{\frac{3}{10}} \,\mathrm{m}^2\,\mathrm{s}^{-1}, \tag{2.301}$$

$$v_{\varpi} = -2.7 \times 10^2 \epsilon^{\frac{4}{5}} \dot{M}_{13}^{\frac{3}{10}} M_1^{-\frac{1}{4}} \varpi^{-\frac{1}{4}} f^{-\frac{7}{10}} \,\mathrm{m}\,\mathrm{s}^{-1}, \tag{2.302}$$

where $\dot{M}_{13} = \dot{M}_{\mathrm{p}}/10^{13}\,\mathrm{Kg}\,\mathrm{s}^{-1}$, $M_1 = M_{\mathrm{p}}/M_{\odot}$, $\varpi_8 = \varpi/10^8\,\mathrm{m}$, and

$$f = 1 - \left(\frac{R_{\mathrm{p}}}{\varpi}\right)^{\frac{1}{2}}. \tag{2.303}$$

If the accreting star has a significant magnetic field, the viscous disc solution can be modified. A high stellar magnetic moment can result in complete disruption of the disc. A magnetic alternative to the viscous disc, with a dynamo generated field, is presented in Chapter 11.

Time-Dependent Viscous Discs

For a time-dependent, axisymmetric viscous disc, the azimuthal momentum and continuity equations can be written

$$\frac{\partial}{\partial t}(\varpi^2\Omega) + v_\varpi \frac{\partial}{\partial \varpi}(\varpi^2\Omega) + v_z \frac{\partial}{\partial z}(\varpi^2\Omega) = \frac{1}{\varpi\rho}\frac{\partial}{\partial \varpi}\left(\rho\nu\varpi^3 \frac{\partial \Omega}{\partial \varpi}\right), \quad (2.304)$$

$$-\frac{\partial \rho}{\partial t} = \frac{1}{\varpi}\frac{\partial}{\partial \varpi}(\varpi\rho v_\varpi) + \frac{\partial}{\partial z}(\rho v_z). \quad (2.305)$$

If Ω remains Keplerian at all heights in the disc, then $\partial\Omega/\partial t = \partial\Omega/\partial z = 0$ and (2.304) gives

$$\varpi\rho v_\varpi \frac{d}{d\varpi}(\varpi^2\Omega_{\mathrm{K}}) = \frac{\partial}{\partial \varpi}\left(\rho\nu\varpi^3 \frac{d\Omega_{\mathrm{K}}}{d\varpi}\right). \quad (2.306)$$

Vertical integration of (2.305) and (2.306) through the disc yields

$$-\frac{\partial \Sigma}{\partial t} = \frac{1}{\varpi}\frac{\partial}{\partial \varpi}(\varpi\Sigma v_\varpi), \quad (2.307)$$

$$\varpi\Sigma v_\varpi \frac{d}{d\varpi}(\varpi^2\Omega_{\mathrm{K}}) = \frac{\partial}{\partial \varpi}\left(\nu\Sigma\varpi^3 \frac{d\Omega_{\mathrm{K}}}{d\varpi}\right), \quad (2.308)$$

where Σ is the surface density function given by (2.274), and $\rho(\varpi, \pm h) = 0$ is used. Elimination of $\varpi\Sigma v_\varpi$ between (2.307) and (2.308), together with the use of (2.270) for Ω_{K}, leads to

$$\frac{\partial \Sigma}{\partial t} = \frac{3}{\varpi}\frac{\partial}{\partial \varpi}\left(\varpi^{\frac{1}{2}}\frac{\partial}{\partial \varpi}\left(\varpi^{\frac{1}{2}}\nu\Sigma\right)\right). \quad (2.309)$$

This is a diffusion equation for Σ. For realistic disc models ν will depend on Σ and hence the diffusion equation will be non-linear.

The viscosity has the effect of spreading material in the disc, at a rate which increases with spatial gradients. If ν and Σ have radial length-scales of $\sim \varpi$, then (2.309) yields a characteristic viscous time-scale of

$$\tau_{\mathrm{V}} \sim \frac{\varpi^2}{\nu}. \quad (2.310)$$

The instantaneous radial velocity follows from (2.270) and (2.308) as

$$v_\varpi(\varpi, t) = -\frac{3}{\varpi^{\frac{1}{2}}\Sigma}\frac{\partial}{\partial \varpi}\left(\varpi^{\frac{1}{2}}\nu\Sigma\right), \quad (2.311)$$

and hence

$$|v_{\varpi}| \sim \frac{\nu}{\varpi}. \tag{2.312}$$

This shows that $\tau_{\rm V}$ may be expressed as

$$\tau_{\rm V} \sim \frac{\varpi}{|v_{\varpi}|}, \tag{2.313}$$

which is also the radial drift time-scale. The viscosity redistributes angular momentum and mass radially through the disc on the time-scale $\tau_{\rm V}$. Density distributions of length-scale $\ell < \varpi$ diffuse more rapidly with larger associated radial drift velocities.

Other time-scales relevant to discs are the dynamical time and the thermal time. The dynamical time-scale associated with Keplerian rotation is

$$\tau_{\phi} \sim \frac{\varpi}{v_{\phi}} \sim \frac{1}{\Omega_{\rm K}}. \tag{2.314}$$

Deviations from vertical hydrostatic equilibrium are smoothed out on the dynamical time-scale

$$\tau_z \sim \frac{h}{c_{\rm S}}. \tag{2.315}$$

Since $c_{\rm S} \sim (h/\varpi)v_{\rm K}$ for a thin viscous disc, it follows that $\tau_z \sim \tau_{\phi}$.

The thermal time-scale $\tau_{\rm th}$ gives the evolution time of perturbations from thermal equilibrium. This can be estimated by noting that the thermal energy density in a perfect gas is $\sim P = \rho c_{\rm S}^2$, while the viscous dissipation rate per unit volume is $\sim \nu\Sigma(\varpi\Omega_{\rm K}')^2/h$. Hence, since $|\varpi\Omega_{\rm K}'| \sim \Omega_{\rm K}$,

$$\tau_{\rm th} \sim \frac{\rho h c_{\rm S}^2}{\nu\Sigma\Omega_{\rm K}^2} \sim \left(\frac{c_{\rm S}}{v_{\rm K}}\right)^2 \frac{\varpi^2}{\nu},$$

so that

$$\tau_{\rm th} \sim \left(\frac{h}{\varpi}\right)^2 \tau_{\rm V}. \tag{2.316}$$

The dynamical time-scale can be related to the viscous time-scale by using the viscosity prescription (2.264), together with $c_{\rm S} \sim (h/\varpi)v_{\rm K}$ from (2.268). Then

$$\tau_{\rm V} \sim \frac{\varpi^2}{\nu} \sim \frac{\varpi^2}{\epsilon c_{\rm S} h} \sim \frac{1}{\epsilon}\left(\frac{\varpi}{h}\right)^2 \frac{\varpi}{v_{\rm K}} \sim \frac{1}{\epsilon}\left(\frac{\varpi}{h}\right)^2 \tau_{\phi},$$

giving

$$\tau_\phi \sim \epsilon \left(\frac{h}{\varpi}\right)^2 \tau_{\rm V} \sim \epsilon \tau_{\rm th}. \tag{2.317}$$

It follows that, since $\epsilon < 1$, the ordering of the time-scales is

$$\tau_\phi \sim \tau_z < \tau_{\rm th} \ll \tau_{\rm V}. \tag{2.318}$$

The main use of time-dependent disc theory is in the analysis of waves and instabilities. The significant differences between the various time-scales, shown by (2.318), means that different types of perturbations can be identified. Consider an axisymmetric perturbation in the surface density function, so

$$\Sigma = \Sigma_0(\varpi) + \delta\Sigma(\varpi, t).$$

Substitution in (2.309) gives

$$\frac{\partial}{\partial t}(\delta\Sigma) = \frac{3}{\varpi}\frac{\partial}{\partial\varpi}\left(\varpi^{\frac{1}{2}}\frac{\partial}{\partial\varpi}\left(\varpi^{\frac{1}{2}}\delta\mu\right)\right),$$

where $\mu = \nu\Sigma$. The viscosity can be expressed in the form $\nu = \nu(\varpi, \Sigma)$, with $\nu \propto \varpi^{15/14}\Sigma^{3/7}$ in the foregoing model. Then $\mu = \mu(\varpi, \Sigma)$, so

$$\delta\mu = \frac{\partial\mu_0}{\partial\Sigma_0}\delta\Sigma$$

and $\delta\Sigma$ can be eliminated from the diffusion equation to give

$$\frac{\partial}{\partial t}(\delta\mu) = \frac{\partial\mu_0}{\partial\Sigma_0}\frac{3}{\varpi}\frac{\partial}{\partial\varpi}\left(\varpi^{\frac{1}{2}}\frac{\partial}{\partial\varpi}\left(\varpi^{\frac{1}{2}}\delta\mu\right)\right). \tag{2.319}$$

The diffusion coefficient is therefore proportional to $\partial\mu_0/\partial\Sigma_0$. For positive $\partial\mu_0/\partial\Sigma_0$ a perturbation decays on the viscous time-scale $\tau_{\rm V}$. However, if $\partial\mu_0/\partial\Sigma_0$ is negative a perturbation in Σ will grow due to viscous instability. More material will be fed into regions that are denser than their surroundings, and the disc will tend to break up into rings on the time-scale $\tau_{\rm V}$. A steady flow therefore requires $\partial\mu_0/\partial\Sigma_0 > 0$.

It is noted that the foregoing standard viscous accretion disc appears to be non-magnetic. However, as will be seen in Chapter 11, the presence of at least a weak magnetic field is believed to be necessary for the generation and maintenance of turbulence in Keplerian discs. Dynamo action then leads to large-scale magnetic fields, and the associated stresses play a part in the radial advection of angular momentum causing the inflow.

2.5. Spin Dynamics

In spin evolution calculations the compact white dwarf, or neutron star, is treated as a rigid body. This is valid provided the dynamical time-scale for adjustments in the stellar structure is short compared to the spin evolution time-scale. The cases considered here satisfy this condition. A strongly magnetic primary star will experience some distortion from spherical symmetry due to non-radial internal magnetic forces.

The angular momentum of a rigid body of volume V rotating with instantaneous angular velocity $\boldsymbol{\omega}$ is

$$\mathbf{L} = \int_V \mathbf{r} \wedge (\boldsymbol{\omega} \wedge \mathbf{r}) \rho dV,$$

where ρ is the density and $\mathbf{r}$ is measured from the centre of mass, O. Expanding the cross product gives

$$\mathbf{L} = \int_V [r^2 \boldsymbol{\omega} - (\mathbf{r} \cdot \boldsymbol{\omega})\mathbf{r}] \rho dV. \tag{2.320}$$

In an inertial coordinate frame $Oxyz$ the components of $\mathbf{L}$ can be written

$$\begin{aligned} L_x &= I_{xx}\omega_x + I_{xy}\omega_y + I_{xz}\omega_z, \\ L_y &= I_{yx}\omega_x + I_{yy}\omega_y + I_{yz}\omega_z, \\ L_z &= I_{zx}\omega_x + I_{zy}\omega_y + I_{zz}\omega_z, \end{aligned} \tag{2.321}$$

where

$$I_{xx} = \int_V (y^2 + z^2) \rho dV, \qquad I_{yy} = \int_V (x^2 + z^2) \rho dV,$$

$$I_{zz} = \int_V (x^2 + y^2) \rho dV \tag{2.322}$$

are moments of inertia, while

$$I_{xy} = I_{yx} = -\int_V xy\rho dV, \qquad I_{xz} = I_{zx} = -\int_V xz\rho dV,$$

$$I_{yz} = I_{zy} = -\int_V yz\rho dV \tag{2.323}$$

are products of inertia. Equation (2.321) can be expressed in Cartesian tensor form as

$$L_i = I_{ij}\omega_j, \tag{2.324}$$

where the repeated index is summed over. The inertia tensor I_{ij} is symmetric and (2.324) shows that, in general, $\mathbf{L}$ and $\boldsymbol{\omega}$ are not parallel.

Since the inertia tensor is symmetric it corresponds to a real Hermitian matrix. Such a matrix has a complete set of eigenvectors which therefore form a basis in which an arbitrary vector can be represented. The matrix elements I_{ij} will depend on the basis used to represent the column vector on which I acts. The simplest form of I results when its eigenvectors are chosen as the basis. This diagonalization process involves the similarity transformation

$$I(\text{diag}) = X^{-1}IX,$$

where the modal matrix $X = (X_1 X_2 X_3)$ with X_i the eigenvectors of I. This rotates the frame $Oxyz$ to be coincident with the orthogonal eigenvectors. The inertia tensor then has the form

$$I = \begin{pmatrix} I_1 & 0 & 0 \\ 0 & I_2 & 0 \\ 0 & 0 & I_3 \end{pmatrix}. \tag{2.325}$$

The diagonal elements are the principal moments of inertia about the eigenvectors of I which have directions denoted by the unit vectors $\mathbf{e}_1$, $\mathbf{e}_2$ and $\mathbf{e}_3$. These orthogonal principal axes are fixed in the body. Relative to this frame, the angular velocity and angular momentum are

$$\boldsymbol{\omega} = \omega_1\mathbf{e}_1 + \omega_2\mathbf{e}_2 + \omega_3\mathbf{e}_3, \tag{2.326}$$

$$\mathbf{L} = I_1\omega_1\mathbf{e}_1 + I_2\omega_2\mathbf{e}_2 + I_3\omega_3\mathbf{e}_3. \tag{2.327}$$

It is noted that $\boldsymbol{\omega}$ and $\mathbf{L}$ are only parallel if the body is spherical or rotates about a principal axis.

The motion of a rigid body is described by the equation

$$\frac{d\mathbf{L}}{dt} = \mathbf{T}, \tag{2.328}$$

where the time derivative is measured in inertial space, and $\mathbf{T}$ is the torque. The principal unit vectors of the body frame have inertial time derivatives given by

$$\frac{d\mathbf{e}_i}{dt} = \boldsymbol{\omega} \wedge \mathbf{e}_i.$$

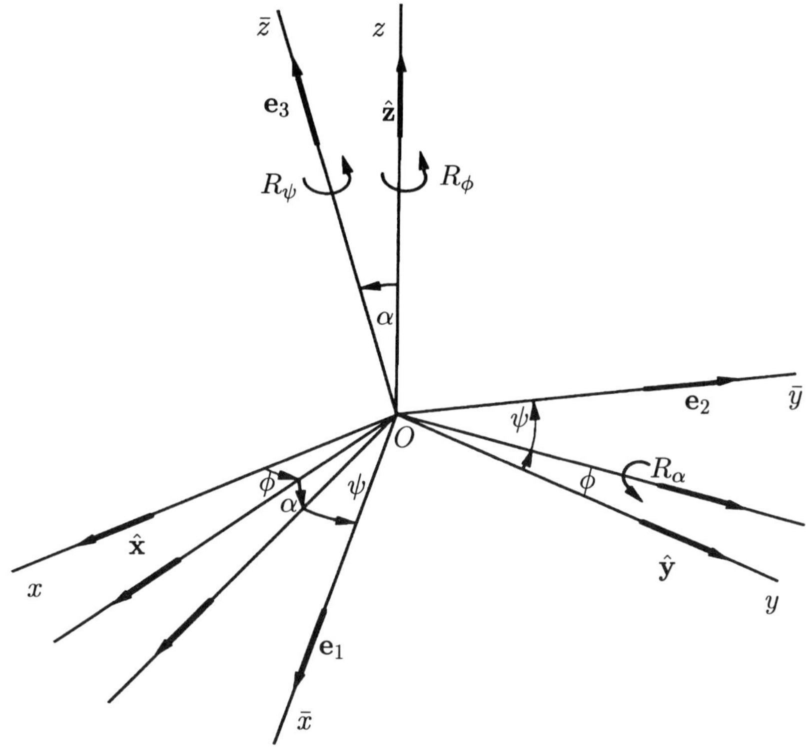

Figure 2.2. The rotation sequence generating an orientation with Euler angles (α, ϕ, ψ).

Denoting the time derivative relative to the body frame by a dot, it follows that

$$\frac{d\mathbf{L}}{dt} = \frac{d}{dt}(L_i \mathbf{e}_i) = \dot{L}_i \mathbf{e}_i + L_i \boldsymbol{\omega} \wedge \mathbf{e}_i,$$

and hence (2.328) can be written as

$$\dot{\mathbf{L}} + \boldsymbol{\omega} \wedge \mathbf{L} = \mathbf{T}. \tag{2.329}$$

The orientation of a rigid body about its centre of mass is specified by three angles. The Euler angles (α, ϕ, ψ) are usually chosen. The angles α and ϕ define the orientation of one axis, say $\mathbf{e}_3$, while ψ defines a rotation of the body about this axis from a standard position. Figure 2.2 shows how the orientation (α, ϕ, ψ) can be attained by three rotations. Firstly, a rotation of ϕ about the z-axis is generated by the matrix

$$R_\phi = \begin{pmatrix} \cos\phi & \sin\phi & 0 \\ -\sin\phi & \cos\phi & 0 \\ 0 & 0 & 1 \end{pmatrix}. \tag{2.330}$$

Secondly, a rotation of α about the new y-axis is generated by

$$R_\alpha = \begin{pmatrix} \cos\alpha & 0 & -\sin\alpha \\ 0 & 1 & 0 \\ \sin\alpha & 0 & \cos\alpha \end{pmatrix}. \tag{2.331}$$

Finally, a rotation of ψ about the new z-axis is given by

$$R_\psi = \begin{pmatrix} \cos\psi & \sin\psi & 0 \\ -\sin\psi & \cos\psi & 0 \\ 0 & 0 & 1 \end{pmatrix}. \tag{2.332}$$

The total rotation is then the matrix product.

$$R = R_\psi R_\alpha R_\phi. \tag{2.333}$$

Equations (2.330)–(2.333) give the elements of the rotation matrix as;

$$\begin{aligned}
R_{11} &= \cos\alpha\cos\phi\cos\psi - \sin\phi\sin\psi, \\
R_{12} &= \cos\alpha\sin\phi\cos\psi + \cos\phi\sin\psi, \\
R_{13} &= -\sin\alpha\cos\psi, \\
R_{21} &= -\cos\alpha\cos\phi\sin\psi - \sin\phi\cos\psi, \\
R_{22} &= -\cos\alpha\sin\phi\sin\psi + \cos\phi\cos\psi, \\
R_{23} &= \sin\alpha\sin\psi, \\
R_{31} &= \sin\alpha\cos\phi, \\
R_{32} &= \sin\alpha\sin\phi, \\
R_{33} &= \cos\alpha.
\end{aligned} \tag{2.334}$$

The components of a vector $\mathbf{V}$ relative to the frame $Oxyz$ are related to those relative to the body frame $O\bar{x}\bar{y}\bar{z}$ by

$$\bar{V}_i = R_{ij}V_j\ , \tag{2.335}$$

where the summation convention is used.

The instantaneous angular velocity can be expressed as the sum of the angular velocities $\dot{\alpha}$, $\dot{\phi}$ and $\dot{\psi}$ about the directions $\hat{\boldsymbol{\alpha}}$, $\hat{\boldsymbol{\phi}}$ and $\hat{\boldsymbol{\psi}}$, so

$$\boldsymbol{\omega} = \dot{\alpha}\hat{\boldsymbol{\alpha}} + \dot{\phi}\hat{\boldsymbol{\phi}} + \dot{\psi}\hat{\boldsymbol{\psi}}. \tag{2.336}$$

It follows from Figure 2.2 that the Euler angular velocities relative to the body frame are

$$\dot{\boldsymbol{\alpha}} = \dot{\alpha}\sin\psi\mathbf{e}_1 + \dot{\alpha}\cos\psi\mathbf{e}_2, \tag{2.337}$$

$$\dot{\boldsymbol{\phi}} = -\dot{\phi}\sin\alpha\cos\psi\mathbf{e}_1 + \dot{\phi}\sin\alpha\sin\psi\mathbf{e}_2 + \dot{\phi}\cos\alpha\mathbf{e}_3, \tag{2.338}$$

$$\dot{\boldsymbol{\psi}} = \dot{\psi}\mathbf{e}_3. \tag{2.339}$$

Equations (2.336)–(2.339) give the components of $\boldsymbol{\omega}$ in the body frame in terms of the Euler angles α, ϕ and ψ as

$$\begin{aligned} \omega_1 &= \dot{\alpha}\sin\psi - \dot{\phi}\sin\alpha\cos\psi, \\ \omega_2 &= \dot{\alpha}\cos\psi + \dot{\phi}\sin\alpha\sin\psi, \\ \omega_3 &= \dot{\phi}\cos\alpha + \dot{\psi}. \end{aligned} \tag{2.340}$$

The torque $\mathbf{T}$ in (2.329) will, in general, be a function of α, ϕ, and ψ, so the components of this equation yield a set of coupled differential equations. The solution of these gives the Euler angles as functions of time, and hence the rotational motion of the body.

In spin stability problems the equation of motion is linearized about a given state. Normal mode solutions can then be sought involving the Euler angle perturbations. In general, the fact that the basic state is dynamical means that independent normal modes may not always exist. However, the high rotation rates occurring in close binary stars can lead to separated modes.

References

Battener, E., 1996. *Astrophysical Fluid Dynamics*, Cambridge University Press, Cambridge.
Campbell, C.G. and Papaloizou, J.C.B., 1983. *Mon. Not. R. Astr. Soc.*, **204**, 433.
Cowling, T.G., 1934. *Mon. Not. R. Astr. Soc.*, **94**, 39.
Cox, J.P. and Giuli, R.T., 1968. *Principles of Stellar Structure*, Vol I, Gordon and Breach.
Eggleton, P.P., 1983. *Astrophys. J.*, **268**, 368.
Faulkner, J., 1971. *Astrophys. J. Lett.*, **170**, L99.
Goldreich, P. and Keeley, D.A., 1977. *Astrophys. J.*, **211**, 934.
Kippenhahn, R. and Weigert, A., 1990. *Stellar Structure and Evolution*, Springer-Verlag.
Kraft, R.P., Mathews, J. and Greenstein, J.L., 1962. *Astrophys. J.*, **136**, 312.
Landau, L. and Lifshitz, E., 1951. *The Classical Theory of Fields*, Addison Wesley.
Lubow, S.H. and Shu, F.H., 1975. *Astrophys. J.*, **198**, 383.
Mestel, L., 1997. *Stellar Magnetic Fields*, Oxford University Press.
Paczyński, B., 1967. *Acta. Astr.*, **17**, 287.
Parker, E.N., 1955. *Astrophys. J.*, **122**, 293.
Parker, E.N., 1979. *Cosmical Magnetic Fields*, Oxford University Press.
Plavec, M. and Kratochvil, P., 1964. *Bull. Astr. Czech.*, **15**, 165.
Roberts, P.H., 1994. In *Lectures on Solar and Planetary Dynamos*, eds. Proctor, M.R.E. and Gilbert, A.D., Cambridge University Press.
Roche, E.N., 1873. *Ann. de l'Acad. Sci. Montpelier.*, **8**, 235.
Shakura, N.I. and Sunyaev, R.A., 1973. *Astron. Astrophys.*, **24**, 337.

Skumanich, A., 1972. *Astrophys. J.*, **171**, 565.
Spitzer, L., 1962. *Physics of Fully Ionized Gases*, Interscience, London.
Verbunt, F. and Zwaan, C., 1981. *Astron. Astrophys.*, **100**, L7.

CHAPTER 3

THE AM HERCULIS STARS

3.1. The Discovery of AM Herculis

Optical observations of AM Herculis extend back to 1890. Berg and Duthie (1977) suggested that AM Her could be the optical counterpart of the high galactic latitude X-ray source 3U 1809 + 50. Further evidence was available from Hearn, Richardson and Clark (1976) which indicated AM Her to be a soft X-ray source. These suggestions were confirmed by Hearn and Richardson (1977). Szkody and Brownlee (1977) found the visual light curve of AM Her to have a broad, deep minimum which repeated on a timescale of 3.1 hr. Linear and circular polarization were observed in the V and I spectral bands by Tapia (1977a), of a strength an order of magnitude larger than previously observed in any object. This suggested the presence of a strong magnetic field, with $B \sim 10^8$ G, assuming the fundamental cyclotron frequency to be observed. Large variations of the polarization were found with a period of 3.1 hr.

Models of AM Her involving an accreting magnetic white dwarf in a binary system were discussed by various authors (e.g. Chanmugam and Wagner, 1977; Michalsky, Stokes and Stokes, 1977; Stockman, 1977). The rotation of the white dwarf was taken to be synchronous with the orbital period, based on the fact that all observed radiations had the same period of intensity variations. The short orbital period of AM Her put it in the category of close binary stars. However, unlike the standard cataclysmic variables, there was no evidence for an accretion disc around the white dwarf. The large linear and circular optical polarization must result from cyclotron radiation emitted by the accreting gas in the strong magnetic field. Highly conducting matter would be channelled by the field, and the 3.1 hr variations were consistent with a localized radiating accretion column changing its orientation to the line of sight due to the white dwarf's rotation.

Matter is lost from the secondary star through the inner Lagrangian point and falls towards the white dwarf primary. The gas stream is ionized by collisions and X-rays from the accreting region. Motion of the conduct-

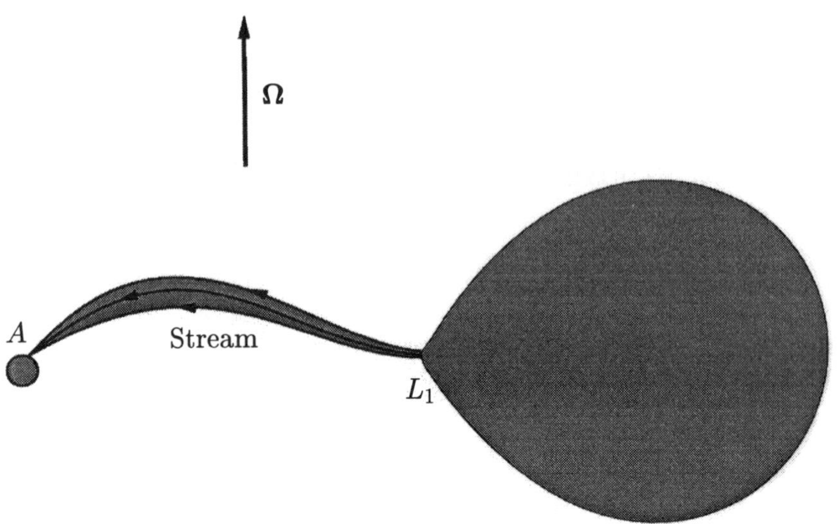

Figure 3.1. An AM Herculis binary, showing the simplest case of single pole accretion. Material lost from the L_1 region of the secondary becomes channelled by the primary's magnetic field and accretes onto its surface through a localized column, A.

ing stream across the primary's magnetic field generates electric currents and a magnetic force is exerted on the material. The infalling matter experiences an increasingly strong magnetic field and ultimately becomes channelled by it onto the surface of the white dwarf. The incoming supersonic stream passes through a standing shock and settles onto the stellar surface through an accretion column. X-rays are emitted in the post-shock flow; most of these are reprocessed by the white dwarf's surface and re-emitted as a softer X-ray component. The basic theory of accretion columns on magnetized white dwarfs was considered by King and Lasota (1979). The shock temperature is given by

$$T_{\rm s} = 3.7 \times 10^8 \left(\frac{M_{\rm p}}{M_\odot}\right) \left(\frac{R_{\rm p}}{10^7 {\rm m}}\right) {\rm K}, \tag{3.1}$$

where $M_{\rm p}$ and $R_{\rm p}$ are the mass and radius of the primary. Electrons in the ionized stream close to the shock spiral around the magnetic field lines and emit strongly polarized cyclotron radiation. The field strength causes this radiation to the emitted at optical and near infra-red wavelengths.

Chanmugam and Wagner (1979) showed that the shocked gas in the accretion column will be optically thick to the lowest cyclotron harmonics, suggesting that the previously measured field strengths were too high since they assumed low harmonics. This was confirmed by Schmidt, Stockman and Margon (1981) who presented spectropolarimetry, obtained during a faint state of AM Her, which showed strong circular polarization and absorption features characteristic of hydrogen in a range of magnetic fields $\sim$ 10–20 MG, significantly lower than previous estimates. A field structure more complicated than that of a centred dipole was also indicated.

Young, Schneider and Shectman (1981) observed AM Her in a low state and measured a 3.1 hr period in emission line velocities. Their observations indicated the secondary star to be an M5 red dwarf. This is consistent with a lobe-filling late main-sequence star.

3.2. Other Systems

Three new members of the class were discovered soon after AM Her; AN UMa (Krzeminski and Serkowski, 1977), VV Pup (Tapia, 1977b) and EF Eri (Tapia, 1979). There are, at present, 48 AM Her systems known, also referred to as the polars. Tables 3.1 and 3.2 list these systems with their orbital periods and, where known, white dwarf surface fields. During episodes of low accretion photospheric Zeeman split absorption lines can be observed from the white dwarf. This method has been used to measure the magnetic fields in AM Her (Schmidt, Stockman and Margon, 1981; Young, Schneider and Shectman, 1981), ST LMi (Schmidt, Stockman and Grandi, 1983), BL Hyi (Wickramasinghe, Visvanathan and Tuohy, 1984), V834 Cen (Beuermann, Thomas and Schwope, 1989; Ferrario *et al*, 1992), EF Eri (Wickramasinghe *et al*, 1990) and MR Ser (Wickramasinghe *et al*, 1991b). The mean field over the observable photosphere is calculated, denoted $\bar{B}_{\rm ph}$ in the tables. Direct measurements of the field at the cyclotron emission regions are possible for those systems showing cyclotron humps in their optical/IR spectra (e.g. VV Pup, V834 Cen, UZ For, MR Ser). The magnetic field at the emission region can be determined from the spacings between the cyclotron harmonics (e.g. Schwope, Beuermann and Thomas, 1990).

Recent observations suggest that the white dwarfs in AM Her systems have magnetic fields more complicated than centred dipoles. Meggit and Wickramasinghe (1989) analysed the linear pulse and polarization data of EF Eri and found three cyclotron emission regions on the white dwarf surface. A quadrupolar component could be consistent with this data. A complex field structure is also indicated in CE Gru (Wickramasinghe *et al*, 1991a) and DP Leo (Cropper and Wickramasinghe, 1993). The cyclotron-determined fields are denoted by $B_{\rm cyc}$ in Tables 3.1 and 3.2, with values in brackets referring to field strengths at second poles, where available.

It is noted that 8 AM Her binaries, representing $\sim$ 17% of the known systems, lie in the $2-3$ hr period gap, usually unoccupied by cataclysmic variables. A possible explanation for this is discussed in Chapter 12.

TABLE 3.1. The AM Herculis binaries

Name	P (hrs)	$M_S/M_\odot$	B_{cyc} (MG)	$\bar{B}_{ph}$
RXJ 0515	7.983	0.77	61	
RXJ 0203	4.591	0.44		
RXJ 1313	4.200	0.40		
EXO 032957	3.806	0.36		
QQ Vul	3.708	0.36	36	
RXJ 2316	3.482	0.33	30	
RXJ 1007	3.466	0.33		
RXJ 0929	3.389	0.33		
RXJ 1940	3.366	0.32		
BY Cam	3.355	0.32	28	
V1500 Cyg	3.351	0.32		
AM Her	3.094	0.30	14	13
RXJ 0501	2.851	0.27	25	
DR V211b	2.662	0.26		
RXJ 1429	2.366	0.23		
QS Tel	2.332	0.22	47 (70)	
RXJ 0531	2.224	0.21	19	
UZ For	2.109	0.20	53 (75)	
EU Cnc	2.090	0.20	42	
HU Aqr	2.084	0.20	35	
AR UMa	1.932	0.19		
WW Hor	1.925	0.18		
AN UMa	1.914	0.18	29	
EK UMa	1.908	0.18	48	
ST LMi	1.898	0.18	12	19
BL Hyi	1.894	0.18		22
MR Ser	1.891	0.18	24	27
RXJ 1802	1.884	0.18		
V2301 Oph	1.883	0.18		7
CE Gru	1.810	0.17		
RXJ 1002	1.766	0.17		
EP Dra	1.744	0.17		
RXJ 0453	1.699	0.16	36	
RXJ 0953	1.699	0.16		
V834 Cen	1.692	0.16	23	22
VV Pup	1.674	0.16	31 (54)	
RXJ 1957	1.647	0.16		
EU UMa	1.502	0.14	43	

TABLE 3.2. The AM Herculis binaries

Name	P (hrs)	$M_{\mathrm{S}}/M_{\odot}$	B_{cyc} (MG)
RXJ 1844	1.501	0.14	
DP Leo	1.497	0.14	31 (59)
RXJ 2315	1.488	0.14	
EF Eri	1.350	0.13	
RXJ 2022	1.334	0.13	67
RXJ 0132	1.334	0.13	68
RXJ 0154	1.334	0.13	
RXJ 1047	1.334	0.13	
RXJ 1015	1.331	0.13	
EV UMa	1.328	0.13	35

The white dwarf masses are not well known, since their determinations require accurate measurements of the radial velocities of both stars and the orbital inclination. A helium white dwarf obeys the mass-radius relation

$$R_{\mathrm{p}} = 7.8 \times 10^{6} \left[\left(\frac{M_{\mathrm{p}}}{M_{\mathrm{C}}} \right)^{-\frac{2}{3}} - \left(\frac{M_{\mathrm{p}}}{M_{\mathrm{C}}} \right)^{\frac{2}{3}} \right]^{\frac{1}{2}} \mathrm{m}, \tag{3.2}$$

where $M_{\mathrm{C}} = 1.44\, M_{\odot}$ (Nauenberg, 1972). Hence, for similar surface field strengths, the primary magnetic moments can vary significantly for a range of M_{p} since they are proportional to R_{p}^{3}.

The secondary in AM Her systems is a low mass red dwarf. Using the condition that it fills its Roche lobe and obeys a main sequence mass-radius relation, its mass can be calculated from the orbital period. Equation (2.226) gives a mass range of $0.13 < M_{\mathrm{s}}/M_{\odot} < 0.46$, with the one higher mass system RXJ 0515 having $M_{\mathrm{s}} = 0.77\, M_{\odot}$. Spectroscopic observations of the secondary are difficult because most of the energy is emitted from the primary. Phase resolved, high resolution red spectroscopy is available for AM Her (Young and Schnieder, 1979), ST LMi, MR Ser, QQ Vul (Mukai and Charles, 1986, 1987; Schwope *et al*, 1991, 1993b), UZ For (Beuermann, Thomas and Schwope, 1988) and V834 Cen (Schwope *et al*, 1993a). The radial velocities can be used for locating the position of the secondary in non-eclipsing systems. Six systems are known to have eclipses of the primary by the secondary: DP Leo (Biermann *et al*, 1985; Bailey *et al*, 1993), WW Hor (Bailey *et al*, 1988), UZ For (Ferrario *et al*, 1989), EP Dra (Remillard *et al*, 1991), HU Aqr (Schwope, Thomas and Beuermann, 1993) and RXJ 0929 (Sekiguchi, Nakada and Bassett, 1994). High resolution observations made during eclipses confirm the standard model of AM Her binaries.

The spectra of AM Her systems, in the UV, optical and infra-red, are rich in emission lines. Phase resolved and high resolution studies of the optical spectra reveal lines with multiple components, each with distinctive radial velocity variations. These lines are thought to arise in the accretion stream. Cowley and Crampton (1977) identified a broad base component, a narrower peak component and a very narrow component in VV Pup. Most of the polars have such components. Schneider and Young (1980) modelled these by assuming the broad component came from near the accretion regions, and the narrower components from further away. A phase shift between broad and narrow components was identified with curvature of the stream. Mukai (1988) suggested that some of the narrow components from QQ Vul, ST LMi and VV Pup arise in an initial part of the stream before it is channelled by the primary's magnetic field. Ferrario *et al* (1989) obtained a reasonable fit to the emission line data of V834 Cen, ST LMi, and UZ For with a model in which field channelling becomes effective at a distance from the L_1 point which is about a third of its distance, A, from the primary. The stream was broadened before converging on the primary. Recent observations of HU Aqr by Schwope, Mantel and Horne (1996), using Doppler tomography, also indicate a partially channelled stream. Analysis of the data suggests that the initial, weakly channelled stream forms a tenuous curtain of material above the orbital plane. The bulk of the stream subsequently lifts out of the orbital plane at a distance from the white dwarf of $\gtrsim 0.3A$.

3.3. Synchronous Rotation of the Primary

Synchronous rotation of the primary with the orbit has always been assumed in AM Her systems, based on the modulation of their emissions at single periods. However, the periodic variations seen in the optical light curves, the X-rays and optical polarization are due to the rotation of the accretion column with the white dwarf. Observations of the relatively faint red dwarf secondary are necessary in order to measure the orbital period. The first direct support for synchronism of the primary came when Young and Schneider (1979) observed absorption lines from the secondary in AM Her. Orbital periods were subsequently measured in other polars, either from eclipses or ellipsoidal variations in the infra-red (e.g. Bailey *et al*, 1985). The primary and orbital angular velocities were found to be the same, to the accuracy limits of the methods employed. For the eclipsing system DP Leo, over a baseline of four years, Biermann *et al* (1985) obtained $|\omega|/\Omega < 2 \times 10^{-6}$, where ω is the synodic angular velocity of the primary and Ω is the orbital angular velocity. Cropper (1988) compiled data for twelve polars and found a tendency for the main accreting pole to lead the motion of the line of stellar centres. However, Bailey *et al* (1993) showed that the longitudes of the magnetic poles in DP Leo and WW Hor can vary by up to 20° over several years.

The white dwarf spin period in V1500 Cyg has been found to be a few percent shorter than the orbital period (Katz, 1991). This may be a result of its recent nova outburst.

3.4. MHD Problems

The synchronous rotation of the primary and channelling of the accretion stream present problems of magnetohydrodynamics. The removal of asynchronous motions of the primary requires a dissipative process, and magnetic interaction with an orbitally synchronized secondary gives a plausible mechanism for this. Asynchronous motions of the primary will induce electric currents in the conducting secondary, and these are dissipated causing the primary to approach synchronism.

The accretion stream interacts with the white dwarf's magnetic field, ultimately being channelled by it. The reaction to the magnetic force exerted on the stream leads to a torque felt by the primary. This accretion torque depends on the geometry of the stream, and hence on the magnetic orientation of the primary relative to the secondary. The torque will be time-dependant for an asynchronous primary.

The dissipative magnetic torque, which removes asynchonous motions, cannot balance the accretion torque to produce a synchronous state, since the former vanishes at corotation. A non-dissipative torque is necessary and interaction of the primary with a magnetized secondary generates such a torque. The secondary's field could be produced by dynamo action. The torque balance must be stable to all perturbations.

Even if a stable synchronous state exists it is not clear that the white dwarf can reach it while accreting material. Overshooting could occur if the dissipation is insufficient to remove the primary's synodic rotational energy over one synodic period, close to synchronism. The magnetic diffusivity of the secondary plays a key role here. These problems are addressed in the next four chapters.

References

Bailey, J.A., Watts, D.J., Sherrington, M.R., Axon, D.J., Giles, A.B., Hanes, D.A., Heathcote, S.R., Hough, J.A., Hughes, S., Jameson, R.F. and McLean, I., 1985. *Mon. Not. R. Astr. Soc.*, **215**, 179.

Bailey, J.A., Wickramasinghe, D.T., Ferrario, L., Hough, J.H. and Cropper, M.S., 1993. *Mon. Not. R. Astr. Soc.*, **261**, L31.

Bailey, J.A., Wickramasinghe, D.T., Hough, J.H. and Cropper, M.S., 1988. *Mon. Not. R. Astr. Soc.*, **234**, 19P.

Berg, R.P. and Duthie, J.G., 1977. *Astrophys. J.*, **211**, 859.
Beuermann, K., Thomas, H.C. and Schwope, A.D., 1988. *Astron. Astrophys.*, **195**, L15.
Beuermann, K., Thomas, H.C. and Schwope, A.D., 1989. *IAU Circ., No. 4775.*
Biermann, P., Schmidt, G.D., Liebert, J., Stockman, H.S., Tapia, S., Kuhr, H., Strittmatter, P.A., West, S. and Lamb, D.Q., 1985. *Astrophys. J.*, **293**, 303.
Chanmugam, G. and Wagner, R.L., 1977. *Astrophys. J. Lett.*, **213**, L13.
Chanmugam, G. and Wagner, R.L., 1979. *Astrophys. J.*, **232**, 895.
Cowley, A.P. and Crampton, D., 1977. *Astrophys. J. Lett.*, **212**, L121.
Cropper, M.S., 1988. *Mon. Not. R. Astr. Soc.*, **231**, 597.
Cropper, M.S. and Wickramasinghe, D.T., 1993. *Mon. Not. R. Astr. Soc.*, **260**, 696.
Ferrario, L., Wickramasinghe, D.T., Bailey, J.A., Hough, J. and Tuohy, I., 1992. *Mon. Not. R. Astr. Soc.*, **256**, 252.
Ferrario, L., Wickramasinghe, D.T., Bailey, J.A., Tuohy, I.R. and Hough, J.H., 1989. *Astrophys. J.*, **337**, 832.
Hearn, D.R. and Richardson, J.A., 1977. *Astrophys. J. Lett.*, **213**, L115.
Hearn, D.R., Richardson, J.A. and Clark, G.W., 1976. *Astrophys. J. Lett.*, **210**, L23.
Katz, J.H., 1991. *Comments. Astrophys.*, **15**, No. 4, 177.
King, A.R. and Lasota, J.P., 1979. *Mon. Not. R. Astr. Soc.*, **188**, 653.
Krzeminski, W. and Serkowski, K., 1977. *Astrophys. J. Lett.*, **216**, L45.
Meggitt, S.M.A. and Wickramasinghe, D.T., 1989. *Mon. Not. R. Astr. Soc.*, **236**, 31.
Michalsky, J.J., Stokes, G.M. and Stokes, R.A., 1977. *Astrophys. J.*, **216**, L35.
Mukai, K., 1988. *Mon. Not. R. Astr. Soc.*, **232**, 175.
Mukai, K. and Charles, P.A., 1986. *Mon. Not. R. Astr. Soc.*, **222**, 1P.
Mukai, K. and Charles, P.A., 1987. *Mon. Not. R. Astr. Soc.*, **226**, 209.
Nauenberg, P., 1972. *Astrophys. J.*, **175**, 417.
Remillard, R.A., Stroozas, B.A., Tapia, S. and Sibler, A., 1991. *Astrophys. J.*, **379**, 715.
Schmidt, G.D., Stockman, H.S. and Grandi, S.A., 1983. *Astrophys. J.*, **271**, 735.
Schmidt, G.D., Stockman, H.S. and Margon, B., 1981. *Astrophys. J. Lett.*, **243**, L157.
Schneider, D.P. and Young, P., 1980. *Astrophys. J.*, **238**, 946.
Schwope, A.D., Beuermann, K., Jordon, S. and Thomas, H.C., 1993b. *Astron. Astrophys.*, **287**, 487.
Schwope, A.D., Beuermann, K. and Thomas, H.C., 1990. *Astron. Astrophys.*, **230**, 120.
Schwope, A.D., Mantel, K. and Horne, K., 1996. *Astron. Astrophys.*, in press.
Schwope, A.D., Thomas, H.C. and Beuermann, K., 1993. *Astron. Astrophys.*, **271**, L25.
Schwope, A.D., Thomas, H.C., Beuermann, K. and Naundorf, C.E., 1991. *Astron. Astrophys.*, **244**, 373.
Schwope, A.D., Thomas, H.C., Beuermann, K. and Reinsch, K., 1993a. *Astron. Astrophys.*, **267**, 103.
Sekiguchi, K., Nakada, Y. and Bassett, B., 1994. *Mon. Not. R. Astr. Soc.*, **266**, L51.
Stockman, H.S., 1977. *Astrophys. J. Lett.*, **218**, L57.
Szkody, P. and Brownlee, D.E., 1977. *Astrophys. J. Lett.*, **212**, L113.
Tapia, S., 1977a. *Astrophys. J. Lett.*, **212**, L125.
Tapia, S., 1977b. *IAU Circ., No. 3054.*
Tapia, S., 1979. *IAU Circ., No. 3327.*
Wickramasinghe, D.T., Archilleos, N., Wu, K. and Boyle, B.J., 1990. *Int. Astr. Union. Circ.*, No. 4962.
Wickramasinghe, D.T., Cropper, M.S., Mason, K.O. and Garlick, M., 1991b. *Mon. Not. R. Astr. Soc.*, **250**, 692.
Wickramasinghe, D.T., Ferrario, L., Cropper, M.S. and Bailey, J., 1991a. *Mon. Not. R. Astr. Soc.*, **251**, 137.
Wickramasinghe, D.T., Visvanathan, N. and Tuohy, I.R., 1984. *Astrophys. J.*, **286**, 328.
Young, P. and Schneider, D.P., 1979. *Astrophys. J.*, **230**, 502.
Young, P., Schneider, D.P. and Shectman, S.A., 1981. *Astrophys. J.*, **245**, 1043.

CHAPTER 4

MAGNETIC COUPLING TO THE SECONDARY

4.1. Induction of Currents in the Secondary

The first problem to consider in the AM Her binaries is how the white dwarf primary can approach synchronous rotation with the orbit. Spin angular momentum must be removed from or added to the primary, depending on whether it is initially over or under-synchronous, for corotation to be approached. A coupling mechanism, with an associated torque, is therefore necessary and magnetic interaction with the secondary star provides this.

The red dwarf secondary is expected to be corotating with the orbit, this being achieved by tidal interaction and viscous dissipation. In general, if the magnetic primary is asynchronous the secondary will experience a time-dependent magnetic field. The induced electric field drives currents in the star which are dissipated due to its electrical resistance. In these short period systems, the secondary has a mass in the range $0.1\,M_{\odot} \lesssim M_{\rm s} \lesssim 0.4\,M_{\odot}$ and so will be largely, if not fully convective. Although there is no rigorous theory of turbulence, it is generally believed (and observationally suggested in the case of the sun) that it enhances the diffusion rate of magnetic fields, compared to ohmic processes. The chaotic motions break up the magnetic field, reducing its length-scale and allowing ohmic dissipation to diffuse it at a greatly increased rate (e.g. Parker, 1979). The mean-field theory related to turbulence was discussed in §2.3. The white dwarf magnetosphere will be tenuous and highly conducting so, whilst it will carry currents, very little dissipation will occur in it. Most dissipation occurs in the diffusive secondary and this results in a decrease in the synodic (i.e. relative to the orbit) rotational energy of the primary.

The large-scale induced magnetic field in the secondary obeys the induction equation

$$\nabla \wedge (\mathbf{v} \wedge \mathbf{B}) - \nabla \wedge (\eta \nabla \wedge \mathbf{B}) = \frac{\partial \mathbf{B}}{\partial t}, \tag{4.1}$$

where η is the turbulent magnetic diffusivity. It is natural to use the orbital frame, in which $\mathbf{v} = \mathbf{0}$ for a synchronized secondary. The induced electric currents will result in a $\mathbf{J} \wedge \mathbf{B}$ magnetic force and associated motions. These motions will be localized to the surface layers where the density of stellar material is lowest and the magnetic force is a larger perturbation to the structure than deeper in the star. Hence, in the main body of the corotating secondary, the $\mathbf{v} \wedge \mathbf{B}$ term in equation (4.1) is relatively small. The effect of this term on the total dissipative torque is greatest for larger degrees of asynchronism, for which the field penetration is lowest. Such motions would tend to enhance the dissipation leading to a stronger torque.

It is noted that the α-effect term $\nabla \wedge (\alpha \mathbf{B})$, discussed in §2.3 for a turbulent medium, is not relevant here since the poloidal field in the secondary has the primary as its source. The primary's magnetic field will not affect a dynamo operating in the secondary provided that the induced $\mathbf{J} \wedge \mathbf{B}$ force is not large enough to significantly perturb the dynamo sources. This should be the case if the dynamo were generating internal fields larger than those induced by the time-varying primary field, requiring internal dynamo fields $\gtrsim 10^2$ G. Such a dynamo process can then be treated separately. It is of interest that dynamo fields of at least this magnitude are required in the maintenance of synchronism, discussed in Chapter 6.

The effect of a corotating white dwarf magnetosphere has been estimated by Chanmugam and Dulk (1983), Lamb *et al* (1983) and Kaburaki (1986). These authors point out that there is a potential difference between field lines of different latitudes, and that this drives currents through the stars with the conducting magnetosphere connecting them. This is an important effect, but present estimates do not account for the slippage of field lines through the diffusive secondary and hence over estimate the torque. This is discussed in §4.5.

The degree of asynchronism can be measured by the ratio $|\omega|/\Omega$, where $\omega = \omega_{\rm in} - \Omega$ is the synodic angular velocity of the primary with $\omega_{\rm in}$ its inertial angular velocity and Ω the orbital angular velocity. The case of high asynchronism ($|\omega|/\Omega \sim 1$) was considered by Papaloizou and Pringle (1978), and Joss, Katz and Rappaport (1979). They estimated the torque due to magnetic dissipation in the surface layers of the secondary. The rate of magnetic energy dissipation in a layer of depth δ is

$$W \simeq \frac{B^2}{2\mu_0}(4\pi R_{\rm s}^2 \delta)\omega.$$

The skin depth $\delta = (2\eta/\omega)^{1/2}$, and taking a dipolar field gives a torque

$$|T_{\rm D}| = \frac{W}{\omega} \simeq \frac{4\pi}{\sqrt{2}} \frac{(B_{\rm p})_0^2 R_{\rm s}^2 R_{\rm p}^6}{\mu_0 D^6} \left(\frac{\eta}{|\omega|}\right)^{\frac{1}{2}}, \tag{4.2}$$

where $R_{\rm s}$ is the mean radius of the secondary, $(B_{\rm p})_0$ and $R_{\rm p}$ the surface polar field and radius of the white dwarf, and D is the orbital separation.

A synchronization time can be defined as

$$t_s = \frac{I|\omega|}{|T_\mathrm{D}|}, \tag{4.3}$$

where I is the moment of inertia of the primary. Use of (4.2) then yields $t_s \sim 10^7$ yrs for $\eta \sim 10^8 \mathrm{m}^2\mathrm{s}^{-1}$ and typical system parameters. However, such formulae cannot be used as $|\omega|$ decreases since the field penetration then increases and $|T_\mathrm{D}|$ must vanish as $|\omega| \to 0$.

The following sections contain the details of the solution of the induction equation found by Campbell (1983). This enables the asynchronous range $0 < |\omega|/\Omega \lesssim 1$ to be investigated. In §4.2 the induction equation is solved to find the magnetic field induced in the secondary. In §4.3 this solution is used to calculate the dissipation in the secondary and hence the torque as a function of the degree of asynchronism. Asymptotic forms for the dissipation torque and associated synchronization times are presented in §4.4, and the results are discussed in §4.5.

4.2. Solution of the Diffusion Equation

Consider a secondary of mass M_s and a primary mass M_p in circular orbits with separation D and period $2\pi/\Omega$. Figure 4.1 shows the orbital frame with the coordinates used in the analysis. The white dwarf is taken to have a centred dipole field with magnetic moment $\mathbf{m}$ having orientation angles (α, β), and a synodic angular velocity $\boldsymbol{\omega}$ parallel to $\boldsymbol{\Omega}$. The region between the stars is treated as a vacuum. As previously explained, although there is likely to be tenuous material in this region which can carry currents, the majority of the dissipation will occur in the diffusive secondary. In general, this happens whether or not the external medium is taken to be a vacuum. The special case of $\alpha = 0$ will be discussed in §4.5.

The primary's magnetic field $\mathbf{B}_\mathrm{p}$ can be expressed in terms of a scalar potential Ψ_m where, relative to the origin O',

$$\Psi_\mathrm{m} = \frac{\mu_0 m}{4\pi r'^3}\mathbf{r}' \cdot \hat{\mathbf{m}}(t), \tag{4.4}$$

with the unit dipole moment being

$$\hat{\mathbf{m}}(t) = \sin\alpha\cos\omega t\,\mathbf{i}' + \sin\alpha\sin\omega t\,\mathbf{j}' + \cos\alpha\,\mathbf{k}. \tag{4.5}$$

It then follows, using identity (A2), that

$$\mathbf{B}_\mathrm{p} = -\nabla\Psi_\mathrm{m} = -\frac{\mu_0 m}{4\pi r'^3}\left[\hat{\mathbf{m}} - 3(\hat{\mathbf{m}}\cdot\hat{\mathbf{r}}')\hat{\mathbf{r}}'\right]. \tag{4.6a,b}$$

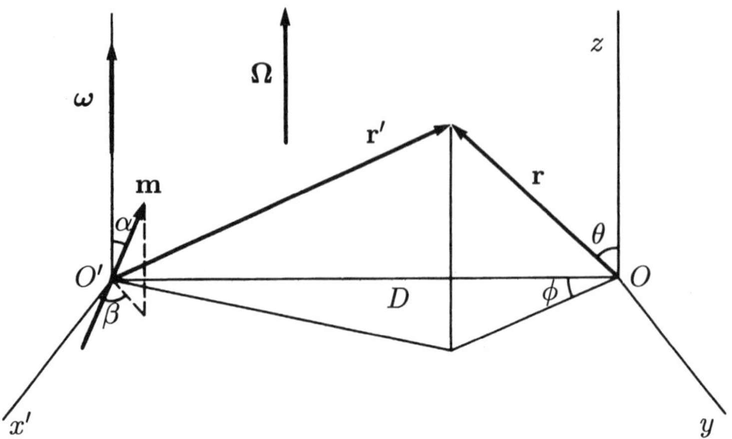

Figure 4.1. The orbital frame, showing the coordinates used in the analysis. The centres of mass of the primary and secondary stars are at O' and O, respectively. Based on Campbell (1983).

The unit vector $\hat{\mathbf{m}}$ can be resolved into components parallel and perpendicular to $\boldsymbol{\omega}$. This enables $\mathbf{B}_\mathrm{p}$ to be expressed as the sum of a time-independent part $\mathbf{B}_0$ and a time-dependent part $\mathbf{B}_t$, so

$$\mathbf{B}_\mathrm{p} = \mathbf{B}_0 + \mathbf{B}_t,$$

where

$$\mathbf{B}_0 = -\frac{\mu_0 m}{4\pi r'^3}\cos\alpha\left(\mathbf{k} - \frac{3z}{r'}\hat{\mathbf{r}}'\right), \tag{4.7}$$

$$\mathbf{B}_t = -\frac{\mu_0 m}{4\pi r'^3}\left[\hat{\mathbf{m}}_\perp - \frac{3}{r'}(\hat{\mathbf{m}}_\perp \cdot \mathbf{r}')\hat{\mathbf{r}}'\right], \tag{4.8}$$

with

$$\hat{\mathbf{m}}_\perp = \sin\alpha\cos\omega t\,\mathbf{i}' + \sin\alpha\sin\omega t\,\mathbf{j}'. \tag{4.9}$$

The time-dependent component $\mathbf{B}_t$ induces currents in the secondary star, which are dissipated at the expense of the synodic rotational energy of the primary. Hence, in the absence of other torques, $|\omega|$ tends to zero. The synchronization time-scale is

$$t_s(\omega) = \left|\frac{\omega}{\dot{\omega}}\right| = \frac{I|\omega|}{|T|}, \tag{4.10}$$

where T is the dissipative torque and I is the moment of inertia of the primary, taken to be spherical. Except very close to synchronism, the timescale t_s greatly exceeds the synodic period so

$$\frac{|\omega| t_s}{2\pi} \gg 1. \tag{4.11}$$

It follows that $\boldsymbol{\omega}$ is essentially constant over a synodic period because the response time to the periodic variations of the torque far exceeds $2\pi/\omega$. It will be shown that $\boldsymbol{\omega}$ remains aligned with $\boldsymbol{\Omega}$ during the synchronization process. The orbital motion acts as a large sink or source of angular momentum, so the compact primary can exchange angular momentum with the orbit without significantly affecting $\boldsymbol{\Omega}$.

The potential Ψ_{m}, given by (4.4), can be expressed relative to the secondary's origin O. The required coordinate transformations are, from Figure 4.1,

$$\begin{aligned} r' &= (r^2 + D^2 - 2rD\sin\theta\cos\phi)^{\frac{1}{2}}, \\ x' &= r\sin\theta\sin\phi, \\ y' &= D - r\sin\theta\cos\phi, \\ z &= r\cos\theta. \end{aligned} \tag{4.12}$$

Equations (4.4), (4.5) and (4.12) give the time-dependent part of Ψ_{m}, to second order in r/D, as

$$\begin{aligned} \Psi_{\mathrm{m}} = & \frac{\mu_0 m \sin\alpha}{4\pi D^3} r P_1^1 (2\cos\phi\sin\omega t + \sin\phi\cos\omega t) \\ & - \frac{3\mu_0 m\sin\alpha}{8\pi D^4} r^2 \left[P_2^0 \sin\omega t - P_2^2 \left(\frac{1}{2}\cos 2\phi\sin\omega t + \frac{1}{3}\sin 2\phi\cos\omega t \right) \right], \end{aligned} \tag{4.13}$$

where $P_l^{|m|}$ are associated Legendre functions, given by (A27) as

$$\begin{aligned} P_1^1 &= \sin\theta, \\ P_2^0 &= \frac{1}{2}(3\cos^2\theta - 1), \\ P_2^2 &= 3\sin^2\theta. \end{aligned} \tag{4.14}$$

On and within the surface of the secondary, where Ψ_{m} will be required, the radial coordinate satisfies $r/D \lesssim 1/4$. The above second order expansion is therefore a reasonable approximation.

The time-varying magnetic field of the primary induces a current density $\mathbf{J}$ in the secondary. These currents are dissipated and the magnetic field obeys the diffusive induction equation

$$\nabla \wedge (\eta \nabla \wedge \mathbf{B}) = -\frac{\partial \mathbf{B}}{\partial t}. \tag{4.15}$$

The magnetic field, being solenoidal, can be expressed as the sum of generalized poloidal and toroidal fields as in (2.143). Since the external field is of a poloidal nature, only that type of field will be generated in the secondary. Hence $\mathbf{B}$ can be expressed as

$$\mathbf{B} = \nabla \wedge [\nabla \wedge (\Phi \hat{\mathbf{r}})]. \tag{4.16}$$

The scalar Φ can be expanded in a set of radial functions and spherical harmonics as in (2.151), where, due to the periodic nature of the external primary field, each harmonic has a time-dependence $\exp(i\omega t)$. By virtue of condition (4.11), ω can be taken as time-independent in the process of solving equation (4.15). In the following expressions, involving general harmonic components, summation over the indices l and m with appropriate coefficients, phases and complex conjugates is implied. The poloidal scalar is

$$\Phi = G_l(r) Y_l^m(\theta, \phi) e^{i\omega t}. \tag{4.17}$$

Equations (4.16) and (4.17) give the magnetic field components as

$$B_r = \frac{l(l+1)}{r^2} G_l Y_l^m e^{i\omega t}, \tag{4.18}$$

$$B_\theta = \frac{1}{r} \frac{dG_l}{dr} \frac{\partial Y_l^m}{\partial \theta} e^{i\omega t}, \tag{4.19}$$

$$B_\phi = \frac{1}{r \sin\theta} \frac{dG_l}{dr} \frac{\partial Y_l^m}{\partial \phi} e^{i\omega t}, \tag{4.20}$$

using the eigenvalue equation (2.153) for spherical harmonics. The components of the curl are

$$(\nabla \wedge \mathbf{B})_r = 0, \tag{4.21}$$

$$(\nabla \wedge \mathbf{B})_\theta = -\frac{1}{r \sin\theta} \left[\frac{d^2 G_l}{dr^2} - \frac{l(l+1)}{r^2} G_l \right] \frac{\partial Y_l^m}{\partial \phi} e^{i\omega t}, \tag{4.22}$$

$$(\nabla \wedge \mathbf{B})_\phi = \frac{1}{r} \left[\frac{d^2 G_l}{dr^2} - \frac{l(l+1)}{r^2} G_l \right] \frac{\partial Y_l^m}{\partial \theta} e^{i\omega t}, \tag{4.23}$$

and the current density is

$$\mathbf{J} = \frac{1}{\mu_0} \nabla \wedge \mathbf{B}. \tag{4.24}$$

Taking $\eta = \eta(r)$ in (4.15), then using (4.18)–(4.23) and equating harmonics, gives the following differential equation for the radial functions,

$$\frac{d^2 G_l}{dr^2} - \left[\frac{i\omega}{\eta} + \frac{l(l+1)}{r^2}\right] G_l = 0. \tag{4.25}$$

Specifying η and solving this equation gives the radial dependence of the poloidal scalar Φ in the secondary star. The solutions in the outer region are found by solving (4.25) with $\eta \to \infty$. The inner and outer forms of Φ must then be matched at the surface of the secondary to satisfy the boundary conditions.

The secondary star will be tidally and rotationally distorted. However, to make the problem tractable, a spherical surface of radius R_s is considered enclosing the same volume as the secondary's Roche lobe. The essential symmetries of the problem are preserved and no additional properties are introduced by using this simplification. The conditions at the secondary's surface are that $\hat{\mathbf{n}} \cdot \mathbf{J} = 0$, where $\hat{\mathbf{n}}$ is the unit normal, and that $\mathbf{B}$ is continuous. The radial current condition is met by equation (4.21), while from (4.18)–(4.20) the continuity of $\mathbf{B}$ requires

$$G_l \text{ and } \frac{dG_l}{dr} \text{ continuous at } r = R_s. \tag{4.26}$$

Consider, first, the free-space solutions of equation (4.25), for which $\eta \to \infty$, so

$$\frac{d^2 G_l}{dr^2} - \frac{l(l+1)}{r^2} G_l = 0. \tag{4.27}$$

The general solution of this equation is

$$G_l = a r^{l+1} + \frac{b}{r^l}, \tag{4.28}$$

where a and b are constants. The solution r^{l+1} corresponds to the poloidal scalar of the primary's field, while the solution r^{-l} corresponds to the poloidal scalar of the outer field $\mathbf{B}_s$ resulting from the induced current source $\mathbf{J}$ in the secondary. Equations (4.6a) and (4.16) show that the magnetic potential and free-space poloidal scalar are related by

$$\Psi_m = -\frac{\partial \Phi}{\partial r}. \tag{4.29}$$

Equations (4.13) and (4.29) give the poloidal scalar of the primary's field as

$$\Phi_p = -\frac{\mu_0 m \sin\alpha}{8\pi D^3} r^2 P_1^1 (2\cos\phi \sin\omega t + \sin\phi \cos\omega t)$$
$$+ \frac{\mu_0 m \sin\alpha}{8\pi D^4} r^3 \left[P_2^0 \sin\omega t - P_2^2 \left(\frac{1}{2}\cos 2\phi \sin\omega t + \frac{1}{3}\sin 2\phi \cos\omega t\right)\right]. \tag{4.30}$$

The poloidal scalar defining the field $\mathbf{B}_\mathrm{s}$ must then take the form

$$\begin{aligned}\Phi_\mathrm{s} =&\frac{1}{r}P_1^1\cos\phi(\alpha_1\sin\omega t+\alpha_2\cos\omega t)+\frac{1}{r}P_1^1\sin\phi(\alpha_3\sin\omega t+\alpha_4\cos\omega t)\\ &+\frac{1}{r^2}P_2^0(\beta_1\sin\omega t+\beta_2\cos\omega t)+\frac{1}{r^2}P_2^2\cos 2\phi(\gamma_1\sin\omega t+\gamma_2\cos\omega t)\\ &+\frac{1}{r^2}P_2^2\sin 2\phi(\gamma_3\sin\omega t+\gamma_4\cos\omega t),\end{aligned} \tag{4.31}$$

where α_i, β_i and γ_i are constants.

Consider, now, the inner solution of (4.25), for which η is finite. Since the secondary is convective, a turbulent origin for η is appropriate. The simple mixing length theory of §2.2.10 suggests that the variation of η through the star is not large so, in the absence of a more rigorous theory of turbulence, η is taken as constant. None of the physical essentials of the problem are lost in making this mathematical simplification. Making the subsitution in (4.25) of

$$G_l=\left(\frac{i\omega}{\eta}\right)^{-\frac{1}{4}}u^{\frac{1}{2}}F_l(u),$$

where $u=i^{3/2}(\omega/\eta)^{1/2}r$, gives

$$\frac{d^2F_l}{du^2}+\frac{1}{u}\frac{dF_l}{du}+\left[1-\frac{(l+\frac{1}{2})^2}{u^2}\right]F_l=0.$$

This is Bessel's equation of order $\pm(l+1/2)$. The solution of negative order is singular at $r=0$, so the required non-singular solution of (4.25) is

$$G_l=i^{\frac{1}{2}}r^{\frac{1}{2}}J_{l+\frac{1}{2}}\left(i^{\frac{3}{2}}\left(\omega/\eta\right)^{\frac{1}{2}}r\right), \tag{4.32}$$

where $J_{l+\frac{1}{2}}$ is a Bessel function of the first kind of order $l+1/2$, given by

$$J_{l+\frac{1}{2}}(u)=(-1)^l\left(\frac{2}{\pi}\right)^{\frac{1}{2}}u^{l+\frac{1}{2}}\frac{d^l}{(udu)^l}\left(\frac{\sin u}{u}\right).$$

The required radial functions are then

$$G_1=-\left(\frac{2ir}{\pi}\right)^{\frac{1}{2}}\frac{1}{u^{\frac{1}{2}}}\left[\cos u-\frac{\sin u}{u}\right], \tag{4.33a}$$

$$G_2=\left(\frac{2ir}{\pi}\right)^{\frac{1}{2}}\frac{1}{u^{\frac{1}{2}}}\left[\left(\frac{3}{u^2}-1\right)\sin u-\frac{3\cos u}{u}\right]. \tag{4.33b}$$

The functions G_l can be expressed in the form

$$G_l(r) = C_l(r)\exp[i\delta_l(r)]. \tag{4.34}$$

The poloidal scalar, given by (4.17), is then

$$\Phi = C_l Y_l^m \exp[i(\omega t + \delta_l)]. \tag{4.35}$$

The boundary conditions, specified by (4.26), require Φ and $\partial\Phi/\partial r$ to be continuous across the stellar surface. Matching a suitable linear combination of the harmonics of Φ, and their radial derivatives, to those of $\Phi_{\rm p} + \Phi_{\rm s}$ given by (4.30) and (4.31), yields the inner poloidal scalar as

$$\begin{aligned}\Phi =& C_1 P_1^1 \cos\phi[A_1 \sin(\omega t + \delta_1 - \delta_{1s}) + A_2 \cos(\omega t + \delta_1 - \delta_{1s})] \\ &- \frac{1}{2} C_1 P_1^1 \sin\phi[A_2 \sin(\omega t + \delta_1 - \delta_{1s}) - A_1 \cos(\omega t + \delta_1 - \delta_{1s})] \\ &+ C_2 P_2^0 [B_1 \sin(\omega t + \delta_2 - \delta_{2s}) + B_2 \cos(\omega t + \delta_2 - \delta_{2s})] \\ &- \frac{1}{2} C_2 P_2^2 \cos 2\phi[B_1 \sin(\omega t + \delta_2 - \delta_{2s}) + B_2 \cos(\omega t + \delta_2 - \delta_{2s})] \\ &+ \frac{1}{3} C_2 P_2^2 \sin 2\phi[B_2 \sin(\omega t + \delta_2 - \delta_{2s}) - B_1 \cos(\omega t + \delta_2 - \delta_{2s})],\end{aligned} \tag{4.36}$$

where the ω-dependent coefficients A_i and B_i are

$$A_1 = -\frac{3\mu_0 m R_{\rm s}^2 \sin\alpha}{4\pi D^3}(C_1 + R_{\rm s}C_1')_s\left[(C_1 + R_{\rm s}C_1')_s^2 + (R_{\rm s}C_1\delta_1')_s^2\right]^{-1}, \tag{4.37}$$

$$A_2 = -\frac{R_{\rm s}C_{1s}\delta_{1s}'}{(C_1 + R_{\rm s}C_1')_s}A_1, \tag{4.38}$$

$$B_1 = \frac{5\mu_0 m R_{\rm s}^3 \sin\alpha}{16\pi D^4}\left(C_2 + \frac{R_{\rm s}}{2}C_2'\right)_s\left[\left(C_2 + \frac{R_{\rm s}}{2}C_2'\right)_s^2 + \left(\frac{R_{\rm s}}{2}C_2\delta_2'\right)_s^2\right]^{-1} \tag{4.39}$$

$$B_2 = -\frac{R_{\rm s}C_{2s}\delta_{2s}'}{(C_2 + \frac{1}{2}R_{\rm s}C_2')_s}B_1, \tag{4.40}$$

with primes denoting differentiation with respect to r, and subscripts s surface values. The functions C_l and δ_l can be found from (4.33) and (4.34).

4.3. The Dissipation Torque

The dissipation of energy in the secondary and the torque acting on the primary can be simply related since, as will be shown, the angle α remains constant during the synchronization process. The electric currents induced in the secondary are dissipated, due to its finite diffusivity, the total dissipation rate being

$$W = \frac{1}{\mu_0} \int \eta (\nabla \wedge \mathbf{B})^2 dV, \tag{4.41}$$

where the integral is over the stellar volume. It follows from (4.15)–(4.17) that

$$\nabla \wedge \mathbf{B} = \frac{i\omega}{\eta} \hat{\mathbf{r}} \wedge \nabla \Phi. \tag{4.42}$$

The dissipation is at the expense of the synodic rotational energy of the primary. Condition (4.11) means that ω will not respond significantly to the periodic time-dependence of the torque. Only its secular variation is then relevant and it is therefore appropriate to consider the time-averaged dissipation

$$\frac{dE}{dt} = -\frac{|\omega|}{2\pi} \int_0^{\frac{2\pi}{\omega}} W dt. \tag{4.43}$$

Using (4.36) and (4.42), performing the angular integrations in (4.41) employing the orthogonality relation (A31), and the time-average in (4.43), gives

$$\frac{dE}{dt} = -\pi \mu_0 \omega^2 \left[\frac{5}{3}(A_1^2 + A_2^2) \int_0^{R_s} \frac{C_1^2}{\eta} dr + \frac{64}{5}(B_1^2 + B_2^2) \int_0^{R_s} \frac{C_2^2}{\eta} dr \right], \tag{4.44}$$

where A_i and B_i are given by (4.37)–(4.40). The radial integrals can be expressed in terms of functions evaluated on the stellar surface. Substituting (4.34) for G_l into (4.25), and equating the real and imaginary parts to zero, gives

$$\frac{d^2 C_l}{dr^2} - \left[\delta_l'^2 + \frac{l(l+1)}{r^2} \right] C_l = 0, \tag{4.45}$$

$$\frac{d}{dr}(C_l^2 \delta_l') - \frac{\omega}{\eta} C_l^2 = 0. \tag{4.46}$$

Equation (4.45) requires C_l to vanish at $r = 0$, so integrating (4.46) from centre to surface gives

$$\int_0^{R_s} \frac{C_l^2}{\eta} dr = \frac{1}{\omega} (C_l^2 \delta_l')_s. \tag{4.47}$$

Using (4.37)–(4.40) and (4.47) in (4.44), and simplifying, gives the dissipation rate as

$$\frac{dE}{dt} = -\frac{5\mu_0 m^2 R_s^3 \sin^2\alpha}{4\pi D^6}\omega f(\omega,\eta), \tag{4.48}$$

where the dimensionless function f is

$$f(\omega,\eta) = \frac{3}{4}(C_1^2\delta_1^{'})_s F_1 + \left(\frac{R_s}{D}\right)^2 (C_2^2\delta_2^{'})_s F_2, \tag{4.49}$$

with

$$F_1 = R_s C_{1s}^2\left[(C_1^2 + R_s C_1 C_1^{'})_s^2 + R_s^2 (C_1^2\delta_1^{'})_s^2\right]^{-1}, \tag{4.50}$$

$$F_2 = R_s C_{2s}^2\left[\left(C_2^2 + \frac{1}{2}R_s C_2 C_2^{'}\right)_s^2 + \frac{1}{4}R_s^2 (C_2^2\delta_2^{'})_s^2\right]^{-1}. \tag{4.51}$$

The real and imaginary parts of (4.33) give the functions C_l in (4.34) which lead to

$$C_{1s}^2 = \frac{\sqrt{2}R_s}{\pi a_s}\left[\cosh a_s + \cos a_s - \frac{2}{a_s}(\sinh a_s + \sin a_s) + \frac{2}{a_s^2}(\cosh a_s - \cos a_s)\right], \tag{4.52}$$

$$C_{2s}^2 = \frac{\sqrt{2}R_s}{\pi a_s}\left[\cosh a_s - \cos a_s - \frac{6}{a_s}(\sinh a_s - \sin a_s) + \frac{18}{a_s^2}(\cosh a_s + \cos a_s) - \frac{36}{a_s^3}(\sinh a_s + \sin a_s) + \frac{36}{a_s^4}(\cosh a_s - \cos a_s)\right], \tag{4.53}$$

where a_s is a dimensionless similarity variable, given by

$$a_s = \left(\frac{2|\omega|}{\eta}\right)^{\frac{1}{2}} R_s. \tag{4.54}$$

It is noted that $a_s \sim (\tau_d/P_{syn})^{1/2}$, where τ_d is the characteristic diffusion time of a magnetic field through the secondary, and $P_{syn} = 2\pi/\omega$ is the synodic rotation period of the primary. The quantities $(C_l C_l^{\prime})_s$ and $(C_l^2\delta_l^{\prime})_s$ occurring in f are found from the real and imaginary parts of (4.33) and their relations to C_l and δ_l in (4.34). Some differentiation and lengthy algebra yields

$$(C_1 C_1')_s = \frac{\sqrt{2}}{\pi}\left[\frac{1}{2}(\sinh a_s - \sin a_s) - \frac{1}{a_s}(\cosh a_s + \cos a_s) + \frac{2}{a_s^2}(\sinh a_s + \sin a_s) - \frac{2}{a_s^3}(\cosh a_s - \cos a_s)\right], \quad (4.55)$$

$$(C_2 C_2')_s = \frac{\sqrt{2}}{\pi}\left[\frac{1}{2}(\sinh a_s + \sin a_s) - \frac{3}{a_s}(\cosh a_s - \cos a_s) + \frac{12}{a_s^2}(\sinh a_s - \sin a_s) - \frac{36}{a_s^3}(\cosh a_s + \cos a_s) + \frac{72}{a_s^4}(\sinh a_s + \sin a_s) - \frac{72}{a_s^5}(\cosh a_s - \cos a_s)\right], \quad (4.56)$$

$$(C_1^2 \delta_1')_s = \frac{1}{\pi\sqrt{2}}\left[\sinh a_s + \sin a_s - \frac{2}{a_s}(\cosh a_s - \cos a_s)\right], \quad (4.57)$$

$$(C_2^2 \delta_2')_s = \frac{1}{\pi\sqrt{2}}\left[\sinh a_s - \sin a_s - \frac{6}{a_s}(\cosh a_s + \cos a_s) + \frac{12}{a_s^2}(\sinh a_s + \sin a_s) - \frac{12}{a^3}(\cosh a_s - \cos a_s)\right]. \quad (4.58)$$

The dissipation rate, given by equation (4.48), can be related to the torque exerted on the primary by its interaction with $\mathbf{B}_{\mathrm{s}}$, the outer field whose source is the current density induced in the secondary. This torque is

$$\mathbf{T} = \mathbf{m} \wedge \mathbf{B}_{\mathrm{s}}(\mathbf{r}_{\mathrm{p}}), \quad (4.59)$$

where, from (4.6a) and (4.29),

$$\mathbf{B}_{\mathrm{s}}(\mathbf{r}_{\mathrm{p}}) = \left[\nabla\left(\frac{\partial \Phi_{\mathrm{s}}}{\partial r}\right)\right]_{\mathbf{r}=\mathbf{r}_{\mathrm{p}}}, \quad (4.60)$$

and the position of the primary $\mathbf{r}_{\mathrm{p}} = D\mathbf{i}$. From Figure 4.1, the polar components of $\mathbf{B}_{\mathrm{s}}(\mathbf{r}_{\mathrm{p}})$ relative to $Oxyz$ can be related to its Cartesian components in $O'x'y'z$ by

$$B_{\mathrm{s}r} = -B_{\mathrm{s}y'}, \quad B_{\mathrm{s}\theta} = -B_{\mathrm{s}z}, \quad B_{\mathrm{s}\phi} = B_{\mathrm{s}x'}.$$

The components of the torque are then

$$T_{x'} = m(\hat{m}_z B_{\mathrm{s}r} - \hat{m}_{y'} B_{\mathrm{s}\theta}), \quad (4.61)$$

$$T_{y'} = m(\hat{m}_{x'} B_{s\theta} + \hat{m}_z B_{s\phi}), \tag{4.62}$$

$$T_z = -m(\hat{m}_{x'} B_{sr} + \hat{m}_{y'} B_{s\phi}), \tag{4.63}$$

at $(r, \theta, \phi) = (D, \pi/2, 0)$. Equations (4.31) and (4.60) give the required field components as

$$B_{sr} = \frac{1}{D^3}\left[2\alpha_1 - \frac{3}{D}(\beta_1 - 6\gamma_1)\right]\sin\omega t + \frac{1}{D^3}\left[2\alpha_2 - \frac{3}{D}(\beta_2 - 6\gamma_2)\right]\cos\omega t, \tag{4.64}$$

$$B_{s\theta} = 0, \tag{4.65}$$

$$B_{s\phi} = -\frac{1}{D^3}\left(\alpha_3 + \frac{12}{D}\gamma_3\right)\sin\omega t - \frac{1}{D^3}\left(\alpha_4 + \frac{12}{D}\gamma_4\right)\cos\omega t. \tag{4.66}$$

Using the components of $\hat{\mathbf{m}}$ in (4.5) together with (4.64)–(4.66) in (4.61)–(4.63), and averaging over a period $2\pi/\omega$, gives

$$\langle T_{x'} \rangle = \langle T_{y'} \rangle = 0, \tag{4.67}$$

$$\langle T_z \rangle = -\frac{m \sin\alpha}{D^3}\left[\alpha_2 - \frac{1}{2}\alpha_3 - \frac{3}{2D}(\beta_2 - 6\gamma_2 + 4\gamma_3)\right]. \tag{4.68}$$

Equation (4.67) shows that there is no tendency for secular changes in the tilt angle α of the dipole moment $\mathbf{m}$.

The quantities α_i, β_i and γ_i in (4.68) can be expressed in terms of A_2 and B_2 given by (4.38) and (4.40), by means of the boundary conditions at $r = R_s$. This gives

$$\alpha_2 = R_s C_{1s} A_2, \quad \alpha_3 = -\frac{1}{2}\alpha_2, \quad \beta_2 = R_s^2 C_{2s} B_2,$$
$$\gamma_2 = -\frac{1}{2}\beta_2, \quad \gamma_3 = \frac{1}{3}\beta_2.$$

Substitution of these in (4.68) yields

$$\langle T_z \rangle = -\frac{5\mu_0 m^2 R_s^3 \sin^2\alpha}{4\pi D^6}\left[\frac{3}{4}(C_1^2 \delta_1')_s F_1 + \left(\frac{R_s}{D}\right)^2 (C_2^2 \delta_2')_s F_2\right], \tag{4.69}$$

where the functions F_1 and F_2 are identical to those given by (4.50) and (4.51). The quantity in square brackets is therefore $f(\omega, \eta)$, appearing in (4.48), and hence the synodic average of the total dissipative torque is

$$\langle \mathbf{T} \rangle = \mathbf{T}_D = -\frac{5\mu_0 m^2 R_s^3 \sin^2\alpha}{4\pi D^6} f(\omega, \eta)\, \mathbf{k}. \tag{4.70}$$

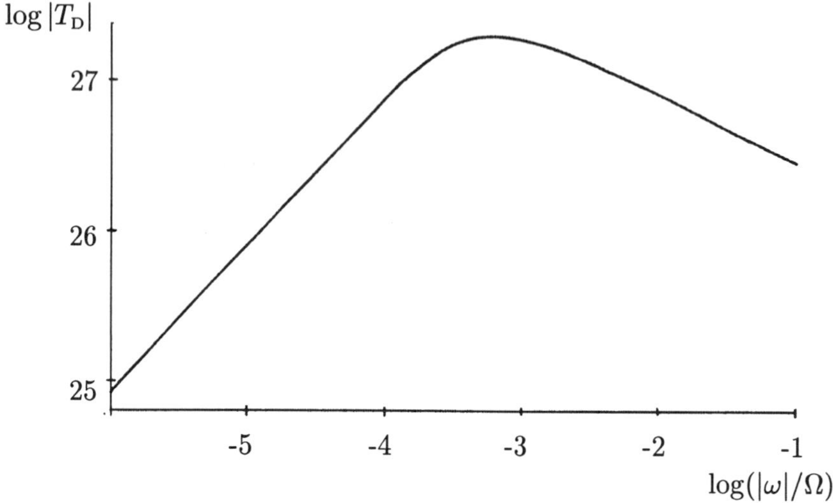

Figure 4.2. The dissipation torque versus degree of asynchronism. Based on Campbell (1983).

The dissipation rate of the synodic rotational energy of the primary, given by (4.48), is therefore related to this torque by

$$\frac{dE}{dt} = \omega T_{\mathrm{D}}. \tag{4.71}$$

The vanishing synodic averages of the horizontal torque components are a consequence of the intrinsic symmetry of the dipole field, and the initial condition of $\boldsymbol{\omega}$ parallel to $\boldsymbol{\Omega}$. Equations (4.8) and (4.9) show that the time-dependent part of the primary's magnetic field has a vanishing z-component in the orbital plane. The outer field $\mathbf{B}_{\mathrm{s}}$, resulting from the current distribution induced in the secondary, also has this property. At the primary $B_{\mathrm{s}\theta} = -B_{\mathrm{s}z} = 0$, so (4.61) and (4.62) give

$$T_{x'} = m \cos \alpha B_{\mathrm{s}r},$$
$$T_{y'} = m \cos \alpha B_{\mathrm{s}\phi}.$$

It then follows from (4.64) and (4.66) that $\langle T_{x'} \rangle = \langle T_{y'} \rangle = 0$.

If $\boldsymbol{\omega}$ were initially misaligned from $\boldsymbol{\Omega}$ part of the dissipation in the secondary would be due to the component of $\boldsymbol{\omega}$ in the orbital plane. This component would be dissipated at a similar rate to the synchronization of ω_z so, in the absence of other torques, $\boldsymbol{\omega}$ would tend to $\boldsymbol{\Omega}$ on the time-scale t_s given by (4.3).

Figure 4.2 is a plot of $\log |T_{\mathrm{D}}|$ against $\log(|\omega|/\Omega)$, where T_{D} is given by (4.70). The parameters used are $P = 1.5\,\mathrm{hrs}$, $M_{\mathrm{s}} = 0.14 M_{\odot}$ and $M_{\mathrm{p}} =$

$0.4M_\odot$, so $R_s = 10^8$ m, and $D = 3.8 \times 10^8$ m. These values are consistent with (2.210), (2.225) and (2.226), with $q = 1.1$. Equation (3.2) gives $R_p = 1.1 \times 10^7$ m. The polar magnetic field is taken as $(B_p)_0 = 1.4 \times 10^3$ tesla, and the dipole moment $m = 2\pi (B_p)_0 R_p^3/\mu_0$. A value $\eta = 5 \times 10^8\ \mathrm{m^2 s^{-1}}$ is used for the magnetic diffusivity. The torque decreases for $|\omega|/\Omega \lesssim 10^{-4}$ since, although the field penetration becomes large, its time variation, and hence the induced current density, becomes small. The torque also decreases for $|\omega|/\Omega \gtrsim 10^{-3}$ because the currents are localized to a skin-depth δ beneath the stellar surface, and then $\delta \to 0$ as $|\omega|$ increases.

4.4. Asymptotic Synchronization Times

The dimensionless quantity a_s, given by (4.54), can be expressed as

$$a_s = \frac{\sqrt{2} R_s \Omega^{\frac{1}{2}}}{\eta^{\frac{1}{2}}} \left(\frac{|\omega|}{\Omega} \right)^{\frac{1}{2}} .$$

Using the orbital and lobe-filling conditions

$$\Omega^2 = \frac{GM}{D^3} \text{ and } \left(\frac{R_s}{D} \right)^3 \frac{M}{M_s} = 0.1,$$

where $M = M_s + M_p$, gives

$$a_s = \left(\frac{2GM_s R_s}{5\eta^2} \right)^{\frac{1}{4}} \left(\frac{|\omega|}{\Omega} \right)^{\frac{1}{2}} . \tag{4.72}$$

For the parameters previously used, this yields

$$a_s = 2.3 \times 10^2 \left(\frac{|\omega|}{\Omega} \right)^{\frac{1}{2}} . \tag{4.73}$$

It follows from (4.54) that

$$a_s \sim \left(\frac{\tau_d}{P_{syn}} \right)^{\frac{1}{2}} \sim \frac{R_s}{\delta}, \tag{4.74}$$

where τ_d is the magnetic diffusion time through the secondary, $P_{syn} = 2\pi/|\omega|$ and $\delta \sim (\eta/\omega)^{1/2}$ is the characteristic penetration depth of the magnetic field.

Asymptotic forms can be found for the dissipation torque $T_D = \langle T \rangle$, for $a_s \gg 1$ and $a_s \ll 1$. These correspond to low and high magnetic field penetration into the secondary, respectively, and can be used to find synchronization times in these regimes.

Low field penetration

For $a_s \gg 1$ the primary's magnetic field penetrates the secondary to a skin-depth $\delta \ll R_{\rm s}$. Equations (4.49)–(4.58) and (4.70) yield

$$T_{\rm D} = \mp \frac{15\sqrt{2}\pi}{8\mu_0} \left[1 + \frac{16}{3}\left(\frac{R_{\rm s}}{D}\right)^2\right] \frac{(B_{\rm p})_0^2 R_{\rm p}^6 R_{\rm s}^2 \sin^2\alpha}{D^6} \left(\frac{\eta}{|\omega|}\right)^{\frac{1}{2}}, \tag{4.75}$$

where the negative and positive signs apply for $\omega > 0$ and $\omega < 0$, respectively. The corresponding synchronization time is

$$t_s = \frac{1.3 \times 10^7 \left(\frac{k_{\rm p}^2}{0.2}\right) \left(\frac{P}{1.5\,{\rm hr}}\right)^{-\frac{3}{2}} \left(\frac{M_{\rm p}}{0.4 M_\odot}\right) \left(\frac{D}{3.8 \times 10^8\,{\rm m}}\right)^6 \left(\frac{|\omega|}{\Omega}\right)^{\frac{3}{2}}}{\left(\frac{R_{\rm s}}{10^8\,{\rm m}}\right)^2 \left(\frac{R_{\rm p}}{1.1 \times 10^7\,{\rm m}}\right)^4 \left(\frac{(B_{\rm p})_0}{14\,{\rm MG}}\right)^2 \left(\frac{\eta}{5 \times 10^8\,{\rm m\,s^{-1}}}\right)^{\frac{1}{2}} N_1} \,{\rm yr},$$

where $k_{\rm p} R_{\rm p}$ is the radius of gyration of the primary, and

$$N_1 = \left[1 + \frac{16}{3}\left(\frac{R_{\rm s}}{D}\right)^2\right] \sin^2\alpha. \tag{4.76a,b}$$

These expressions hold to good accuracy for $|\omega|/\Omega \gtrsim 10^{-2}$. It is seen that for $|\omega|/\Omega \sim 1$, t_s is less than the lifetime of the system, given by $M_{\rm s}/|\dot{M}_{\rm s}| \sim 10^9$yrs. Equation (4.75) agrees with (4.2), except for a numerical factor which derives from taking the field structure into account here.

High field penetration

For $a_s \ll 1$ the field penetration is essentially complete and Taylor expansion of f about $a_s = 0$ yields

$$T_{\rm D} = \mp \frac{\pi}{12\mu_0} \left[1 + \frac{48}{35}\left(\frac{R_{\rm s}}{D}\right)^2\right] \frac{(B_{\rm p})_0^2 R_{\rm p}^6 R_{\rm s}^5 \sin^2\alpha}{D^6} \frac{|\omega|}{\eta}. \tag{4.77}$$

It follows that $|\omega|$ decreases exponentially on a time-scale

$$t_s = \frac{80 \left(\frac{k_{\rm p}^2}{0.2}\right) \left(\frac{M_{\rm p}}{0.4 M_\odot}\right) \left(\frac{D}{3.8 \times 10^8\,{\rm m}}\right)^6 \left(\frac{\eta}{5 \times 10^8\,{\rm m^2\,s^{-1}}}\right)}{\left(\frac{(B_{\rm p})_0}{14\,{\rm MG}}\right)^2 \left(\frac{R_{\rm p}}{1.1 \times 10^7\,{\rm m}}\right)^4 \left(\frac{R_{\rm s}}{10^8\,{\rm m}}\right)^5 N_2 \sin^2\alpha} \,{\rm yr}, \tag{4.78a}$$

where

$$N_2 = \left[1 + \frac{48}{35}\left(\frac{R_{\rm s}}{D}\right)^2\right]. \tag{4.78b}$$

4.5. Discussion

The foregoing calculation took the medium between the stars to be a vacuum. Consequently, the magnetic field must have an explicit time dependence for currents to be induced in the secondary. In the special case of $\alpha = 0$ the primary star's angular velocity and magnetic moment are parallel to $\boldsymbol{\Omega}$ and hence the time-dependent field vanishes, in accordance with (4.8) and (4.9). It follows that there is no current induction in this case so the dissipation torque, given by (4.70), vanishes.

If the white dwarf has a corotating magnetosphere and is asynchronous with the orbit, then there will be a shear across the surface of a synchronized secondary. This would lead to the creation of toroidal magnetic field from the poloidal component by the ω-effect, discussed in §2.3.1. Electric currents would flow in the secondary and through the magnetosphere and the surface layers of the primary. A torque would result on the primary through the $B_p B_\phi$ stresses, this being finite even in the case $\alpha = 0$. Such an effect needs to be calculated. However, apart from the difference in α-dependence, this process has similarities to the vacuum case calculated here. In both cases most dissipation occurs in the diffusive secondary. In the corotating magnetosphere case there is a synodic frequency $\omega = \omega_{\text{in}} - \Omega$ associated with the primary's asynchronism, even for $\alpha = 0$ where there is no explicit time dependence. In the orbital frame, magnetospheric material has angular velocity ω while material in the main body of the secondary is stationary. As in the vacuum case, the torque will vanish as ω tends to zero. The ω-dependence of the radial distribution of the field in the secondary should be similar in both cases, with smaller penetration for higher ω. Hence the torque curve for the magnetosphere case should be qualitatively similar to that of the vacuum case, shown in Figure 4.2. For a general α value, both mechanisms will be operable.

The primary's magnetic field was taken as dipolar here, while there is some evidence of a quadrupolar component in the field. However, such higher multipole fields fall more rapidly with distance and, since the torque depends on B^2, these would only make small contributions.

It is noted that the process of synchronizing the primary involves angular momentum and energy exchange with the orbit. The distribution of $\mathbf{J} \wedge \mathbf{B}$ in the secondary leads to a non-central force on its centre of mass, and hence to an orbital torque about the centre of mass of the system. This enables an under-synchronous primary to gain inertial spin energy from the orbital energy. The magnetic stellar torques and the orbital torque sum to zero, as in the case of non-dissipative magnetic torques considered in §6.6.

Magnetic dissipation in the secondary explains how the white dwarf can attain synchronism in the absence of other significant torques. However, the presence of the accretion torque requires the problems of the maintenance and attainment of synchronism to be carefully considered. These problems are addressed in the next three chapters.

References

Campbell, C.G., 1983. *Mon. Not. R. Astr. Soc.*, **205**, 1031.
Chanmugan, G. and Dulk, G.A., 1983. *Cataclysmic Variables and Related Objects*, Reidel, Holland.
Joss, P.C., Katz, J.I. and Rappaport, S.A., 1979. *Astrophys. J.*, **230**, 176.
Kaburaki, O., 1985. *Astrophys. Sp. Sci.*, **119**, 85.
Lamb, F.K., Aly, J.J., Cook, M.C. and Lamb, D.Q., 1983. *Astrophys. J.*, **274**, L71.
Papaloizou, J. and Pringle, J.E., 1978. *Astron. Astrophys.*, **70**, L65.
Parker, E.N., 1979. *Cosmical Magnetic Fields*, Oxford University Press.

CHAPTER 5

THE ACCRETION TORQUE

5.1. The Effect of Magnetic Force

Accretion discs do not form in the AM Herculis binaries. The effect of a strongly magnetic accretor on a disc is considered in Chapter 9. If the disc's magnetic diffusivity η is due to turbulence, or buoyancy, disruption occurs where the azimuthal magnetic force starts to dominate that due to viscosity. Thermal equilibrium breaks down in such regions, which also tend to be viscously unstable. The radius inside which disruption occurs is given by

$$\varpi_{\mathrm{m}} = \frac{5.5 \times 10^9 \, \gamma^{\frac{8}{29}} \left(\frac{(B_{\mathrm{p}})_0}{2 \times 10^3 \, \mathrm{T}}\right)^{\frac{16}{29}} \left(\frac{R_{\mathrm{p}}}{8.7 \times 10^6 \, \mathrm{m}}\right)^{\frac{48}{29}} (1 - \xi_{\mathrm{R}})^{\frac{8}{29}}}{\left(\frac{\epsilon}{0.1}\right)^{\frac{36}{145}} \left(\frac{\dot{M}_{\mathrm{p}}}{10^{-10} \, M_{\odot} \, \mathrm{yr}^{-1}}\right)^{\frac{46}{145}} \left(\frac{M_{\mathrm{p}}}{0.6 \, M_{\odot}}\right)^{\frac{1}{29}}} \, \mathrm{m},$$

where $\gamma < 1$, $(B_{\mathrm{p}})_0$, M_{p} and R_{p} are the surface polar field, mass and radius of the primary, ϵ is the ratio of the rms turbulent speed to the sound speed and

$$\xi_{\mathrm{R}} = \frac{\Omega_{\mathrm{p}}}{\Omega_{\mathrm{K}}(\varpi_{\mathrm{m}})}$$

is the ratio of the stellar rotation rate to that of the inner edge of the disc. Since the mean radius of the primary's Roche lobe is typically $R_1 \sim 10^8$ m, it is seen that $\varpi_{\mathrm{m}} > R_1$ so discs would not be expected in AM Her stars.

Matter lost from the L_1 region of the secondary interacts with the magnetic field of the primary. The ratio of the magnetic energy density to the thermal energy density can be estimated in this region. Assuming the white dwarf to have a dipolar magnetic field, and taking its distance from L_1 as $r_{1\mathrm{p}} \sim D/2$, gives

$$B_{L_1} \simeq \left(\frac{2R_{\mathrm{p}}}{D}\right)^3 (B_{\mathrm{p}})_0. \tag{5.1}$$

The density of material leaving the L_1 region can be estimated from the mass loss rate

$$\dot{M}_{\rm s} = -\pi \rho_{L_1} c_{\rm s} H^2, \tag{5.2}$$

where $c_{\rm s}$ is the isothermal sound speed and H is the density scale height perpendicular to the line of stellar centres, just beneath L_1. This height is given by (2.235) so, with $f = 1$,

$$H = \left(\frac{c_{\rm s}}{\Omega D}\right) R_{\rm s}. \tag{5.3}$$

Use of (5.2) and (5.3) yields the density of material in the vicinity of L_1 as

$$\rho_{L_1} \simeq \frac{|\dot{M}|_{\rm s}\Omega^2}{\pi c_{\rm s}^3}\left(\frac{D}{R_{\rm s}}\right)^2. \tag{5.4}$$

The ratio of magnetic to thermal energy density is then given by (5.1) and (5.4) as

$$\frac{E_{\rm m}}{E_{\rm th}} \simeq \frac{B_{L_1}^2}{\mu_0 \rho_{L_1} c_{\rm s}^2} \simeq \frac{\pi}{\mu_0}\left(\frac{2R_{\rm p}}{D}\right)^6 \left(\frac{R_{\rm s}}{D}\right)^2 \frac{(B_{\rm p})_0^2 c_{\rm s}}{\dot{M}_{\rm s}\Omega^2}. \tag{5.5}$$

This ratio is most sensitive to the white dwarf radius $R_{\rm p}$, and hence to its mass $M_{\rm p}$. For $M_{\rm p} = 0.6M_\odot$, and typical system parameters, the ratio is ~ 1.

Recent observations indicate that effective channelling of material starts to occur at some distance from the L_1 region (Wu and Wickramasinghe, 1993; Schwope, Mantel and Horne, 1996). Equation (5.5) suggests that the white dwarf's magnetic field should affect the gas near L_1. However, the stream rapidly becomes supersonic, so its kinetic energy density soon dominates the magnetic energy density. Closer to the white dwarf this situation is reversed and matter is effectively channelled by its field.

A steady flow of highly conducting material obeys the induction equation

$$\nabla \wedge (\mathbf{v} \wedge \mathbf{B}) = \mathbf{0},$$

which is satisfied by

$$\mathbf{v} = \kappa(\mathbf{r})\mathbf{B}. \tag{5.6}$$

In the strongly channelled region the local distortion of the magnetic field, $\mathbf{B}'$, caused by the flow is small. Matter then flows very nearly parallel to the unperturbed field, $\mathbf{B}$, and experiences a force per unit mass

$$\mathbf{F} = \frac{1}{\mu_0 \rho}(\nabla \wedge \mathbf{B}') \wedge \mathbf{B}. \tag{5.7}$$

In general, the channelled part of the accretion stream will lie out of the orbital plane. There is a magnetic torque $\mathbf{r} \wedge \mathbf{F}$ exerted on material and the reaction to this is transmitted to the white dwarf to which the field is attached. A torque also results from the angular momentum flux of matter flowing onto the accretion pole. Some angular momentum exchange between the stream and the orbit can occur, due to gravitational interaction with the secondary star.

After leaving the L_1 region matter will flow in the orbital plane, slightly leading the line of stellar centres. As the magnetic force $\mathbf{F}$ strengthens, the stream will be diverted out of the orbital plane into the strongly channelled region. This region gives the dominant contribution to the magnetic torque exerted on the white dwarf. Most angular momentum exchange with the orbit occurs in the flow region near L_1, where the gravitational coupling is largest.

The accretion torque is required for the synchronous and asynchronous cases. In the synchronous case the magnetic field lines are stationary in the orbital frame and so, for steady mass loss from L_1, the accretion torque is time-independent. In the asynchronous case the geometry of the channelled stream changes, and the accretion torque will be time-dependent. In general, the torque will have components perpendicular and parallel to the orbital plane. The ratio of these components, as a function of magnetic orientation, is of central interest.

In §5.2 the angular momentum transfer is calculated due to a field-channelled flow from the L_1 region onto a synchronized white dwarf, for an arbitrary orientation. Section 5.3 considers the resulting accretion torque and also applies this to the case of small asynchronism. Partial field channelling is considered in §5.4, and the results are discussed in §5.5.

5.2. Angular Momentum Transfer

The precise nature of the white dwarf's magnetic field is unknown. It has recently been suggested (e.g. Meggitt and Wickramasinghe, 1989) that the field has a quadrupolar component, in addition to a dipole. The stream might then split and lead to two accretion poles, as appear to be observed in some systems. However, to investigate the effect of field channelling out of the orbital plane, a dipole geometry was considered by Campbell (1986). By asssuming that matter links to the primary's magnetic field near L_1, an upper bound can be found for the angular momentum exchanged with the orbit. By considering the distribution of magnetic torque along the stream, cases of partial field channelling can then be gauged.

Figure 5.1 shows the orbital frame, with the primary's dipole moment having orientation (α, β). Material is lost from the lobe-filling secondary

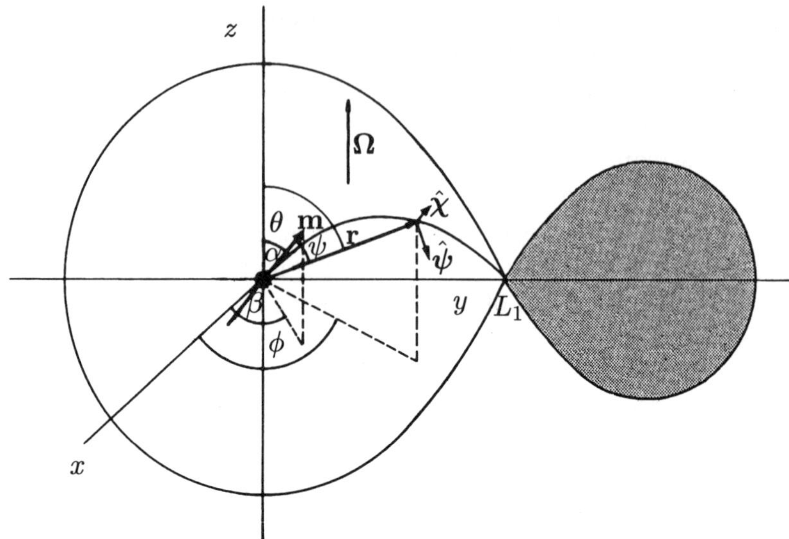

Figure 5.1. The orbital frame, showing the rotated spherical polar coordinate system (r, ψ, χ). Based on Campbell (1986).

through the L_1 region, with speed $\sim c_{\rm s} \ll \Omega D$, at a steady rate $\dot{M}_{\rm s}$ and is assumed to be immediately channelled by the primary's field. The case of a synchronized white dwarf is considered in this section, so the lines of $\mathbf{B}$ are stationary in the orbital frame. The accretion stream is centred on the field line passing through L_1. For a thin, steady stream the accretion torque is found by calculating the specific angular momentum imparted to the primary along this line and multiplying it by $\dot{M}_{\rm p} = -\dot{M}_{\rm s}$.

For a general orientation of the dipole moment $\mathbf{m}$ the unperturbed magnetic field $\mathbf{B}$ will be a function of all three coordinates if described in the polar system (r, θ, ϕ). Since $\mathbf{B}$ is rotationally symmetric about $\mathbf{m}$ the natural coordinates to use are those of a spherical polar system (r, ψ, χ) measured with respect to a Cartesian frame having orientation (α, β) in the orbital frame. Hence ψ is measured from $\hat{\mathbf{m}}$ and χ from a vector $\hat{\mathbf{m}}_\perp$ where, relative to the orbital frame,

$$\hat{\mathbf{m}} = \sin\alpha\cos\beta\,\mathbf{i} + \sin\alpha\sin\beta\,\mathbf{j} + \cos\alpha\,\mathbf{k}, \tag{5.8}$$

and

$$\hat{\mathbf{m}}_\perp = \cos\alpha\cos\beta\,\mathbf{i} + \cos\alpha\sin\beta\,\mathbf{j} - \sin\alpha\,\mathbf{k}. \tag{5.9}$$

In this system the unperturbed field $\mathbf{B}$ will only depend on r and ψ and the equation of a field line will take its simplest form. The field $\mathbf{B}$ can be derived from a magnetic scalar potential $\Psi_{\mathrm{m}}(r,\psi)$ through

$$\mathbf{B} = -\nabla\Psi_{\mathrm{m}}. \tag{5.10}$$

By expressing the gradient operator in the two polar coordinate systems, it follows from (5.10) that the unit vectors $\hat{\boldsymbol{\psi}}$ and $\hat{\boldsymbol{\chi}}$ are related to $\hat{\boldsymbol{\theta}}$ and $\hat{\boldsymbol{\phi}}$ by

$$\hat{\boldsymbol{\psi}} = \frac{\partial\psi}{\partial\theta}\hat{\boldsymbol{\theta}} + \frac{1}{\sin\theta}\frac{\partial\psi}{\partial\phi}\hat{\boldsymbol{\phi}}, \tag{5.11}$$

$$\hat{\boldsymbol{\chi}} = -\frac{1}{\sin\theta}\frac{\partial\psi}{\partial\phi}\hat{\boldsymbol{\theta}} + \frac{\partial\psi}{\partial\theta}\hat{\boldsymbol{\phi}}, \tag{5.12}$$

and hence

$$\left(\frac{\partial\psi}{\partial\theta}\right)^2 + \frac{1}{\sin^2\theta}\left(\frac{\partial\psi}{\partial\phi}\right)^2 = 1. \tag{5.13}$$

Using the fact that $\cos\psi = \hat{\mathbf{m}}\cdot\hat{\mathbf{r}}$ and $\sin\chi = -\hat{\mathbf{m}}_{\perp}\cdot\hat{\boldsymbol{\chi}}$, together with (5.8), (5.9) and (5.12), it can be shown that ψ and χ are related to θ and ϕ through

$$\cos\psi = \sin\alpha\sin\theta\cos(\phi-\beta) + \cos\alpha\cos\theta, \tag{5.14}$$

$$\sin\psi\sin\chi = \sin\theta\sin(\phi-\beta). \tag{5.15}$$

The field line passing through L_1 lies in the plane containing $\hat{\mathbf{m}}$ and $\mathbf{j}$. By its definition, the unit vector $\hat{\boldsymbol{\chi}}$ is constant over the plane of a field line and for the required plane is, from (5.8),

$$\hat{\boldsymbol{\chi}} = \frac{1}{(1-\sin^2\alpha\sin^2\beta)^{\frac{1}{2}}}\hat{\mathbf{m}}\wedge\mathbf{j}. \tag{5.16}$$

Since $\sin\chi = -\hat{\mathbf{m}}_{\perp}\cdot\hat{\boldsymbol{\chi}}$, (5.8), (5.9) and (5.16) give the constant value of χ over the required plane from

$$\sin\chi = \frac{\cos\beta}{(1-\sin^2\alpha\sin^2\beta)^{\frac{1}{2}}}. \tag{5.17}$$

The magnetic potential for a dipole field is

$$\Psi_{\mathrm{m}} = \frac{\mu_0 m}{4\pi r^2}\cos\psi, \tag{5.18}$$

so (5.10) gives the magnetic field components as

$$B_r = \frac{\mu_0 m}{2\pi r^3}\cos\psi, \tag{5.19}$$

$$B_\psi = \frac{\mu_0 m}{4\pi r^3}\sin\psi. \tag{5.20}$$

The equation of the relevant field line is found by integrating the relation

$$\frac{1}{r}\frac{dr}{d\psi} = \frac{B_r}{B_\psi}$$

and applying the condition that the line passes through L_1. This gives

$$r = \frac{A\sin^2\psi}{1-\sin^2\alpha\sin^2\beta}, \tag{5.21}$$

where A is the distance of the L_1 point from the centre of the primary.

For gas temperatures of $\sim 10^4$ K, the thermal pressure gradient along the stream has negligible effect on the speed of material since $v/c_\mathrm{s} \gg 1$. The equation of motion for the accretion stream can be written

$$\ddot{\mathbf{r}} = -\nabla\Psi - 2\boldsymbol{\Omega}\wedge\dot{\mathbf{r}} + \mathbf{F}, \tag{5.22}$$

where $\Psi(r,\psi,\chi)$ is taken to be the Roche potential. The specific magnetic force $\mathbf{F}$, given by (5.7), is perpendicular to $\dot{\mathbf{r}}$ and so does no work on the material. The Roche potential $\Psi(r,\psi,\chi)$ is found by using (2.199) in the coordinates (r,θ,ϕ) and expressing the angular terms as functions of ψ and χ. The required transformations are obtained by using the z-component of $\hat{\mathbf{m}}_\perp$ expressed in (α,β) and in (θ,ϕ,ψ,χ), together with (5.14) and (5.15). The result is

$$\begin{aligned}\Psi = -\frac{GM_\mathrm{p}}{r} - \frac{GM_\mathrm{s}}{D}\left[1 - 2\frac{r}{D}Q_1 + \left(\frac{r}{D}\right)^2\right]^{-\frac{1}{2}} \\ + \frac{GM_\mathrm{s}}{D^2}rQ_1 - \frac{1}{2}\Omega^2 r^2 Q_2,\end{aligned} \tag{5.23}$$

where

$$Q_1 = \sin\theta\sin\phi = \sin\alpha\sin\beta\cos\psi + \sin\psi(\cos\beta\sin\chi + \cos\alpha\sin\beta\cos\chi),$$
$$Q_2 = \sin^2\theta = 1 - (\cos\alpha\cos\psi - \sin\alpha\sin\psi\cos\chi)^2.$$

The angular momentum equation follows by taking the cross product of (5.22) with $\mathbf{r}$, and noting that $\dot{\mathbf{r}}$ lies in the plane of the relevant field line, so

$$\frac{d}{dt}(r^2\dot{\psi})\hat{\boldsymbol{\chi}} = -\mathbf{r}\wedge\nabla\Psi - 2\mathbf{r}\wedge(\boldsymbol{\Omega}\wedge\dot{\mathbf{r}}) + \mathbf{r}\wedge\mathbf{F}. \tag{5.24}$$

The left-hand side of this equation represents the rate of change of specific angular momentum of material, and it follows that the ψ-components on the right-hand side must sum to zero. Equation (5.24) therefore splits into two parts; one representing the rate of change of material angular momentum due to torques acting about the $\hat{\boldsymbol{\chi}}$ direction, and the other representing the balance of torques about the $\hat{\boldsymbol{\psi}}$ direction. The Coriolis term can be split into its ψ and χ-components by expressing $\boldsymbol{\Omega}$ as $\Omega_\chi \hat{\boldsymbol{\chi}} + \boldsymbol{\Omega}_\parallel$, where $\boldsymbol{\Omega}_\parallel$ is the component of $\boldsymbol{\Omega}$ parallel to the plane of the stream. Equation (5.24) then gives

$$\frac{1}{\sin\psi}\frac{\partial \Psi}{\partial \chi}\hat{\boldsymbol{\psi}} - 2\mathbf{r} \wedge (\boldsymbol{\Omega}_\parallel \wedge \dot{\mathbf{r}}) - rF_\chi \hat{\boldsymbol{\psi}} = \mathbf{0}, \tag{5.25}$$

$$\frac{d}{dt}(r^2\dot{\psi}) = -\frac{\partial \Psi}{\partial \psi} - \frac{d}{dt}(r^2\Omega_\chi) + rF_\psi. \tag{5.26}$$

The above equations must be integrated along the stream to calculate the total specific angular momentum delivered to the white dwarf. This total consists of the integral of the magnetic torque, per unit mass, plus the specific angular momentum of material at the accretion pole. The magnetic torque can be expressed as a rate of advection of specific angular momentum by

$$rF_\psi \hat{\boldsymbol{\chi}} - rF_\chi \hat{\boldsymbol{\psi}} = -\frac{d\mathbf{L}}{dt}, \tag{5.27}$$

where, for a steady flow along the path specified by (5.21), $d/dt = \dot{\psi} d/d\psi$. Equations (5.25) and (5.26) therefore become

$$\left(\frac{d\mathbf{L}}{d\psi}\right)_\psi \hat{\boldsymbol{\psi}} = \left[2r\left(r\Omega_r - \frac{dr}{d\psi}\Omega_\psi\right) + \frac{1}{\dot{\psi}\sin\psi}\frac{\partial \Psi}{\partial \chi}\right]\hat{\boldsymbol{\psi}}, \tag{5.28}$$

$$\frac{d}{d\psi}(r^2\dot{\psi}) + \frac{dL_\chi}{d\psi} = -\frac{d}{d\psi}(r^2\Omega_\chi) - \frac{1}{\dot{\psi}}\frac{\partial \Psi}{\partial \psi}. \tag{5.29}$$

Integrating these equations along the stream gives

$$\Delta\mathbf{L}_\parallel = \int_{\psi_1}^{\psi_a}\left[2r\left(r\Omega_r - \frac{dr}{d\psi}\Omega_\psi\right) + \frac{1}{\dot{\psi}\sin\psi}\frac{\partial \Psi}{\partial \chi}\right]\hat{\boldsymbol{\psi}}\, d\psi, \tag{5.30}$$

$$R_\mathrm{p}^2\dot{\psi}_a + \Delta L_\chi = (A^2 - R_\mathrm{p}^2)\Omega_\chi - \int_{\psi_1}^{\psi_a}\frac{1}{\dot{\psi}}\frac{\partial \Psi}{\partial \psi}\, d\psi, \tag{5.31}$$

where ψ_1 is the value of ψ at L_1 and ψ_a is its value at the accreting pole. Equation (5.21) gives these limits as

$$\psi_1 = \cos^{-1}(\sin\alpha\sin\beta), \tag{5.32}$$

$$\psi_a = \sin^{-1}\left[\frac{R_\mathrm{p}(1-\sin^2\alpha\sin^2\beta)}{A}\right]^{\frac{1}{2}}. \tag{5.33}$$

The angular velocity of material, $\dot{\psi}$, can be found from (5.22). Taking the scalar product of this with $\dot{\mathbf{r}}$, integrating with respect to time using the condition of essentially vanishing speed (i.e. $c_\mathrm{s} \ll \Omega D$) at L_1, and noting that r is a function of ψ along the stream, gives

$$\dot{\psi} = \pm\frac{[2(\Psi_1-\Psi)]^{\frac{1}{2}}}{\left[r^2+(dr/d\psi)^2\right]^{\frac{1}{2}}}, \tag{5.34}$$

where the negative sign applies for $0 < \beta < \pi$ and the positive sign for $\pi < \beta < 2\pi$. It is noted that in the strong field limit the magnetic force makes no explicit contribution to $\dot{\psi}$, since (5.7) shows $\mathbf{F}$ is perpendicular to $\mathbf{B}$ and hence to $\dot{\mathbf{r}}$. However, $\mathbf{F}$ determines $dr/d\psi$ since it fixes the shape of the path $r(\psi)$ by acting as a perpendicular constraint force to the stream. Equations (5.17), (5.23) and (5.32) give the Roche potential in the plane of the stream as

$$\begin{aligned}\Psi(r,\psi) &= -\frac{GM_\mathrm{p}}{r} - \frac{GM_\mathrm{s}}{D}\left[1-2\frac{r}{D}\cos(\psi_1-\psi)+\left(\frac{r}{D}\right)^2\right]^{-\frac{1}{2}}\\ &+\frac{GM_\mathrm{s}}{D^2}r\cos(\psi_1-\psi)-\frac{1}{2}\Omega^2r^2\left[\frac{\sin^2\alpha\cos^2\beta+\cos^2\alpha\cos^2(\psi_1-\psi)}{1-\sin^2\alpha\sin^2\beta}\right].\end{aligned} \tag{5.35}$$

The unit vectors $\hat{\boldsymbol{\psi}}$ and $\hat{\boldsymbol{\chi}}$ must be expressed in terms of their components in the orbital frame $Oxyz$, in order to obtain the components of the specific angular momentum transfer along these axes. Equations (5.8) and (5.16) give the Cartesian components of $\hat{\boldsymbol{\chi}}$. The components of $\hat{\boldsymbol{\psi}}$ are then found using $\hat{\boldsymbol{\psi}} = \hat{\boldsymbol{\chi}}\wedge\hat{\mathbf{r}}$ and (5.14), (5.17) and (5.32), together with the definition of Q_1 in (5.23). The results are

$$\hat{\boldsymbol{\psi}} = -\frac{\sin\alpha\cos\beta\cos(\psi_1-\psi)}{(1-\sin^2\alpha\sin\beta)^{\frac{1}{2}}}\mathbf{i}+\sin(\psi_1-\psi)\mathbf{j}-\frac{\cos\alpha\cos(\psi_1-\psi)}{(1-\sin^2\alpha\sin^2\beta)^{\frac{1}{2}}}\mathbf{k}, \tag{5.36}$$

$$\hat{\boldsymbol{\chi}} = -\frac{\cos\alpha}{(1-\sin^2\alpha\sin^2\beta)^{\frac{1}{2}}}\mathbf{i}+\frac{\sin\alpha\cos\beta}{(1-\sin^2\alpha\sin^2\beta)^{\frac{1}{2}}}\mathbf{k}. \tag{5.37}$$

Using these relations together with (5.30) and (5.31) gives the components of the specific angular momentum transfer as

$$\begin{aligned}\Delta J_x = &-\frac{(A^2 - R_{\mathrm{p}}^2)\Omega_\chi \cos\alpha}{(1-\sin^2\alpha\sin^2\beta)^{\frac{1}{2}}} \\ &- \frac{2\sin\alpha\cos\beta}{(1-\sin^2\alpha\sin^2\beta)^{\frac{1}{2}}}\int_{\psi_1}^{\psi_a} r\left(r\Omega_r - \frac{dr}{d\psi}\Omega_\psi\right)\cos(\psi_1-\psi)d\psi \qquad (5.38)\\ &+ \frac{1}{(1-\sin^2\alpha\sin^2\beta)^{\frac{1}{2}}}\int_{\psi_1}^{\psi_a}\frac{1}{\dot\psi}\left(\cos\alpha\frac{\partial\Psi}{\partial\psi} - \frac{\sin\alpha\cos\beta\cos(\psi_1-\psi)}{\sin\psi}\frac{\partial\Psi}{\partial\chi}\right)d\psi,\end{aligned}$$

$$\begin{aligned}\Delta J_y = &\,2\int_{\psi_1}^{\psi_a} r\left(r\Omega_r - \frac{dr}{d\psi}\Omega_\psi\right)\sin(\psi_1-\psi)d\psi \\ &+ \int_{\psi_1}^{\psi_a}\frac{\sin(\psi_1-\psi)}{\dot\psi\sin\psi}\frac{\partial\Psi}{\partial\chi}d\psi, \qquad (5.39)\end{aligned}$$

$$\begin{aligned}\Delta J_z = &\,\frac{(A^2 - R_{\mathrm{p}}^2)\Omega_\chi \sin\alpha\cos\beta}{(1-\sin^2\alpha\sin^2\beta)^{\frac{1}{2}}} \\ &- \frac{2\cos\alpha}{(1-\sin^2\alpha\sin^2\beta)^{\frac{1}{2}}}\int_{\psi_1}^{\psi_a} r\left(r\Omega_r - \frac{dr}{d\psi}\Omega_\psi\right)\cos(\psi_1-\psi)d\psi \qquad (5.40)\\ &- \frac{1}{(1-\sin^2\alpha\sin^2\beta)^{\frac{1}{2}}}\int_{\psi_1}^{\psi_a}\frac{1}{\dot\psi}\left(\sin\alpha\cos\beta\frac{\partial\Psi}{\partial\psi} + \frac{\cos\alpha\cos(\psi_1-\psi)}{\sin\psi}\frac{\partial\Psi}{\partial\chi}\right)d\psi.\end{aligned}$$

Equations (5.21), (5.36) and (5.37) give

$$r\Omega_r - \frac{dr}{d\psi}\Omega_\psi = \frac{A\Omega\cos\alpha\sin\psi\,[\sin\psi\sin(\psi_1-\psi) + 2\cos\psi\cos(\psi_1-\psi)]}{(1-\sin^2\alpha\sin^2\beta)^{\frac{3}{2}}}. \qquad (5.41)$$

The required derivatives of the Roche potential are obtained from (5.17), (5.23) and (5.32) as

$$\frac{\partial\Psi}{\partial\psi} = -\frac{GM_{\mathrm{s}}}{D^2}r\left[\frac{M}{M_{\mathrm{s}}}\frac{r}{D}\frac{\cos^2\alpha\cos(\psi_1-\psi)}{(1-\sin^2\alpha\sin^2\beta)} + W^{-\frac{3}{2}} - 1\right]\sin(\psi_1-\psi),$$

where

$$W = 1 - 2\frac{r}{D}\cos(\psi_1-\psi) + \left(\frac{r}{D}\right)^2, \qquad (5.42\text{a,b})$$

and

$$\frac{1}{\sin\psi}\frac{\partial\Psi}{\partial\chi} = \Omega^2 r^2\frac{\sin\alpha\cos\alpha\cos\beta}{1-\sin^2\alpha\sin^2\beta}\sin(\psi_1-\psi), \qquad (5.43)$$

with $M = M_{\rm s} + M_{\rm p}$. It is noted that (5.43) is purely centrifugal.

Substituting (5.21), (5.34), (5.35) and (5.41)–(5.43) in (5.38)–(5.40), and evaluating the Coriolis integrals, gives the components of the specific angular momentum transfer as

$$\Delta J_x = A^2 \Omega f_x(\alpha, \beta), \tag{5.44}$$

$$\Delta J_y = A^2 \Omega f_y(\alpha, \beta), \tag{5.45}$$

$$\Delta J_z + R_{\rm p}^2 \Omega \sin^2 \theta_a = A^2 \Omega f_z(\alpha, \beta). \tag{5.46}$$

The second term on the left-hand side of (5.46) represents the additional specific angular momentum of material about the z-axis at the accreting pole when measured in the inertial frame. The dimensionless functions $f_i(\alpha, \beta)$ are

$$f_x(\alpha, \beta) = \left(\frac{R_{\rm p}}{A}\right)^2 \frac{\sin 2\alpha \cos \beta \sin^2(\psi_1 - \psi_a)}{2(1 - \sin^2 \alpha \sin^2 \beta)} - \int_{\psi_a}^{\psi_1} K_x(\alpha, \beta, \psi) d\psi, \tag{5.47}$$

$$f_y(\alpha, \beta) = \frac{\cos \alpha}{(1 - \sin^2 \alpha \sin^2 \beta)^{\frac{5}{2}}} J(\alpha, \beta) + \int_{\psi_a}^{\psi_1} K_y(\alpha, \beta, \psi) d\psi, \tag{5.48}$$

$$f_z(\alpha, \beta) = 1 + \int_{\psi_a}^{\psi_1} K_z(\alpha, \beta, \psi) d\psi, \tag{5.49}$$

where

$$K_x = \pm \frac{\cos \alpha \sin^3 \psi \sin(\psi_1 - \psi)(\sin^2 \psi + 4\cos^2 \psi)^{\frac{1}{2}}}{(1 - \sin^2 \alpha \sin^2 \beta)^{\frac{5}{2}} \sqrt{2Q}} \bar{W}, \tag{5.50}$$

$$\begin{aligned} J = &\frac{3}{4}(\psi_a - \psi_1) + \frac{3}{8}\sin(\psi_1 - \psi_a)\cos(\psi_1 + \psi_a) \\ &+ \frac{1}{4}(\sin^3 \psi_1 \cos \psi_1 - \sin^3 \psi_a \cos \psi_a) \\ &+ \sin^4 \psi_a \sin(\psi_1 - \psi_a)\cos(\psi_1 - \psi_a), \end{aligned} \tag{5.51}$$

$$K_y = \pm \frac{A}{D} \frac{\sin \alpha \cos \alpha \cos \beta \sin^2(\psi_1 - \psi) \sin^5 \psi (\sin^2 \psi + 4\cos^2 \psi)^{\frac{1}{2}}}{(1 - \sin^2 \alpha \sin^2 \beta)^4 \sqrt{2Q}}, \tag{5.52}$$

$$K_z = \pm \frac{M_{\rm s}}{M} \frac{\sin \alpha \cos \beta \sin(\psi_1 - \psi) \sin^3 \psi (\sin^2 \psi + 4\cos^2 \psi)^{\frac{1}{2}} (W^{-\frac{3}{2}} - 1)}{(1 - \sin^2 \alpha \sin^2 \beta)^{\frac{5}{2}} \sqrt{2Q}}, \tag{5.53}$$

with W given by (5.42b) and

$$\bar{W} = \frac{A}{D}\frac{\sin^2\psi\cos(\psi_1-\psi)}{(1-\sin^2\alpha\sin^2\beta)} + \frac{M_{\rm s}}{M}\left(W^{-\frac{3}{2}}-1\right), \tag{5.54}$$

$$\begin{aligned} Q = & \frac{M_{\rm p}}{M}\left(\frac{D}{r}-\frac{D}{A}\right) + \frac{M_{\rm s}}{M}\left[\frac{A}{D}-\frac{r}{D}\cos(\psi_1-\psi)\right] \\ & -\frac{M_{\rm s}}{M}\left[\left(1-\frac{A}{D}\right)^{-1} - W^{-\frac{1}{2}}\right] \\ & -\frac{1}{2}\left[\left(\frac{A}{D}\right)^2 - \left(\frac{r}{D}\right)^2\left(\frac{\sin^2\alpha\cos^2\beta+\cos^2\alpha\cos^2(\psi_1-\psi)}{1-\sin^2\alpha\sin^2\beta}\right)\right]. \end{aligned} \tag{5.55}$$

In equations (5.50), (5.52) and (5.53) the positive signs apply for $0<\beta<\pi$ and the negative signs for $\pi<\beta<2\pi$. The integrals in (5.47)–(5.49) are evaluated numerically and the ratio A/D is found from (2.221) with $r_{1\rm p}=A$.

5.3. Synchronous and Asynchronous Cases

5.3.1. THE SYNCHRONOUS CASE

Since the primary accretes matter at a steady rate $\dot{M}_{\rm p}$, the components of the accretion torque are

$$T_x = A^2\Omega\dot{M}_{\rm p}f_x(\alpha,\beta), \tag{5.56}$$

$$T_y = A^2\Omega\dot{M}_{\rm p}f_y(\alpha,\beta), \tag{5.57}$$

$$T_z = A^2\Omega\dot{M}_{\rm p}f_z(\alpha,\beta). \tag{5.58}$$

Because matter is lost from L_1 with a speed $\sim c_{\rm s} \ll \Omega D$, it follows from energy conservation that the functions $f_i(\alpha,\beta)$ can only be evaluated at orientations for which the channelled stream lies entirely within the primary's Roche lobe. For a given value of α, a critical angle β_c will exist at which the primary's field line through L_1 is tangent to its Roche lobe at that point. Hence, for total field channelling, the calculation of the accretion torque can only be performed for $\beta_c \le \beta \le \pi-\beta_c$ and $\beta_c+\pi \le \beta \le 2\pi-\beta_c$. An expression can be derived for $\beta_c(\alpha)$ by performing a Taylor expansion of Ψ, given by (5.35), along the field line defined by (5.21). When $\beta=\beta_c$ it follows that $(d^2\Psi/d\psi^2)_{L_1} = 0$, giving

$$\beta_c(\alpha) = \sin^{-1}\left[\frac{\frac{M_s}{M}\left(\frac{A}{D}\right)^2\left[\left(1-\frac{A}{D}\right)^{-3}-1\right]+\left(\frac{A}{D}\right)^3\cos^2\alpha}{\left(\frac{M_s}{M}\left(\frac{A}{D}\right)^2\bar{Q}+4\left(\frac{A}{D}\right)^3+8\frac{M_p}{M}\right)\sin^2\alpha}\right]^{\frac{1}{2}}, \quad (5.59)$$

where

$$\bar{Q} = \left(1+8\frac{A}{D}\right)\left(1-\frac{A}{D}\right)^{-3} - 1.$$

As the inclination of the magnetic moment to the orbital plane increases (i.e. as α decreases) β_c increases and hence the range of β for which the fully-channelled calculation can be performed is reduced. A minimum value of α is reached when $\beta_c = \pi/2$ and so it follows from (5.59) that

$$\alpha_{\min} = \sin^{-1}\left[\frac{\frac{M_s}{M}\left(\frac{A}{D}\right)^2\left[\left(1-\frac{A}{D}\right)^{-3}-1\right]+\left(\frac{A}{D}\right)^3}{\frac{M_s}{M}\left(\frac{A}{D}\right)^2\bar{Q}+5\left(\frac{A}{D}\right)^3+8\frac{M_p}{M}}\right]^{\frac{1}{2}}. \quad (5.60)$$

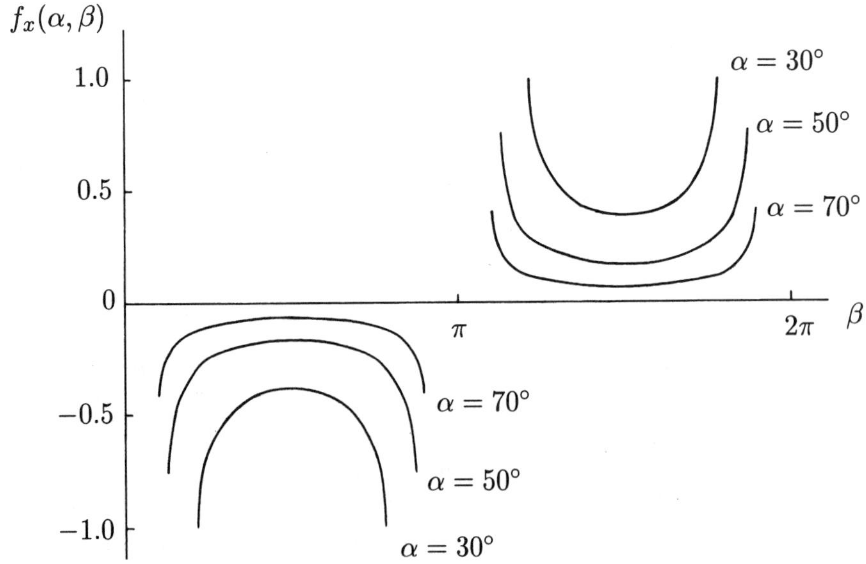

Figure 5.2. The function $f_x(\alpha, \beta)$. Based on Campbell (1986).

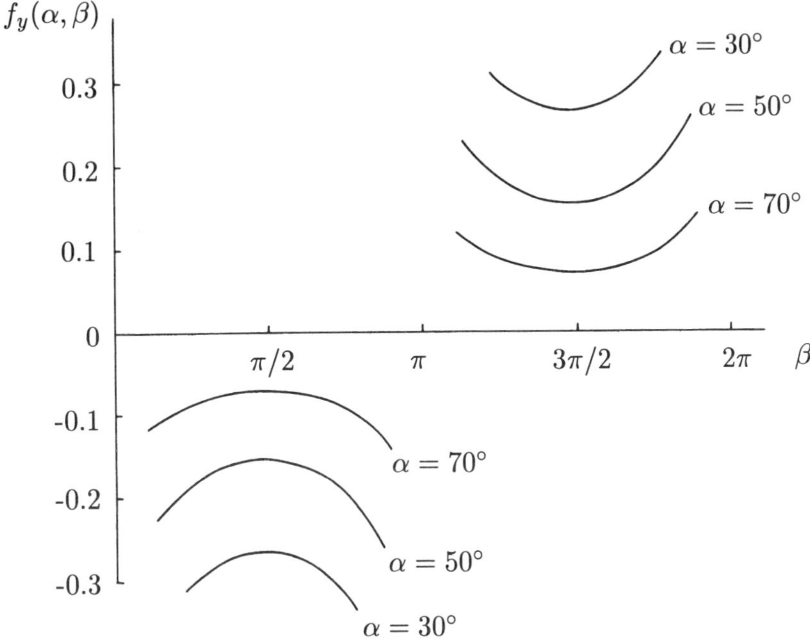

Figure 5.3. The function $f_y(\alpha, \beta)$. Based on Campbell (1986).

Taking $P = 2.0$ hrs, (2.226) gives a lobe-filling secondary with $M_s = 0.2\,M_\odot$. Using $M_p = 0.6\,M_\odot$, (2.222) yields $A/D = 0.61$ and (5.60) gives $\alpha_{\min} = 19°$. The functions $f_i(\alpha, \beta)$ were calculated for three values of α. The results are shown in Figures 5.2–5.4 and Tables 5.1–5.3.

Figure 5.1 illustrates that if $\alpha_{\min} \leq \alpha \leq \pi/2$ then the accreting magnetic pole lies above the orbital plane for $\beta_c \leq \beta \leq \pi - \beta_c$ and beneath it for $\beta_c + \pi \leq \beta \leq 2\pi - \beta_c$. The situation is reversed for $\pi/2 < \alpha \leq \pi - \alpha_{\min}$. Figures 5.2 and 5.3 show that $f_x(\alpha, \beta)$ and $f_y(\alpha, \beta)$ are negative when the accreting pole is above the orbital plane. For given values of α and β, it is seen that the magnitude of f_x is generally larger than that of f_y. Equations (5.38), (5.44) and (5.47) show that Coriolis, centrifugal and gravitational terms contribute to $f_x(\alpha, \beta)$, although the Coriolis term is small since $R_p/A \ll 1$. The torque component T_x therefore involves gravitational coupling of the stream to the orbit. Equations (5.39) and (5.43) show that T_y does not involve any gravitational coupling, since $\partial\Psi/\partial\chi$ is purely centrifugal in the required χ-plane. It can be shown that the Coriolis term gives the major contribution to T_y. It is noted that for $\alpha = \pi/2$, corresponding to the accreting pole lying in the orbital plane, equations (5.47) and (5.48) show that f_x and f_y vanish, as expected.

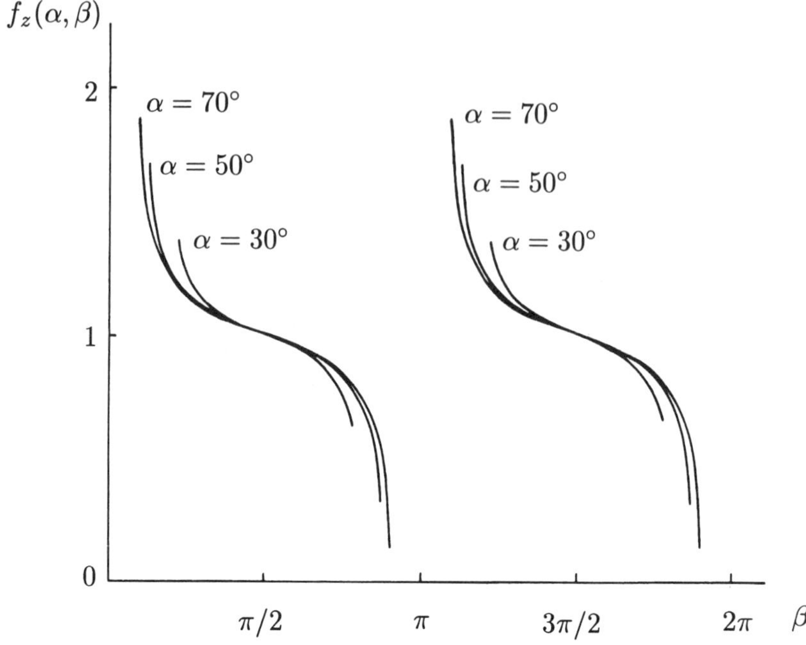

Figure 5.4. The dimensionless function $f_z(\alpha, \beta)$. Based on Campbell (1986).

Figure 5.4 shows the variations of $f_z(\alpha, \beta)$. Equation (5.49) illustrates that all the orientation dependence of f_z is contained in the integral of K_z. This integral arose from the last term in (5.40) and is purely gravitational since the centrifugal part of the term in $\partial\Psi/\partial\psi$ cancels the term in $\partial\Psi/\partial\chi$. In an inertial frame centred on the white dwarf material at L_1 has specific angular momentum $A^2\Omega$. For $\beta \neq \pi/2$ or $3\pi/2$, the orbit adds to or subtracts from this and the integrated resultant is delivered to the white dwarf through the total magnetic torque. For $\beta = \pi/2$ or $3\pi/2$ the gravitational coupling term K_z is zero and specific angular momentum $A^2\Omega$ is delivered to the primary about the z-axis, so that $f_z = 1$. For $\beta_c \leq \beta < \pi/2$ and $\beta_c + \pi \leq \beta < 3\pi/2$ the accreting magnetic pole lags the motion of the line of stellar centres and the stream gains angular momentum about the z-axis from the orbit, resulting in $f_z > 1$. For $\pi/2 < \beta \leq \pi - \beta_c$ and $3\pi/2 < \beta \leq 2\pi - \beta_c$ the accreting pole leads the line of stellar centres and the stream loses angular momentum to the orbit, so $f_z < 1$. It is seen from Figure 5.4 that for significant ranges of β centred on $\beta = \pi/2$ or $3\pi/2$, f_z is not sensitive to changes in α at fixed β. This is because for these ranges of β the gravitational coupling to the orbit, through the integral of K_z, is weak.

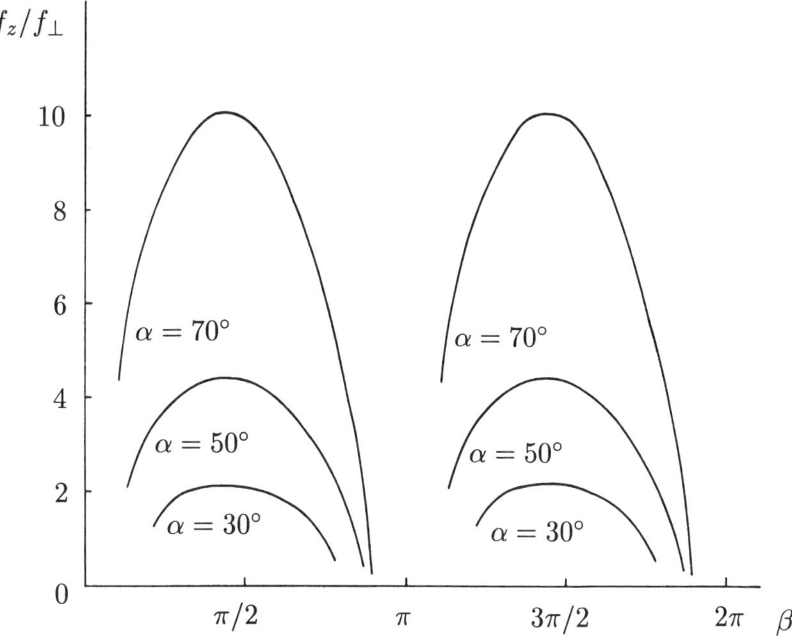

Figure 5.5. The ratio $f_z(\alpha,\beta)/f_\perp(\alpha,\beta)$. Based on Campbell (1986).

Figure 5.5 shows the ratio of the vertical to the horizontal component of the accretion torque, where $f_\perp = (f_x^2 + f_y^2)^{1/2}$. It is seen that for $\alpha = 30°$, corresponding to a high inclination of the accreting pole to the orbital plane, f_z and $f_\perp$ are comparable over the range of β.

TABLE 5.1. $\alpha = 30°$, $\beta_c = 39.3°$

β (deg)	f_x	f_y	f_z	$f_z/f_\perp$
β_c	-1.022	-0.312	1.370	1.282
50	-0.568	-0.293	1.168	1.827
60	-0.465	-0.279	1.106	2.038
70	-0.414	-0.270	1.064	2.154
90	-0.380	-0.265	1.000	2.157
100	-0.388	-0.268	0.969	2.054
110	-0.414	-0.276	0.936	1.881
120	-0.465	-0.289	0.894	1.632
130	-0.568	-0.308	0.832	1.288
180-β_c	-1.022	-0.334	0.630	0.586

TABLE 5.2. $\alpha = 50°, \beta_c = 24°$

β_c (deg)	f_x	f_y	f_z	$f_z/f_\perp$
β_c	-0.765	-0.226	1.680	2.107
30	-0.412	-0.212	1.338	2.888
40	-0.285	-0.191	1.204	3.509
50	-0.229	-0.176	1.137	3.942
60	-0.197	-0.165	1.092	4.242
70	-0.180	-0.158	1.057	4.417
90	-0.167	-0.154	1.000	4.398
110	-0.180	-0.162	0.943	3.903
120	-0.198	-0.171	0.908	3.478
130	-0.229	-0.185	0.863	2.935
140	-0.285	-0.205	0.796	2.265
150	-0.412	-0.234	0.662	1.397
180-β_c	-0.765	-0.258	0.320	0.396

TABLE 5.3. $\alpha = 70°, \beta_c = 19°$

β (deg)	f_x	f_y	f_z	$f_z/f_\perp$
β_c	-0.411	-0.119	1.872	4.381
30	-0.160	-0.102	1.299	6.844
40	-0.117	-0.091	1.192	8.050
50	-0.095	-0.083	1.131	8.960
60	-0.083	-0.077	1.088	9.608
70	-0.076	-0.074	1.055	9.992
90	-0.071	-0.071	1.000	9.952
110	-0.076	-0.075	0.945	8.849
120	-0.083	-0.080	0.912	7.913
130	-0.095	-0.087	0.869	6.732
140	-0.177	-0.098	0.808	5.304
150	-0.160	-0.113	0.071	3.575
180-β_c	-0.411	-0.141	0.128	0.294

It is noted that the critical angle β_c results from the condition of strong field channelling at L_1. This restriction can be removed by allowing material to become field-channelled at some distance from L_1. Provided this distance is a small fraction of A, the curves in Figures 5.2–5.5 can be extended to all β values and will be slightly less steep at their ends.

5.3.2. THE ASYNCHRONOUS CASE

If the white dwarf primary is asynchronous with the orbit then its magnetic field geometry will vary with time in the orbital frame and matter will be channelled along moving field lines. Whatever the path taken by material, the transfer time from L_1 to the accretion pole is given by $\Delta t \sim P$, where P is the orbital period. It follows that two relevant regimes of asynchronism can be distinguished, defined by $P_{\text{syn}} \lesssim P$ and $P_{\text{syn}} \gg P$, where P_{syn} is the synodic period of the asynchronism.

If $P_{\text{syn}} \lesssim P$ then in the time during which a given flux tube is in contact with the L_1 region it can only be partly filled with material. As a result matter will be accreted onto the primary in discontinuous streams, reaching it at different times and along different paths. The analysis of the previous section can therefore not be used to calculate the time-dependent accretion torque in this case.

If $P_{\text{syn}} \gg P$, or equivalently $\omega/\Omega \ll 1$ where $\omega = 2\pi/P_{\text{syn}}$, then a given magnetic flux tube of the primary will completely fill with material during its time of contact with the L_1 region. A flux tube will correspondingly rapidly empty as soon as it becomes disconnected from the L_1 source. Hence, at a given time, the primary will experience a torque due to the single continous stream along the flux tube connecting it to L_1. It then follows that the expressions for the components of $\mathbf{T}$ obtained in §5.3.1 for time-independent orientations (α, β) can be used to obtain the variation of $\mathbf{T}$ for time-dependent (α, β), provided that $\omega/\Omega \ll 1$. For the present, the case is considered in which $\boldsymbol{\omega}$ is parallel to $\boldsymbol{\Omega}$ so α is constant and $\beta = \omega t$. The time-dependent accretion torque is then simply obtained by making this substitution for β in (5.56)–(5.58).

When considering the long term effect of the accretion torque on the primary, and comparing it with the dissipation torque given by (4.70), a synodic average is appropriate. This can be found by exploiting certain symmetry properties possessed by the functions $f_i(\alpha, \beta)$ defined by (5.47)–(5.49). First, consider magnetic orientations of the primary whose corresponding accretion streams are mirror images about the orbital plane $z = 0$. Figure 5.1 shows that, for fixed α, such paths are characterized by orientations β and $\beta + \pi$, while mirror points on them about $z = 0$ have angular positions ψ and $\pi - \psi$. It follows from (5.42b), (5.50), (5.54), and (5.55) that

$$K_x(\alpha, \beta + \pi, \pi - \psi) = K_x(\alpha, \beta, \psi).$$

Defining the integral appearing in (5.47) for $f_x(\alpha, \beta)$ as

$$I_x(\alpha, \beta) = \int_{\psi_a}^{\psi_1} K_x(\alpha, \beta, \psi) d\psi,$$

it is then simple to show that

$$I_x(\alpha, \beta + \pi) = -I_x(\alpha, \beta). \tag{5.61}$$

In a similar way it can be shown that the integral in (5.48) for $f_y(\alpha, \beta)$ satisfies

$$I_y(\alpha, \beta + \pi) = -I_y(\alpha, \beta) \tag{5.62}$$

for mirror paths about the orbital plane. Since the Coriolis terms in (5.47) and (5.48) also possess the above antisymmetric property, it follows that

$$f_x(\alpha, \beta + \pi) = -f_x(\alpha, \beta), \tag{5.63}$$

$$f_y(\alpha, \beta + \pi) = -f_y(\alpha, \beta). \tag{5.64}$$

Consider, now, the $x = 0$ plane. Figure 5.1 shows that for $\beta_c \leq \beta \leq \pi - \beta_c$ mirror paths about this plane are characterized by β and $\pi - \beta$, while mirror points have the same values of ψ. Equations (5.53)–(5.55) give

$$K_z(\alpha, \pi - \beta, \psi) = -K_z(\alpha, \beta, \psi).$$

Writing the integral in (5.49) for $f_z(\alpha, \beta)$ as $I_z(\alpha, \beta)$, it can be shown that

$$I_z(\alpha, \pi - \beta) = -I_z(\alpha, \beta). \tag{5.65}$$

The corresponding result for $\beta_c + \pi \leq \beta \leq 2\pi - \beta_c$ is

$$I_z(\alpha, 3\pi - \beta) = -I_z(\alpha, \beta). \tag{5.66}$$

Since the asynchronous motion considered here has constant α and $\beta = \omega t$, the average value of an accretion torque component over one synodic period is given by

$$\langle T_i \rangle = \frac{1}{2\pi} \int_0^{2\pi} T_i(\alpha, \beta) d\beta. \tag{5.67}$$

Equations (5.49), (5.56)–(5.58) and (5.63)–(5.67) then show that

$$\langle T_x \rangle = \langle T_y \rangle = 0, \tag{5.68}$$

$$\langle T_z \rangle = A^2 \Omega \dot{M}_{\rm p}. \tag{5.69}$$

Hence the average value of the accretion torque for this type of asynchronism is

$$\langle \mathbf{T} \rangle = A^2 \Omega \dot{M}_{\rm p} \, \mathbf{k}. \tag{5.70}$$

5.4. Partial Field Channelling

Observations suggest that effective field-channelling of the stream occurs at some distance from the L_1 region (e.g. Liebert and Stockman 1984; Wu and Wickramasinghe, 1993; Schwope, Mantel and Horne, 1996). After leaving L_1 the stream will lie in the orbital plane, leading the line of stellar centres, and cause large local distortions of the primary's magnetic field. As the field strength increases its distortion will decrease and effective channelling, in general out of the orbital plane, will ultimately occur.

The dominant contribution to the accretion torque should arise from the channelled region, since this is where the generated magnetic stresses are greatest. An estimate of the accretion torque in the case of a partially channelled flow can therefore be made by considering the angular momentum transfer due to a section of the totally channelled stream lying inside a radius $R_{\rm m}(\alpha, \beta)$, measured from the white dwarf. This will give an estimate of the relative magnitudes of the accretion torque components for varying degrees of field channelling.

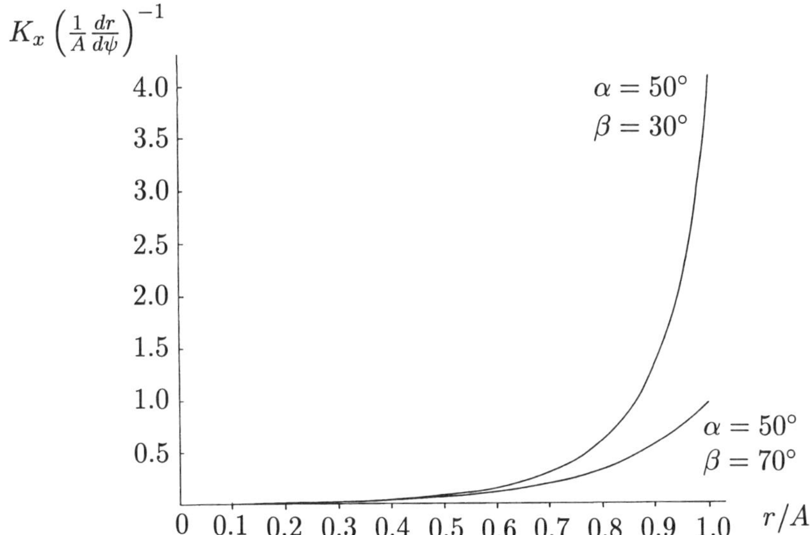

Figure 5.6. The variation of $K_x A(dr/d\psi)^{-1}$ with r/A along primary field lines passing through L_1. Based on Campbell (1986).

The dominant contribution to $f_x(\alpha, \beta)$ arises from the second term on the right hand side of (5.47), involving the integral of K_x along the stream. The major contribution to $f_y(\alpha, \beta)$ comes from the first term on the right hand side of (5.48). This term arose from the Coriolis integral in (5.39) and hence the variation of its integrand, denoted by $K(\alpha, \beta, \psi)$, is required along the stream together with the variations of K_x and K_z. Although

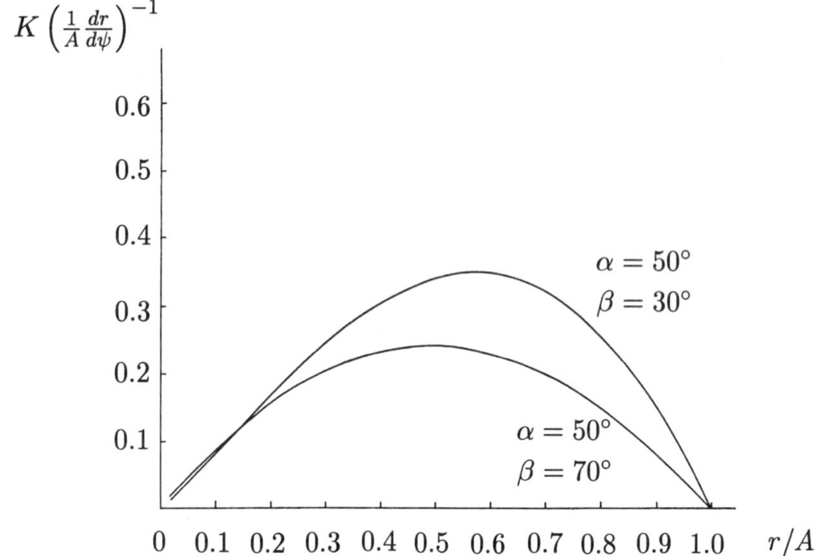

Figure 5.7. The variation of $KA(dr/d\psi)^{-1}$ with r/A along primary field lines passing through L_1. Based on Campbell (1986).

these integrands were expressed as funcitons of ψ, in the present context the coordinate r/A must be used in their integration. It follows that the variation of $K_x(\alpha, \beta, \psi(r))A(dr/d\psi)^{-1}$ with r/A is required to estimate the contribution made between $r = R_\mathrm{p}$ and $r = R_\mathrm{m}$, and likewise for the other two integrands.

Figures 5.6–5.8 show the variations of the relevant functions with r/A. Consider $R_\mathrm{m} = 0.5A$; Figure 5.6 illustrates that the section of the stream having $r < R_\mathrm{m}$, and hence $r/A < 0.5$, contributes only a few per cent to the total line integral of $K_xA(dr/d\psi)^{-1}$. It therefore follows that, for such a partially channelled stream, the values of f_x will typically be reduced by an order of magnitude from those shown in Tables 5.1–5.3.

Figure 5.7 shows that the section of the stream having $r < R_\mathrm{m}$ typically contributes $\sim$ 50% to the total line integral of $KA(dr/d\psi)^{-1}$. Hence, for $R_\mathrm{m} = 0.5A$, the values of f_y shown in Tables 5.1–5.3 will only be approximately halved. The magnitude of f_y will then generally exceed that of f_x, this being the reverse of the case of a totally channelled stream.

For $R_\mathrm{m} = 0.5A$, Figure 5.8 illustrates that the integral on the right hand side of (5.49), for $f_z(\alpha, \beta)$, will typically be reduced by at least an order of magnitude from its value for a totally field-controlled flow. The deviations of $f_z(\alpha, \beta)$ from unity will therefore be significantly reduced from those shown in Tables 5.1–5.3. This is a result of the fact that most of the exchange of the z-component of angular momentum with the orbit occurs near L_1, due

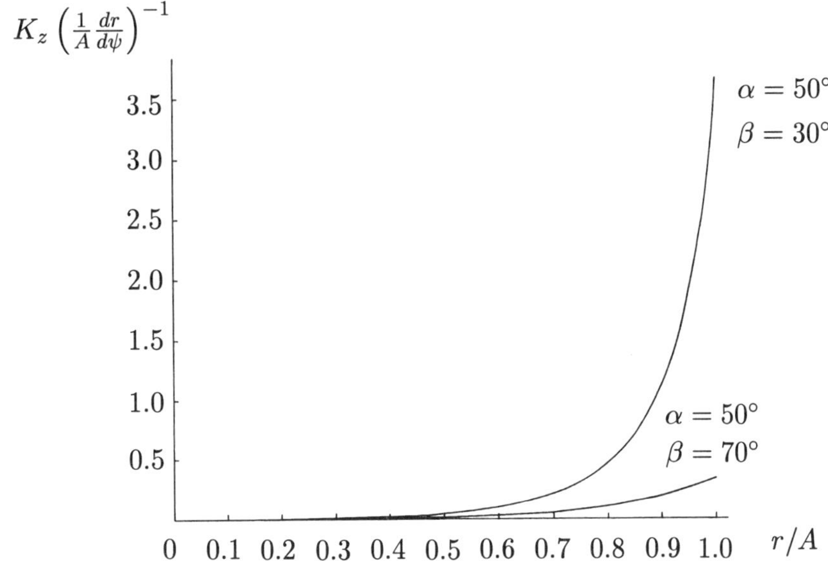

Figure 5.8. The variation of $K_z A(dr/d\psi)^{-1}$ with r/A along primary field lines passing through L_1. Based on Campbell (1986).

to gravitational coupling to the secondary.

5.5. Discussion

The conclusions of the foregoing calculation can be summarized. For a stream channelled from the L_1 region, and high inclinations of the white dwarf dipole moment, the horizontal components of the accretion torque can be comparable to the vertical component. For slow asynchronous motions, with the primary's angular velocity parallel to $\mathbf{\Omega}$, the horizontal torque components average to zero over a synodic period, while the vertical component averages to the conserved form $A^2\Omega\dot{M}_\text{p}$.

For a partially channelled stream the magnitudes of the horizontal torque components are reduced. For a channelling radius of $0.5A$ the horizontal torque component perpendicular to the line of stellar centres becomes a small fraction ($\sim 10^{-2}$) of the vertical component. However, the torque component along the line of stellar centres can remain significant for some magnetic orientations, since it does not involve gravitational coupling to the orbit, being principally due to Coriolis torque. The distribution of this torque is fairly uniform along the stream so, even for a channelling radius as small as $\sim 0.2A$, T_y can be a significant fraction of T_z for channelling well out of the orbital plane. Recent observations of HU Aqr by Schwope, Mantel and Horne (1996), using Doppler tomography, indicate the existence of an accretion curtain of tenuous material above the initial stream. Closer to the white dwarf the whole stream lifts out of the orbital plane. This suggests that the field linkage process is continuous, beginning near L_1 but only becoming strongly effective closer to the primary. The observations indicate a channelling radius of at least $0.3A$, so the T_y component of the accretion torque would be significant.

A major future problem is the calculation of the effect of the primary's magnetic field on the structure of the accretion stream. The channelling radius $R_{\rm m}(\alpha, \beta)$ could then be predicted. This is a formidable problem, with similar difficulties to obtaining a self-consistent poloidal field structure for magnetic wind flows (see §12.2.4). However, since observations suggest that $R_{\rm m}$ is not very small, the conclusions reached here regarding the significance of T_y should be valid. These results have important consequences for the maintenance of synchronism, which are discussed in the next chapter.

References

Campbell, C.G., 1986. *Mon. Not. R. Astr. Soc.*, **219**, 589.
Liebert, J. and Stockman, H.S., 1984. *Cataclysmic Variables and Low Mass X-Ray Binaries*, Reidel, Dordrecht, Holland.
Meggitt, S.M.A. and Wickramasinghe, D.T., 1989. *Mon. Not. R. Astr. Soc.*, **236**, 31.
Schwope, A.D., Mantel, K. and Horne, K., 1996. *Astron. Astrophys.*, in press.
Wu, K. and Wickramasinghe, D.T., 1993. *Mon. Not. R. Astr. Soc.*, **260**, 141.

CHAPTER 6

THE MAINTENANCE OF SYNCHRONISM

6.1. Non-Dissipative Torques

Before the attainment of synchronism can be considered, the nature of the corotating state must be investigated. As the white dwarf approaches corotation the dissipative torque, given by (4.70), tends to zero. However, the accretion torque, given by (5.56)–(5.58), remains finite and so for a spherical primary a non-dissipative torque must act to cancel it. In this case the total torque on the synchronous white dwarf vanishes and its angular velocity and angular momentum are conserved, being parallel to $\boldsymbol{\Omega}$. If the primary is non-spherical, in general, its inertial angular velocity $\boldsymbol{\omega}_{\mathrm{in}}$ and angular momentum $\mathbf{L}$ will not be parallel. Since synchronism requires $\boldsymbol{\omega}_{\mathrm{in}} = \boldsymbol{\Omega}$ then $\mathbf{L}$ will precess about $\boldsymbol{\omega}_{\mathrm{in}}$ and a finite torque will act. This torque must have the required form to generate the precession in $\mathbf{L}$. It will be seen that the accretion torque alone cannot satisfy this condition and hence a non-dissipative torque is again needed.

Early considerations of the maintenance of synchronism pointed out that a non-dissipative magnetostatic torque would result if the primary's field was excluded from the secondary. However, this can only happen if the decay time of a magnetic field in the secondary far exceeds the lifetime of the binary system. The decay times given by (2.163) correspond to the dissipation of poloidal modes. The longest-lived principal mode has $l = j = 1$ and has a decay time of

$$\tau_{\mathrm{d}} = \frac{R_{\mathrm{s}}^2}{\pi^2 \eta},$$

where $\eta = 1/\mu_0 \sigma$ with σ the conductivity. The decay time is maximized by taking an Ohmic diffusivity, given by (2.165) as

$$\eta_{\mathrm{ohm}} = 5.2 \times 10^7 \ln \Lambda \, T^{-\frac{3}{2}} \, \mathrm{m}^2 \, \mathrm{s}^{-1},$$

where T is the temperature and $\ln \Lambda$ a shielding factor of typical magnitude $5 \lesssim \ln \Lambda \lesssim 20$. Taking a mean temperature of 10^6 K, together with $\ln \Lambda = 6$

and $R_s = 2.3 \times 10^8$ m gives $\tau_d = 5.4 \times 10^8$ yrs. The binary lifetime is measured by the mass transfer time-scale $\tau_M \sim M_s/|\dot{M}_s|$, which is typically $\sim 4 \times 10^9$ yrs for an AM Her system. Hence $\tau_d < \tau_M$ which invalidates a magnetostatic approach, since the primary's field will not be excluded from the secondary. In fact, as previously noted, the low mass secondaries in AM Her systems will be largley convective and η is then believed to be greatly enhanced, so shortening τ_d by several orders of magnitude and making a magnetostatic torque even less valid.

Two mechanisms have been considered to generate non-dissipative torques. Firstly, it was pointed out by Joss, Katz and Rappaport (1979) that the secondary could contain an intrinsic magnetic field and that this would generate a torque on the primary. The existence of such a field is very likely since turbulence and rapid rotation are present and these can lead to dynamo action, as seen in §2.3. Secondly, Katz (1989) proposed that the white dwarf could be distorted by a $\mathbf{J} \wedge \mathbf{B}$ force, where $\mathbf{J}$ is the current density source of its field. Tidal interaction with the secondary's gravitational field then leads to a torque on the primary. These mechanisms are investigated in detail in this chapter.

Sections 6.2 and 6.3 consider balances between a non-dissipative magnetic torque and the accretion torque, in two and three-dimensions. Quadrupolar fields are discussed in §6.4. In §6.5 the three-dimensional balance between a tidal torque, exerted on a magnetically distorted primary, and a magnetic torque is analysed. Orbital torques and their evolutionary effects are considered in §6.6, while the results are discussed in §6.7.

6.2. Accretion Versus Magnetism in Two-Dimensions

The simplest case, in which both stars have dipolar fields with moments $\mathbf{m}_p$ and $\mathbf{m}_s$ lying in the orbital plane, was considered in Campbell (1985). Figure 6.1 shows the orientation angles used to define this situation. The vector $\mathbf{m}_s$ is fixed in the synchronous secondary star by dynamo processes, while $\mathbf{m}_p$ can rotate with the white dwarf. The primary is taken to be spherical. Since the accretion stream is centred in the orbital plane, the torque components in the x and y directions, given by (5.56) and (5.57), vanish. If matter is field-channelled some distance from L_1 then $f_z(\pi/2, \beta)$ is close to unity and the accretion torque can be written

$$\mathbf{T}_a = A^2 \Omega \dot{M}_p \mathbf{k}, \tag{6.1}$$

where $\mathbf{k}$ is a unit vector in the z-direction.

The magnetic torque exerted on the primary by the field of the secondary is

$$\mathbf{T}_m = -m_p B_{sp} \sin(\beta - \bar{\delta}) \mathbf{k}, \tag{6.2}$$

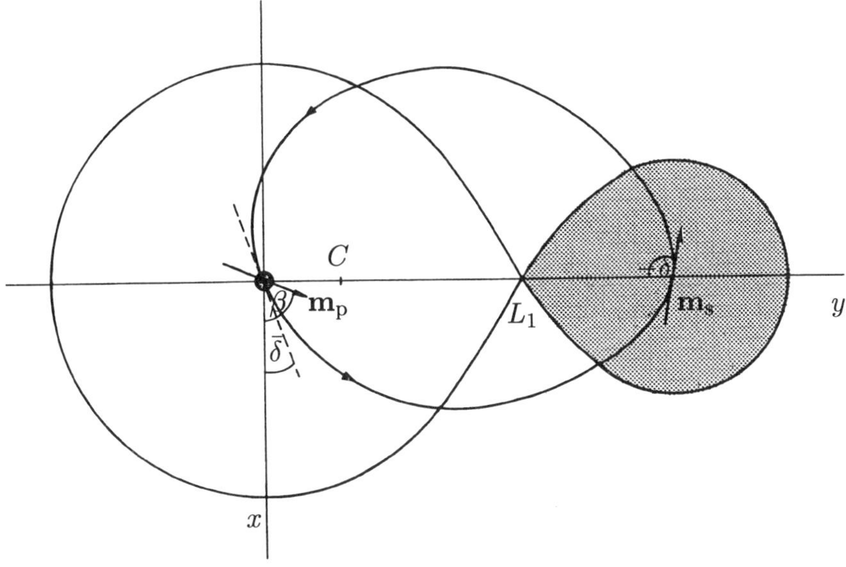

Figure 6.1. The dipole orientations in the orbital plane. Based on Campbell (1985).

where $\mathbf{B}_{\mathrm{sp}}$ is the secondary's magnetic field at the position of the primary, and $\bar{\delta}$ is the tangent angle shown in Figure 6.1. The components of $\mathbf{B}_{\mathrm{sp}}$ are

$$(B_{\mathrm{sp}})_x = -\frac{\mu_0 m_{\mathrm{s}}}{4\pi D^3}\sin\delta, \tag{6.3}$$

$$(B_{\mathrm{sp}})_y = -\frac{\mu_0 m_{\mathrm{s}}}{2\pi D^3}\cos\delta. \tag{6.4}$$

From the definition of $\bar{\delta}$, together with (6.3) and (6.4), it follows that

$$\sin\delta = \frac{2\cos\bar{\delta}}{(3\cos^2\bar{\delta}+1)^{\frac{1}{2}}}, \tag{6.5}$$

$$\cos\delta = -\frac{\sin\bar{\delta}}{(3\cos^2\bar{\delta}+1)^{\frac{1}{2}}}. \tag{6.6}$$

Equations (6.3)–(6.6) give the magnitude of $\mathbf{B}_{\mathrm{sp}}$ as

$$B_{\mathrm{sp}} = \frac{\mu_0 m_{\mathrm{s}}}{2\pi D^3}\frac{1}{(3\cos^2\bar{\delta}+1)^{\frac{1}{2}}}. \tag{6.7}$$

The sum of the accretion and magnetic torques acting on the primary is

$$\mathbf{T} = [A^2 \Omega \dot{M}_{\rm p} - m_{\rm p} B_{\rm sp} \sin(\beta - \bar{\delta})]\mathbf{k}, \tag{6.8}$$

where its dipole moment is related to its radius and polar field by

$$m_{\rm p} = \frac{2\pi}{\mu_0} R_{\rm p}^3 (B_{\rm p})_0, \tag{6.9}$$

with an equivalent expression for $m_{\rm s}$. For a synchronous primary at orientation $\beta = \beta_s$ the torque $\mathbf{T}$ must vanish. Equations (6.7)–(6.9) give this condition as

$$\frac{\sin(\beta_s - \bar{\delta})}{(3\cos^2\bar{\delta} + 1)^{\frac{1}{2}}} = \frac{10\mu_0 A^2 \dot{M}_{\rm p} M}{(B_{\rm s})_0 (B_{\rm p})_0 R_{\rm p}^3 M_{\rm s} P}, \tag{6.10}$$

where $M = M_{\rm s} + M_{\rm p}$, $(B_{\rm s})_0$ is the secondary's polar magnetic field and P is the orbital period. The lobe-filling condition

$$\frac{M}{M_{\rm s}} \left(\frac{R_{\rm s}}{D}\right)^3 = 0.1,$$

due Paczyński (1967), has been employed, where $R_{\rm s}$ is the mean radius of the secondary. The torque balance condition (6.10) requires $\sin(\beta_s - \bar{\delta}) > 0$ and hence

$$\bar{\delta} < \beta_s < \bar{\delta} + \pi. \tag{6.11}$$

The linear stability of the synchronous state requires the sign of a torque perturbation due to a perturbation of β about β_s to be opposite to that of $\beta' = \beta - \beta_s$. Hence

$$\left(\frac{dT}{d\beta}\right)_{\beta_s} < 0 \tag{6.12}$$

is necessary for stability. Applying this condition to (6.8) gives $\cos(\beta_s - \bar{\delta}) > 0$ and hence

$$\bar{\delta} < \beta_s < \bar{\delta} + \frac{\pi}{2} \quad \text{or} \quad \bar{\delta} + \frac{3\pi}{2} < \beta_s < \bar{\delta} + 2\pi. \tag{6.13}$$

A stable synchronous state must satisfy (6.11) and (6.13) and so requires

$$\bar{\delta} < \beta_s < \bar{\delta} + \frac{\pi}{2}. \tag{6.14}$$

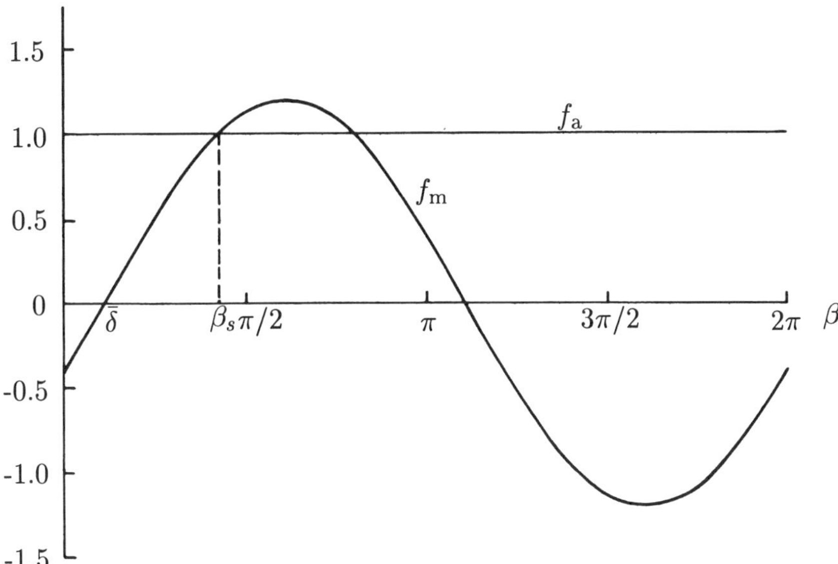

Figure 6.2. The synchronous torque balance in two dimensions, normalized by $A^2\Omega\dot{M}_p$.

It is clear from (6.10) that for a given value of β_s the condition of vanishing torque can be satisfied by choosing B_{sp} and δ independently, subject to the stability condition (6.14). For a given set of parameters there is only one stable synchronous orientation β_s, as illustrated in Figure 6.2. Consider, as an example, an orbital period $P = 1.5$ hrs, giving a lobe-filling secondary with $M_s = 0.14\,M_\odot$. For $M_p = 0.6\,M_\odot$, it follows that $D = 4.2 \times 10^8$ m and $A = 2.7 \times 10^8$ m, while $R_p = 8.7 \times 10^6$ m. The accretion rate is taken as $\dot{M}_p = 10^{-10}\,M_\odot\mathrm{yr}^{-1}$. For $(B_p)_0 = 3.1 \times 10^3$ tesla, $(B_s)_0 = 6.4 \times 10^{-3}$ tesla and $\bar{\delta} = 20°$ (i.e. $\delta = -100°$), equation (6.10) gives $\beta_s = 76°$, which satisfies the stability condition (6.14).

It is of interest to calculate the period of small oscillations of the primary about the synchronous state. An expression for this period can be found by equating the rate of change of angular momentum of the primary to the restoring torque resulting from a small angular displacement about β_s. Expressing the primary's moment of inertia as $I = k_p^2 M_p R_p^2$, (6.8) and (6.10) then give the oscillation period as

$$P_0 = \left[\frac{2\pi k_p^2 M_p R_p^2 P}{A^2 \dot{M}_p}\frac{\sin(\beta_s - \bar{\delta})}{\cos(\beta_s - \bar{\delta})}\right]^{\frac{1}{2}}, \tag{6.15}$$

where $0.1 < k_p^2 < 0.2$. Taking $k_p^2 = 0.18$, corresponding to $M_p = 0.6M_\odot$, and the foregoing parameters, yields $P_0 = 42$ yrs.

It is noted from (6.2) that for an asynchronous primary with $\beta = \omega t$ the average of $\mathbf{T}_\mathrm{m}$ over a synodic period is zero. Hence such a torque does not play a part in the approach to synchronism.

6.3. Accretion Versus Magnetism in Three-Dimensions

6.3.1. ORIENTATIONS WITH ZERO TORQUE

The general case, in which the secondary's magnetic moment has orientation (δ, γ), was considered by Campbell (1989) and is shown in Figure 6.3. A spherical primary is considered, so synchronism at (α_s, β_s) requires zero torque, and such a state must be stable to all angular perturbations.

The components of the secondary's magnetic field at the position of the primary are

$$(B_\mathrm{sp})_x = -\frac{\mu_0 m_\mathrm{s}}{4\pi D^3} \sin\gamma \sin\delta, \tag{6.16}$$

$$(B_\mathrm{sp})_y = -\frac{\mu_0 m_\mathrm{s}}{2\pi D^3} \sin\gamma \cos\delta, \tag{6.17}$$

$$(B_\mathrm{sp})_z = -\frac{\mu_0 m_\mathrm{s}}{4\pi D^3} \cos\gamma. \tag{6.18}$$

The magnetic torque on the primary is

$$\mathbf{T}_\mathrm{m} = \mathbf{m}_\mathrm{p} \wedge \mathbf{B}_\mathrm{sp}, \tag{6.19}$$

where the unit vector $\hat{\mathbf{m}}_\mathrm{p}$ is

$$\hat{\mathbf{m}}_\mathrm{p} = \sin\alpha \cos\beta\, \mathbf{i} + \sin\alpha \sin\beta\, \mathbf{j} + \cos\alpha\, \mathbf{k}. \tag{6.20}$$

Equations (6.16)–(6.20) give the components of the magnetic torque as

$$T_{\mathrm{m}x} = \frac{\mu_0 m_\mathrm{p} m_\mathrm{s}}{4\pi D^3} (2\sin\gamma \cos\delta \cos\alpha - \cos\gamma \sin\alpha \sin\beta), \tag{6.21}$$

$$T_{\mathrm{m}y} = \frac{\mu_0 m_\mathrm{p} m_\mathrm{s}}{4\pi D^3} (\cos\gamma \sin\alpha \cos\beta - \sin\gamma \sin\delta \cos\alpha), \tag{6.22}$$

$$T_{\mathrm{m}z} = \frac{\mu_0 m_\mathrm{p} m_\mathrm{s}}{4\pi D^3} (\sin\delta \sin\beta - 2\cos\delta \cos\beta) \sin\alpha \sin\gamma. \tag{6.23}$$

The components of the accretion torque, $T_\mathrm{a}(\alpha, \beta)$ are given by (5.56)–(5.58). If matter becomes field-channelled at some distance from L_1 it was

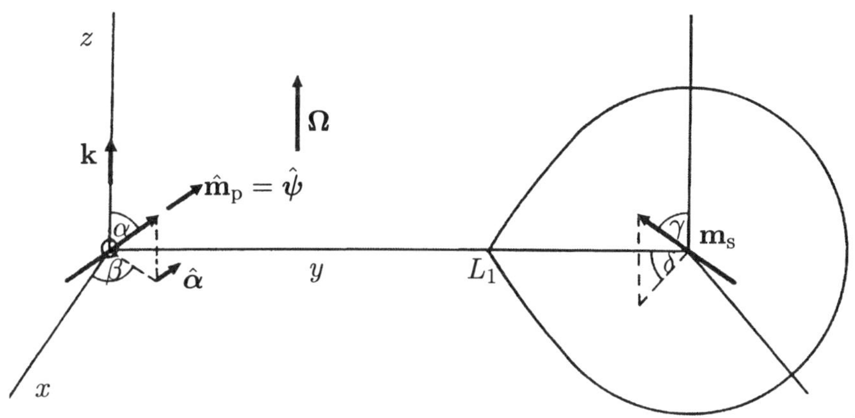

Figure 6.3. The dipole orientations in three dimensions. Based on Campbell (1989).

seen that $T_{\mathrm{a}z}$ becomes effectively independent of (α, β). The horizontal components then only vary by $\sim$ 10% as β is varied over an interval of $\pi/2$. The simplest form for the accretion torque containing its essential properties is then

$$T_{\mathrm{a}x} = A^2\Omega\dot{M}_\mathrm{p}N_x\cos\alpha, \tag{6.24}$$

$$T_{\mathrm{a}y} = A^2\Omega\dot{M}_\mathrm{p}N_y\cos\alpha, \tag{6.25}$$

$$T_{\mathrm{a}z} = A^2\Omega\dot{M}_\mathrm{p}, \tag{6.26}$$

where the constants N_x and N_y are negative for $0 < \beta < \pi$ and positive for $\pi < \beta < 2\pi$, and their magnitudes depend on the channelling radius.

The necessary condition for a synchronous state is that $\mathbf{T}_\mathrm{m}$ and $\mathbf{T}_\mathrm{a}$ cancel. Equations (6.21)–(6.26) then give

$$-\cos\gamma\hat{m}_{\mathrm{p}y} + (2\sin\gamma\cos\delta + QN_x)\hat{m}_{\mathrm{p}z} = 0, \tag{6.27}$$

$$\cos\gamma\hat{m}_{\mathrm{p}x} + (QN_y - \sin\gamma\sin\delta)\hat{m}_{\mathrm{p}z} = 0, \tag{6.28}$$

$$-2\sin\gamma\cos\delta\hat{m}_{\mathrm{p}x} + \sin\gamma\sin\delta\hat{m}_{\mathrm{p}y} = -Q, \tag{6.29}$$

where the characteristic ratio of the accretion and magnetic torques is

$$Q = \frac{4\pi D^3 A^2\Omega\dot{M}_\mathrm{p}}{\mu_0 m_\mathrm{p} m_\mathrm{s}}. \tag{6.30}$$

Solving (6.27)–(6.29) yields the synchronous orientation of the primary in terms of the components of its dipole moment unit vector;

$$\hat{m}_{\mathrm{p}x} = \sin\alpha_s \cos\beta_s = \frac{QN_y - \sin\gamma\sin\delta}{\sin\gamma(N_x\sin\delta + 2N_y\cos\delta)}, \tag{6.31}$$

$$\hat{m}_{\mathrm{p}y} = \sin\alpha_s \sin\beta_s = -\frac{QN_x + 2\sin\gamma\cos\delta}{\sin\gamma(N_x\sin\delta + 2N_y\cos\delta)}, \tag{6.32}$$

$$\hat{m}_{\mathrm{p}z} = \cos\alpha_s = -\frac{1}{\tan\gamma(N_x\sin\delta + 2N_y\cos\delta)}. \tag{6.33}$$

However, since $\hat{\mathbf{m}}_{\mathrm{p}}$ is a unit vector, the condition $\hat{\mathbf{m}}_{\mathrm{p}} \cdot \hat{\mathbf{m}}_{\mathrm{p}} = 1$ must be satisfied. Equations (6.31)–(6.33) then yield a quadratic equation for Q which has solutions

$$Q = \frac{-\sin\gamma(2N_x\cos\delta - N_y\sin\delta) \pm \sqrt{P}}{N_x^2 + N_y^2}, \tag{6.34}$$

where

$$P = (N_x^2 + N_y^2 - 1)\sin^2\gamma(N_x\sin\delta + 2N_y\cos\delta)^2 - (N_x^2 + N_y^2)\cos^2\gamma.$$

It follows from (6.30) and (6.34) that, for given orbital parameters, the orientation and surface polar strength of the secondary's magnetic field cannot be chosen independently if the primary is to experience zero torque. This restriction did not occur in the simple two-dimensional case.

The values of N_x and N_y depend on where material lost from the secondary becomes field-channelled. As seen in §5.4, N_x involves gravitational coupling to the orbit and is more sensitive to the extent of the field-channelling region than N_y, which is due to the more evenly distributed Coriolis torque. If field channelling is far from L_1 (i.e. $\gtrsim A/3$) then $N_x^2 + N_y^2 < 1$ and it follows from (6.34) that Q becomes complex. Hence, in such a situation, an orientation cannot be found at which the magnetic and accretion torques cancel. However, if $Q \ll 1$ the magnetic torque dominates and synchronous states are essentially possible.

6.3.2. STABILITY OF THE SYNCHRONOUS STATE

Having found a synchronous state, its stability to small disturbances must be established. A general perturbation consists of a twist about an axis through the primary. Subsequently the star's rotational motion is defined by an instantaneous angular velocity $\boldsymbol{\omega}$ where, relative to the orbital frame,

$$\boldsymbol{\omega} = \dot{\alpha}\hat{\boldsymbol{\alpha}} + \dot{\beta}\mathbf{k} + \dot{\psi}\hat{\boldsymbol{\psi}}. \tag{6.35}$$

It follows from Figure 6.3 that the components of $\boldsymbol{\omega}$ are

$$\omega_x = -\dot{\alpha}\sin\beta + \dot{\psi}\sin\alpha\cos\beta, \tag{6.36}$$

$$\omega_y = \dot{\alpha}\cos\beta + \dot{\psi}\sin\alpha\sin\beta, \tag{6.37}$$

$$\omega_z = \dot{\beta} + \dot{\psi}\cos\alpha. \tag{6.38}$$

Equations (6.21)–(6.26) give the components of the torque acting on the white dwarf as

$$T_x = -C_{\mathrm{m}}\cos\gamma\sin\alpha\sin\beta + (C_{\mathrm{a}}N_x + 2C_{\mathrm{m}}\sin\gamma\cos\delta)\cos\alpha, \tag{6.39}$$

$$T_y = C_{\mathrm{m}}\cos\gamma\sin\alpha\cos\beta + (C_{\mathrm{a}}N_y - C_{\mathrm{m}}\sin\gamma\sin\delta)\cos\alpha, \tag{6.40}$$

$$T_z = C_{\mathrm{m}}\sin\gamma\sin\alpha(\sin\delta\sin\beta - 2\cos\delta\cos\beta) + C_{\mathrm{a}}, \tag{6.41}$$

where

$$C_{\mathrm{m}} = \frac{\mu_0 m_{\mathrm{p}} m_{\mathrm{s}}}{4\pi D^3}, \qquad C_{\mathrm{a}} = A^2\Omega\dot{M}_{\mathrm{p}}. \tag{6.42a,b}$$

The equation describing the evolution of a perturbation in $\boldsymbol{\omega}$, relative to the orbital frame, is

$$I(\dot{\boldsymbol{\omega}}' + \Omega\mathbf{k}\wedge\boldsymbol{\omega}') = \mathbf{T}', \tag{6.43}$$

where I is the moment of inertia of the primary. Expanding (6.36)–(6.41) to first order about the synchronous state gives the components of (6.43) as

$$\begin{aligned}&-\sin\beta_s\ddot{\alpha}' - \Omega\cos\beta_s\dot{\alpha}'\\ &+ [\bar{C}_{\mathrm{m}}(\cos\gamma\cos\alpha_s\sin\beta_s + 2\sin\gamma\cos\delta\sin\alpha_s) + \bar{C}_{\mathrm{a}}N_x\sin\alpha_s]\alpha'\\ &+ \bar{C}_{\mathrm{m}}\cos\gamma\sin\alpha_s\cos\beta_s\beta' + \sin\alpha_s\cos\beta_s\ddot{\psi}' - \Omega\sin\alpha_s\sin\beta_s\dot{\psi}' = 0,\end{aligned} \tag{6.44}$$

$$\begin{aligned}&\cos\beta_s\ddot{\alpha}' - \Omega\sin\beta_s\dot{\alpha}'\\ &- [\bar{C}_{\mathrm{m}}(\cos\gamma\cos\alpha_s\cos\beta_s + \sin\gamma\sin\delta\sin\alpha_s) - \bar{C}_{\mathrm{a}}N_y\sin\alpha_s]\alpha'\\ &+ \bar{C}_{\mathrm{m}}\cos\gamma\sin\alpha_s\sin\beta_s\beta' + \sin\alpha_s\sin\beta_s\ddot{\psi}' + \Omega\sin\alpha_s\cos\beta_s\dot{\psi}' = 0,\end{aligned} \tag{6.45}$$

$$\begin{aligned}&\bar{C}_{\mathrm{m}}\sin\gamma\cos\alpha_s(2\cos\delta\cos\beta_s - \sin\delta\sin\beta_s)\alpha' + \ddot{\beta}'\\ &- \bar{C}_{\mathrm{m}}\sin\gamma\sin\alpha_s(2\cos\delta\sin\beta_s + \sin\delta\cos\beta_s)\beta' + \cos\alpha_s\ddot{\psi}' = 0,\end{aligned} \tag{6.46}$$

where $\bar{C}_{\rm m} = C_{\rm m}/I$, $\bar{C}_{\rm a} = C_{\rm a}/I$ and subscript s denotes the synchronous state.

The angular perturbations can be written as the product of an exponential time dependence and constant amplitudes, so

$$\alpha' = a_1 \exp(i\sigma t), \quad \beta' = a_2 \exp(i\sigma t), \quad \psi' = a_3 \exp(i\sigma t). \tag{6.47}$$

Substitution of these in (6.44)–(6.46) gives a system of equations for the amplitudes and the vanishing of the determinant of their coefficients, for consistency, yields the following characteristic equation for σ,

$$\sigma^6 + C_1\sigma^4 + C_2\sigma^2 + C_3\sigma = 0, \tag{6.48}$$

where

$$\begin{aligned} C_1 =& \bar{C}_{\rm m} \sin\gamma \sin\alpha_s (2\cos\delta \sin\beta_s + \sin\delta\cos\beta_s) \\ &+ \bar{C}_{\rm m} \frac{\cos\gamma}{\cos\alpha_s}(1+\cos^2\alpha_s) - \Omega^2, \end{aligned} \tag{6.49}$$

$$\begin{aligned} C_2 =& \bar{C}_{\rm m}^2 \frac{\sin\gamma\cos\gamma\sin\alpha_s}{\cos\alpha_s}(2\cos\delta\sin\beta_s + \sin\delta\cos\beta_s) + \bar{C}_{\rm m}^2\cos^2\gamma \\ &- \bar{C}_{\rm m}\Omega^2 \sin\gamma\sin\alpha_s(2\cos\delta\sin\beta_s + \sin\delta\cos\beta_s), \end{aligned} \tag{6.50}$$

$$C_3 = -i\Omega\bar{C}_{\rm m}^2 Q\cos\gamma\cos\alpha_s. \tag{6.51}$$

Consider, first, the case with accretion absent, so that Q vanishes and hence $C_3 = 0$. Equation (6.48) then has the solution $\sigma^2 = 0$, which results from the fact that the torques are independent of the Euler angle ψ. The two finite frequencies are given by the roots of

$$\sigma^4 + C_1\sigma^2 + C_2 = 0. \tag{6.52}$$

Solving this equation for σ^2, using (6.20) and (6.27)–(6.29) with $Q = 0$, gives

$$\sigma^2 = -\bar{C}_{\rm m}\frac{\cos\gamma}{\cos\alpha_s} + \frac{\Omega^2}{2}\left[1 \pm \left(1 - 4\frac{\bar{C}_{\rm m}}{\Omega^2}\cos\gamma\cos\alpha_s\right)^{\frac{1}{2}}\right]. \tag{6.53}$$

If no rotation were present, in this case, (6.53) gives the single frequency

$$\sigma_{\rm m}^2 = -\bar{C}_{\rm m}\frac{\cos\gamma}{\cos\alpha_s}. \tag{6.54}$$

From equations (6.27)–(6.29), with $Q = 0$, it follows that two values of $\cos\gamma/\cos\alpha_s$ are possible, having the same magnitudes but different signs. The negative value corresponds to $\mathbf{m}_\mathrm{p}$ aligned with the secondary's field in the synchronous state, and all perturbations result in stable oscillations with frequency σ_m.

Equations (6.35) and (6.47) give a general perturbation in the angular velocity as

$$\boldsymbol{\omega}' = i\sigma(a_1\hat{\boldsymbol{\alpha}}_s + a_2\mathbf{k} + a_3\hat{\boldsymbol{\psi}}_s)\exp(i\sigma t), \tag{6.55}$$

where subscripts s denote the synchronous state. Equations (6.27)–(6.29), (6.46), (6.47) and (6.54) give

$$a_2\cos\alpha_s + a_3 = 0.$$

Eliminating a_3 in (6.55) yields $\boldsymbol{\omega}'$ for non-rotating magnetic modes as

$$\boldsymbol{\omega}' = i\sigma_\mathrm{m}[a_1\hat{\boldsymbol{\alpha}}_s + a_2\{\mathbf{k} - (\mathbf{k}\cdot\hat{\boldsymbol{\psi}}_s)\hat{\boldsymbol{\psi}}_s\}]\exp(i\sigma_\mathrm{m}t). \tag{6.56}$$

It follows that $\hat{\boldsymbol{\psi}}_s \cdot \boldsymbol{\omega}' = 0$ and hence the angular velocity of these modes is normal to the synchronous direction of $\mathbf{m}_\mathrm{p}$. Any displacement from equilibrium will result in such an oscillation and so the state is stable.

The presence of rotation results in two distinct types of mode, whose frequencies are given by (6.53). The dimensionless quantity $\bar{C}_\mathrm{m}/\Omega^2$ can be written as

$$\frac{\bar{C}_\mathrm{m}}{\Omega^2} = 6.5\times10^{-11}\frac{\left(\frac{(B_\mathrm{p})_0}{20\,\mathrm{MG}}\right)\left(\frac{(B_\mathrm{s})_0}{10^2\,\mathrm{G}}\right)\left(\frac{R_\mathrm{p}}{8.7\times10^6\,\mathrm{m}}\right)\left(\frac{M_\mathrm{s}}{0.2M_\odot}\right)\left(\frac{P}{2.1\,\mathrm{hr}}\right)^2}{\left(\frac{k_\mathrm{p}^2}{0.2}\right)\left(\frac{M_\mathrm{p}}{0.6M_\odot}\right)\left(\frac{M}{0.8M_\odot}\right)}. \tag{6.57}$$

Equation (6.54) gives $\sigma_\mathrm{m}^2 \sim \bar{C}_\mathrm{m}$ and hence it follows that $\sigma_\mathrm{m}/\Omega \sim 10^{-5}$, so the frequency of magnetic oscillations is much less than that of the orbital motion. Expansion of the square root term in (6.53) then yields the two mode frequencies, to high accuracy, as

$$\sigma_+^2 = \Omega^2, \tag{6.58}$$

$$\sigma_-^2 = -\bar{C}_\mathrm{m}\frac{\cos\gamma}{\cos\alpha_s}\sin^2\alpha_s. \tag{6.59}$$

Substituting (6.58) in (6.44) and (6.45), using (6.54) and dropping terms in $(\sigma_\mathrm{m}/\Omega)^2$, gives

$$a_1 = i\sin\alpha_s a_3.$$

Equation (6.46) yields

$$a_2 = -\cos\alpha_2 a_3.$$

Hence the angular velocity perturbation for the σ_+ mode is

$$\boldsymbol{\omega}' = i\Omega a_3[i\sin\alpha_s\hat{\boldsymbol{\alpha}}_s + \{\hat{\boldsymbol{\psi}}_s - (\hat{\boldsymbol{\Omega}}\cdot\hat{\boldsymbol{\psi}}_s)\hat{\boldsymbol{\Omega}}\}]\exp(i\Omega t). \quad (6.60)$$

It follows that $\boldsymbol{\omega}'\cdot\hat{\boldsymbol{\Omega}} = 0$ so $\boldsymbol{\omega}'$ lies in the $z = 0$ plane and rotates with frequency Ω in the orbital frame. This is an inertial mode corresponding to perturbation twists normal to $\boldsymbol{\Omega}$. In this mode the magnetic torque is ineffective since it produces responses on a time-scale $2\pi/\sigma_{\rm m} \gg P$, where P is the orbital period.

The angular velocity perturbation for the σ_- mode is found by using (6.59) for σ_- in (6.44) and (6.45), remembering that $\sigma_{\rm m}/\Omega \ll 1$. This yields

$$a_1 = a_3 = 0.$$

The mode therefore has

$$\boldsymbol{\omega}' = i\sigma_- a_2\exp(i\sigma_- t)\hat{\boldsymbol{\Omega}}. \quad (6.61)$$

This is the purely magnetic mode, given by (6.56), modified by rotation which changes its frequency to $\sigma_{\rm m}\sin\alpha_s$ and aligns it with the $\boldsymbol{\Omega}$ axis. The existence of this mode can be seen directly from the equations of motion. Using (6.27)–(6.29) with $Q = 0$ and (6.54), linear combinations of (6.44)–(6.46) give

$$\ddot{\alpha}' + \sigma_{\rm m}^2\alpha' = -\Omega\sin\alpha_s\dot{\psi}', \quad (6.62)$$

$$\ddot{\beta}' + \sigma_{\rm m}^2\sin^2\alpha_s\beta' = -\cos\alpha_s\ddot{\psi}', \quad (6.63)$$

$$-\sigma_{\rm m}^2\sin\alpha_s\cos\alpha_s\beta' + \sin\alpha_s\ddot{\psi}' = \Omega\dot{\alpha}'. \quad (6.64)$$

If $\psi' \ll \beta'$ then (6.63) implies that β' varies harmonically with frequency $\sigma = \sigma_{\rm m}\sin\alpha_s$. With $\psi' \ll \beta'$, (6.64) yields $\alpha' \sim (\sigma_{\rm m}/\Omega)\beta'$ and hence (6.62) gives $\psi' \sim (\sigma_{\rm m}/\Omega)^2\beta' \ll \beta'$ consistently. Because $\sigma_{\rm m}/\Omega \sim 10^{-5}$, for realistic secondary star fields, it follows that the σ_- mode exists to high accuracy.

For the synchronous state to exist the magnetic torque must balance the accretion torque, or dominate it, since there is no three-dimensional orientation at which the latter vanishes. It follows that $Q = \bar{C}_{\rm a}/\bar{C}_{\rm m} \lesssim 1$ is necessary for synchronism and hence that $\bar{C}_{\rm a}/\Omega^2 \sim (\sigma_{\rm m}/\Omega)^2 \ll 1$. It is therefore clear that the additional terms due to accretion, occurring in the coefficients of α' in (6.44) and (6.45), do not affect the inertial mode given by (6.60). Similar, the rotationally aligned mode still exists, except that accretion modifies its frequency. The frequency can be obtained directly from (6.46) which yields a harmonic equation for β' when α' and ψ' essentially vanish. Use of (6.31) and (6.32) then gives

$$\sigma^2 = \bar{C}_{\mathrm{m}} \left[\frac{\sin\gamma(\sin^2\delta + 4\cos^2\delta) + Q(2N_x\cos\delta - N_y\sin\delta)}{N_x\sin\delta + 2N_y\cos\delta} \right]. \tag{6.65}$$

A stable synchronous state requires $\sigma^2 > 0$.

6.3.3. STABLE SYNCHRONOUS ORIENTATIONS

The values of N_x and N_y, which occur in the horizontal components of the accretion torque, depend on where matter becomes field-channelled. The values used here are $K_x = \pm 1.05$, $K_y = \pm 0.30$, where the negative and positive signs apply for $0 < \beta < \pi$ and $\pi < \beta < 2\pi$, respectively.

The simplest way of determining the synchronous state is to choose (γ, δ) in (6.34) to calculate Q, and then find (α_s, β_s) from (6.31)–(6.33). The sign of σ^2 follows from (6.65). A sample of the results are shown in Table 6.1. The values of β_s illustrate that stable synchronous states, in which the magnetic and accretion torques balance, only occur with the accreting pole lagging the motion of the line of stellar centres.

Observations of AM Her systems suggest some tendency for the accreting pole to lead the line of stellar centres (Cropper, 1988). The foregoing analysis shows that, for dipolar fields, a balance between magnetic and accretion torques cannot produce such observed orientations. The fact that only certain combinations of secondary field strength and orientation allow cancellation of the accretion torque reduces the likelihood of such a balance. Field-channelling far from L_1 makes a magneto-accretion balance in three-dimensions impossible since then $N_x^2 + N_y^2 < 1$ and real values of Q, given by (6.34), do not result.

However, the above restrictions are essentially lifted if the magnetic torque dominates (i.e. $Q \ll 1$). Equation (6.30) can be expressed as

$$Q = 0.67 \frac{\left(\dfrac{A}{3.3 \times 10^8\,\mathrm{m}}\right)^2 \left(\dfrac{\dot{M}_{\mathrm{p}}}{10^{-10}\,M_\odot\,\mathrm{yr}^{-1}}\right) \left(\dfrac{M}{0.8\,M_\odot}\right)}{\left(\dfrac{(B_{\mathrm{p}})_0}{20\,\mathrm{MG}}\right) \left(\dfrac{(B_{\mathrm{s}})_0}{10^2\,\mathrm{G}}\right) \left(\dfrac{R_{\mathrm{p}}}{8.7 \times 10^6\,\mathrm{m}}\right)^3 \left(\dfrac{M_{\mathrm{s}}}{0.2\,M_\odot}\right) \left(\dfrac{P}{2.1\,\mathrm{hr}}\right)}. \tag{6.66}$$

It follows that $Q \sim 10^{-2}$ for $(B_{\mathrm{s}})_0 = 6 \times 10^3\,\mathrm{G}$ and then the magnetic torque alone could produce essentially stable synchronous states with, for appropriate (γ, δ), the accretion pole leading the line of stellar centres.

TABLE 6.1. Stable synchronous orientations for magneto-accretion balance

γ	δ	Q	α_s	β_s
70	100	0.45	113.0	190.3
70	100	0.59	113.0	201.6
70	275	0.24	111.5	5.7
70	275	0.52	111.5	26.2
75	115	0.87	112.6	118.6
75	115	1.01	112.6	203.3
75	295	0.87	112.6	8.6
75	295	1.01	112.6	23.3
80	105	0.96	101.9	216.6
80	300	1.23	106.8	32.9
85	110	1.11	96.4	218.8
85	280	0.88	95.4	39.1
95	120	1.30	81.7	218.2
95	295	1.21	82.8	38.6
100	125	1.30	70.0	208.9
100	280	0.84	79.1	37.1
105	115	0.87	67.4	188.6
105	115	1.01	67.4	203.3
105	295	0.87	67.4	8.6
105	295	1.01	67.4	23.3
110	100	0.45	67.0	190.3
110	100	0.59	67.0	201.6
110	280	0.45	67.0	10.3
110	280	0.59	67.0	21.6

6.4. Quadrupolar Magnetic Fields

Recent observations suggest that the white dwarf may have a quadrupolar component to its magnetic field. The interaction of this with a dipolar secondary field would generate a torque which has a characteristic magnitude, relative to that of the dipole-dipole torque, of $(B_q/B_p)(R_\mathrm{p}/D)$, where B_q and B_p are the polar field strengths. Since $R_\mathrm{p}/D \ll 1$, the situation $B_q \sim B_p$ generally results in a relatively small quadrupole-dipole torque. However, for some orientations this torque would not be ignorable (see below); also the situation $B_q > B_p$ could make the torque significant.

The role of a quadrupole field was considered by Wu and Wickramasinghe (1993). They took the secondary to have a dipole moment $\mathbf{m}_\mathrm{s}$ perpendicular to the orbital plane, and the primary to have a dipole and quadrupole field with an angle θ between their symmetry axes. The primary's dipole moment was taken to be antiparallel to $\mathbf{m}_\mathrm{s}$ so the dipole-dipole torque vanished. The z-component of the quadrupole-dipole torque is

$$T_z = -\frac{m_\mathrm{s} B_q}{2}\left(\frac{R_\mathrm{p}}{D}\right)^4 \sin 2\theta \sin\phi, \tag{6.67}$$

where ϕ is the azimuth of the quadrupole symmetry axis, measured from the x-axis of Figure 6.3. A small tilt of $\mathbf{m}_\mathrm{p}$ results in zero horizontal torque. It was shown that T_z has a magnitude sufficient to cancel the z-component of the accretion torque, though detailed orientations were not found. This calculation requires the horizontal components of the accretion torque to essentially vanish. As shown in §5.4, the x-component of the accretion torque is ignorable if matter becomes field-channelled far from L_1. However, the y-component is significant unless channelling only occurs very close to the white dwarf. Hence, this quadrupole analysis requires such channelling.

6.5. Gravity versus Magnetism

6.5.1. THE SYNCHRONOUS STATE

It was pointed out by Joss, Katz and Rappaport (1979) that a small distortion of the white dwarf would produce a significant gravitational torque due to interaction with the secondary. Katz (1989) showed that an internal toroidal magnetic field, with an associated poloidal current, can produce significant distortions of the white dwarf. He found the difference in the principal moments of inertia of a prolate body as

$$I_3 - I_\perp = -\frac{2\pi}{3\mu_0}\frac{B_0^2 R_\mathrm{p}^6}{GM_\mathrm{p}}, \tag{6.68}$$

where I_3 is the moment about the symmetry axis. The resulting gravitational torque was estimated and can be at least comparable to the accretion torque for $B_0 \sim 10^7$ G. Internal fields of this magnitude, or even somewhat larger, are reasonable given that the observed surface fields are $\sim$ 10–40 MG.

The detailed consequences of a distorted white dwarf were considered in Campbell (1990). The primary is taken to have a magnetic moment $\mathbf{m}_\mathrm{p}$, as shown in Figure 6.3, but to be a spheroidal body with its symmetry axis coincident with $\mathbf{m}_\mathrm{p}$. The moments of inertia about $\mathbf{m}_\mathrm{p}$ and orthogonal axes in the magnetic equatorial plane are denoted by I_3 and $I_\perp$, respectively. Treating the secondary as a spherical gravitational source, the torque it exerts on the primary is

$$\mathbf{T}_{\mathrm{g}} = \frac{3GM_{\mathrm{s}}(I_3 - I_\perp)}{D^3} \sin\alpha \sin\beta \mathbf{j} \wedge \hat{\mathbf{m}}_{\mathrm{p}}, \tag{6.69}$$

where $\hat{\mathbf{m}}_{\mathrm{p}}$ is given by (6.20). This expression follows from (A41) in the Appendix, since $\cos\theta = \sin\alpha \sin\beta$, $\hat{\mathbf{r}} = \mathbf{j}$ and $\mathbf{e}_3 = \hat{\mathbf{m}}_{\mathrm{p}}$.

If the secondary has a magnetic moment $\mathbf{m}_{\mathrm{s}}$, as shown in Figure 6.3, the primary will experience a magnetic torque given by (6.19). It will also experience the accretion torque given by (5.56)–(5.58).

Since the primary is non-spherical, its inertial angular velocity $\boldsymbol{\omega}$ (with the subscript now dropped) and angular momentum $\mathbf{L}$ will not in general be parallel. It is therefore convenient to work in the body frame coincident with the principal axes, thereby diagonalizing the inertia tensor. The principal unit vectors are chosen so $\mathbf{e}_3 = \hat{\mathbf{m}}_{\mathrm{p}}$ and hence $\mathbf{e}_1$ and $\mathbf{e}_2$ lie in the primary's magnetic equatorial plane. The Euler angles describing the orientation of the body frame $O\bar{x}\bar{y}\bar{z}$ relative to the orbital frame $Oxyz$ are (α, β, ψ). The transformation of the components of a vector $\mathbf{V}$ in the orbital frame to those in the body frame is

$$\bar{V}_i = R_{ij} V_j, \tag{6.70a}$$

where from (2.334), with β replacing ϕ,

$$\begin{aligned}
R_{11} &= \cos\alpha \cos\beta \cos\psi - \sin\beta \sin\psi, \\
R_{12} &= \cos\alpha \sin\beta \cos\psi + \cos\beta \sin\psi, \\
R_{13} &= -\sin\alpha \cos\psi, \\
R_{21} &= -\cos\alpha \cos\beta \sin\psi - \sin\beta \cos\psi, \\
R_{22} &= -\cos\alpha \sin\beta \sin\psi + \cos\beta \cos\psi, \\
R_{23} &= \sin\alpha \sin\psi, \\
R_{31} &= \sin\alpha \cos\beta, \\
R_{32} &= \sin\alpha \sin\beta, \\
R_{33} &= \cos\alpha.
\end{aligned} \tag{6.70b}$$

In the synchronous state the angular velocity of the primary is $\boldsymbol{\omega} = \boldsymbol{\Omega}$, a constant vector. In general, such a state cannot be one of zero torque since then $\mathbf{L}$ would be conserved and $\boldsymbol{\omega}$ would precess. Hence the synchronous orientation (α_s, β_s) must be such that a torque acts to cancel the precession of $\boldsymbol{\omega}$, so resulting in a precession of $\mathbf{L}$.

Denoting the time derivative of a vector relative to inertial axes by d/dt, and that along the body axes by a dot, in the body frame $\mathbf{L}$ obeys the equation

$$\dot{\mathbf{L}} + \boldsymbol{\omega} \wedge \mathbf{L} = \mathbf{T}, \tag{6.71}$$

where $\mathbf{T}$ is the total torque acting on the primary. The angular momentum in this frame is

$$\mathbf{L} = I_{\perp}(\omega_1 \mathbf{e}_1 + \omega_2 \mathbf{e}_2) + I_3 \omega_3 \mathbf{e}_3. \tag{6.72}$$

In the synchronous state $\boldsymbol{\omega} = \boldsymbol{\Omega}$ and hence

$$\left(\frac{d\boldsymbol{\omega}}{dt}\right)_s = \dot{\boldsymbol{\omega}}_s = \mathbf{0},$$

so it follows from (6.72) that $\dot{\mathbf{L}}_s = \mathbf{0}$. Equation (6.71) then gives

$$\mathbf{T}_s = \boldsymbol{\Omega} \wedge \mathbf{L}_s \tag{6.73}$$

for the synchronous torque. Equations (6.70), (6.72) and (6.73) yield the components of this torque along the body axes as

$$T_{s1} = \Omega^2 (I_3 - I_{\perp}) \sin\alpha_s \cos\alpha_s \sin\psi_s, \tag{6.74}$$

$$T_{s2} = \Omega^2 (I_3 - I_{\perp}) \sin\alpha_s \cos\alpha_s \cos\psi_s, \tag{6.75}$$

$$T_{s3} = 0. \tag{6.76}$$

In general $\mathbf{T}$ will be the sum of the magnetic, gravitational, accretion and dissipative torques. It follows from (6.19) and (6.69) that $\mathbf{T}_{\mathrm{m}}$ and $\mathbf{T}_{\mathrm{g}}$ satisfy (6.76), since $\hat{\mathbf{m}}_{\mathrm{p}} = \mathbf{e}_3$. However, the accretion torque has a finite component in the $\mathbf{e}_3$ direction and so must be dominated by $\mathbf{T}_{\mathrm{m}}$ and $\mathbf{T}_{\mathrm{g}}$ if such a distorted primary is to be synchronous. From (6.66), this requires a surface secondary field of $\gtrsim 10^3$ G. The accretion torque is, accordingly, excluded in the following analysis and the consistency of this assumption is subsequently checked. The effect of the dissipation is considered in Chapter 7.

The components of $\mathbf{T}_{\mathrm{m}}$ in the body frame are found from (6.21)–(6.23) and (6.70), which give

$$T_{\mathrm{m}1} = \frac{\mu_0 m_{\mathrm{p}} m_{\mathrm{s}}}{4\pi D^3} \left[\sin\gamma \cos\psi_s (2\cos\delta\cos\beta_s - \sin\delta\sin\beta_s) \right. \\ \left. + \sin\psi_s \{\cos\gamma\sin\alpha_s - \sin\gamma\cos\alpha_s(\sin\delta\cos\beta_s + 2\cos\delta\sin\beta_s)\}\right], \tag{6.77}$$

$$T_{\mathrm{m}2} = -\frac{\mu_0 m_{\mathrm{p}} m_{\mathrm{s}}}{4\pi D^3} \left[\sin\gamma \sin\psi_s (2\cos\delta\cos\beta_s - \sin\delta\sin\beta_s) \right. \\ \left. - \cos\psi_s \{\cos\gamma\sin\alpha_s - \sin\gamma\cos\alpha_s(\sin\delta\cos\beta_s + 2\cos\delta\sin\beta_s)\}\right], \tag{6.78}$$

$$T_{\mathrm{m}3} = 0. \tag{6.79}$$

The components of $\mathbf{T}_{\mathrm{g}}$ are given by (6.20), (6.69) and (6.70) as

$$T_{\mathrm{g}1} = \frac{3GM_{\mathrm{s}}(I_3 - I_\perp)}{D^3} \sin\alpha_s \sin\beta_s(\cos\beta_s \cos\psi_s - \cos\alpha_s \sin\beta_s \sin\psi_s), \tag{6.80}$$

$$T_{\mathrm{g}2} = -\frac{3GM_{\mathrm{s}}(I_3 - I_\perp)}{D^3} \sin\alpha_s \sin\beta_s(\cos\beta_s \sin\psi_s + \cos\alpha_s \sin\beta_s \cos\psi_s), \tag{6.81}$$

$$T_{\mathrm{g}3} = 0. \tag{6.82}$$

It follows from the synchronous torque conditions (6.74) and (6.75) that

$$T_{s1} \sin\psi_s + T_{s2} \cos\psi_s = \Omega^2(I_3 - I_\perp) \sin\alpha_s \cos\alpha_s,$$

$$T_{s1} \cos\psi_s - T_{s2} \sin\psi_s = 0.$$

Equations (6.77), (6.78), (6.80) and (6.81) then give

$$\cos\beta_s = -\frac{\sin\gamma \sin\delta}{\left[\frac{\Omega^2}{\bar{C}_{\mathrm{m}}} \frac{(I_3 - I_\perp)}{I_\perp} - \frac{\cos\gamma}{\cos\alpha_s}\right] \sin\alpha_s}, \tag{6.83}$$

$$\frac{\bar{C}_{\mathrm{g}}}{\bar{C}_{\mathrm{m}}} = \frac{\sin\gamma(\sin\delta \sin\beta_s - 2\cos\delta \cos\beta_s)}{\sin\alpha_s \sin\beta_s \cos\beta_s}, \tag{6.84}$$

where

$$\bar{C}_{\mathrm{m}} = \frac{\mu_0 m_{\mathrm{p}} m_{\mathrm{s}}}{4\pi D^3 I_\perp}, \tag{6.85}$$

$$\bar{C}_{\mathrm{g}} = \frac{3GM_{\mathrm{s}}(I_3 - I_\perp)}{D^3 I_\perp}. \tag{6.86}$$

Hence, for given Ω, $\bar{C}_{\mathrm{m}}$ and distortion $(I_3 - I_\perp)/I_\perp$, synchronous states (α_s, β_s) can be found from (6.83) by specifying (γ, δ) and varying α. Equation (6.84) then fixes $\bar{C}_{\mathrm{g}}/\bar{C}_{\mathrm{m}}$.

The quantities $\bar{C}_{\mathrm{m}}$ and $\bar{C}_{\mathrm{g}}$ can be written as

$$\bar{C}_{\mathrm{m}} = \frac{\pi}{10\mu_0} \frac{M_{\mathrm{s}} R_{\mathrm{p}}^3 (B_{\mathrm{p}})_0 (B_{\mathrm{s}})_0}{M I_\perp}, \tag{6.87}$$

$$\bar{C}_{\rm g} = \frac{3\Omega^2 M_{\rm s}}{M} \frac{(I_3 - I_\perp)}{I_\perp}, \tag{6.88}$$

where $M = M_{\rm s} + M_{\rm p}$. For a lobe-filling secondary, choosing Ω fixes $M_{\rm s}$. Then, since $\bar{C}_{\rm m}$ is specified in (6.83), (6.84) fixes $\bar{C}_g$ and hence $M_{\rm p}$. For a given white dwarf polar field $(B_{\rm p})_0$, substitution in (6.87) yields the secondary polar field $(B_{\rm s})_0$.

6.5.2. STABILITY OF THE SYNCHRONOUS STATE

In order to test the linear stability of the synchronous state normal modes of oscillation are sought, though it is not clear *a priori* that such modes exist here, since the basic state is dynamical.

The equation describing the evolution of small perturbations in $\mathbf{L}$ about the synchronous state, expressed in the body frame, is

$$\dot{\mathbf{L}}' + \mathbf{\Omega} \wedge \mathbf{L}' = \mathbf{T}'_{\rm m} + \mathbf{T}'_{\rm g}, \tag{6.89}$$

where

$$\mathbf{L}' = I_\perp(\omega'_1 \mathbf{e}_1 + \omega'_2 \mathbf{e}_2) + I_3 \omega'_3 \mathbf{e}_3. \tag{6.90}$$

Perturbations must be calculated in an inertial frame $OXYZ$ and then expressed in the body frame. To linear order, only the synchronous body frame is required. In an inertial frame $\boldsymbol{\omega}$ has components

$$\begin{aligned} \omega_X &= \dot{\psi} \sin\alpha \cos\phi - \dot{\alpha} \sin\phi, \\ \omega_Y &= \dot{\psi} \sin\alpha \sin\phi + \dot{\alpha} \cos\phi, \\ \omega_Z &= \dot{\phi} + \dot{\psi} \cos\alpha, \end{aligned}$$

where $\phi = \beta + \Omega t$. Perturbation about the synchronous state $\dot{\alpha} = \dot{\psi} = 0$, $\dot{\phi} = \Omega$, and use of (6.70) with ϕ replacing β, gives

$$\omega'_1 = \dot{\alpha}' \sin\psi_s - \dot{\beta}' \sin\alpha_s \cos\psi_s, \tag{6.91}$$

$$\omega'_2 = \dot{\alpha}' \cos\psi_s + \dot{\beta}' \sin\alpha_s \sin\psi_s, \tag{6.92}$$

$$\omega'_3 = \dot{\beta}' \cos\alpha_s + \dot{\psi}'. \tag{6.93}$$

From (6.19) and (6.69), the perturbations in $\mathbf{T}_{\rm m}$ and $\mathbf{T}_{\rm g}$ are

$$\mathbf{T}'_{\rm m} = m_{\rm p} \mathbf{e}'_3 \wedge \mathbf{B}_{\rm sp}, \tag{6.94}$$

$$\mathbf{T}'_{\rm g} = \frac{3GM_{\rm s}(I_3 - I_\perp)}{D^3} (S'\mathbf{j} \wedge \mathbf{e}_{3s} + S_s \mathbf{j} \wedge \mathbf{e}'_3), \tag{6.95}$$

where $S = \sin\alpha \sin\beta$.

The components of the equation of motion (6.89) are found using (6.16)–(6.18), (6.20), (6.70) and (6.90)–(6.95). Suitable linear combinations of the $\bar{x}$ and $\bar{y}$-components are taken, using multiples of $\sin\psi_s$ and $\cos\psi_s$, to generate equations independent of ψ_s. The resulting equations, together with the $\bar{z}$-component, are

$$\begin{aligned}&I_\perp\ddot{\alpha}' + (I_3 - I_\perp)\Omega\sin\alpha_s\cos\alpha_s\dot{\beta}' + I_3\Omega\sin\alpha_s\dot{\psi}'\\&- C_{\rm m}[\sin\gamma\sin\alpha_s(\sin\delta\cos\beta_s + 2\cos\delta\sin\beta_s) + \cos\gamma\cos\alpha_s]\alpha'\\&+ C_{\rm g}[(\cos^2\alpha_s - \sin^2\alpha_s)\sin^2\beta_s\alpha' + \sin\alpha_s\cos\alpha_s\sin\beta_s\cos\beta_s\beta'] = 0, \quad (6.96)\end{aligned}$$

$$\begin{aligned}&I_\perp\Omega\cos\alpha_s\dot{\alpha}' + I_\perp\sin\alpha_s\ddot{\beta}'\\&- C_{\rm m}[\sin\gamma\sin^2\alpha_s(\sin\delta\cos\beta_s + 2\cos\delta\sin\beta_s) + \cos\gamma\sin\alpha_s\cos\alpha_s]\beta'\\&+ C_{\rm g}[\cos\alpha_s\sin\beta_s\cos\beta_s\alpha' + (\cos^2\beta_s - \sin^2\alpha_s\sin^2\beta_s)\sin\alpha_s\beta'] = 0, \quad (6.97)\end{aligned}$$

$$\begin{aligned}&- I_\perp\Omega\sin\alpha_s\dot{\alpha}' + I_3\cos\alpha_s\ddot{\beta}' + I_3\ddot{\psi} - C_{\rm m}\sin\gamma(\sin\delta\sin\beta_s - 2\cos\delta\cos\beta_s)\alpha'\\&- C_{\rm m}[\sin\gamma\sin\alpha_s\cos\alpha_s(\sin\delta\cos\beta_s + 2\cos\delta\sin\beta_s) - \cos\gamma\sin^2\alpha_s]\beta'\\&+ C_{\rm g}[\sin\alpha_s\sin\beta_s\cos\beta_s\alpha' - \sin^2\alpha_s\cos\alpha_s\sin^2\beta_s\beta'] = 0, \qquad (6.98)\end{aligned}$$

where $C_{\rm m} = I_\perp\bar{C}_{\rm m}$ and $C_{\rm g} = I_\perp\bar{C}_{\rm g}$.

The angular perturbations can be written as the product of a constant amplitude and an exponential time dependence, so

$$\alpha' = a\exp(i\sigma t), \quad \beta' = b\exp(i\sigma t), \quad \psi' = c\exp(i\sigma t). \qquad (6.99)$$

Substitution in (6.96)–(6.98) yields a linear set of equations for a, b and c. Vanishing of the determinant of their coefficients, for consistency, gives the following characteristic equation for σ,

$$\sigma^6 + C_1\sigma^4 + C_2\sigma^2 + C_3\sigma = 0, \qquad (6.100)$$

where

$$C_1 = -\Omega^2 + 2\bar{C}_{\rm m}W - \bar{C}_{\rm g}[\cos^2\beta_s + \sin^2\beta_s(\cos^2\alpha_s - 2\sin^2\alpha_s)], \quad (6.101)$$

$$\begin{aligned}C_2 = &-\Omega^2\bar{C}_{\rm m}\sin\alpha_s\sin\gamma(\sin\delta\cos\beta_s + 2\cos\delta\sin\beta_s)\\&- \Omega^2\bar{C}_{\rm g}\sin^2\alpha_s(\sin^2\beta_s - \cos^2\beta_s) + \bar{C}_{\rm m}^2W^2\\&- \bar{C}_{\rm g}^2\sin^2\beta_s[\cos^2\alpha_s\cos^2\beta_s - (\cos^2\alpha_s - \sin^2\alpha_s)(\cos^2\beta_s - \sin^2\alpha_s\sin^2\beta_s)],\\&- \bar{C}_{\rm m}\bar{C}_{\rm g}W[\cos^2\beta_s - \sin^2\alpha_s\sin^2\beta_s + \sin^2\beta_s(\cos^2\alpha_s - \sin^2\alpha_s)], \quad (6.102)\end{aligned}$$

$$C_3 = -i\frac{(I_3 - I_\perp)}{I_\perp}\Omega^2 \bar{C}_{\rm g} \sin^2\alpha_s \cos^2\alpha_s \sin\beta_s \cos\beta_s, \tag{6.103}$$

with

$$W = \sin\gamma \sin\alpha_s(\sin\delta\cos\beta_s + 2\cos\delta\sin\beta_s) + \cos\gamma\cos\alpha_s. \tag{6.104}$$

It follows from (6.84) for the synchronous state that $\bar{C}_{\rm m} \sim \bar{C}_{\rm g}$. The ratio $\bar{C}_{\rm m}/\Omega^2$ can be expressed, from (6.87), as

$$\frac{\bar{C}_{\rm m}}{\Omega^2} = 6.5\times 10^{-10}\frac{\left(\frac{(B_{\rm p})_0}{20\,{\rm MG}}\right)\left(\frac{(B_{\rm s})_0}{10^3\,{\rm G}}\right)\left(\frac{R_{\rm p}}{8.7\times 10^6\,{\rm m}}\right)^3\left(\frac{M_{\rm s}}{0.2M_\odot}\right)\left(\frac{P}{2.1\,{\rm hr}}\right)}{\left(\frac{M}{0.8M_\odot}\right)\left(\frac{I_\perp}{1.8\times 10^{43}\,{\rm Kg\,m^2}}\right)}. \tag{6.105}$$

The quantity $\bar{C}_{\rm m}$ is of order $\sigma_{\rm m}^2$, where $\tau_{\rm m} = 2\pi/\sigma_{\rm m}$ is the characteristic response time of the primary to the torque perturbations $\mathbf{T}'_{\rm m}$ and $\mathbf{T}'_{\rm g}$. Consequently

$$\bar{C}_{\rm m}/\Omega^2 \sim \bar{C}_{\rm g}/\Omega^2 \ll 1.$$

Limits can be placed on the range of the solutions of (6.100) by noting that, if the inertial terms dominate the torque terms in the equations of motion, then $\sigma \sim \Omega$, otherwise $\sigma \sim \sigma_{\rm m}$. Equations (6.101)–(6.104) then show that the linear term in (6.100) is always negligible, due to the above small ratios, so the characteristic equation becomes

$$\sigma^4 + C_1\sigma^2 + C_2 = 0. \tag{6.106}$$

The $\sigma^2 = 0$ solution, which results from the ψ-independence of the torques, has been cancelled.

The solutions of (6.106) give the normal mode frequencies as

$$\sigma^2 = \frac{1}{2}(-C_1 \pm \Omega^2\sqrt{U}), \tag{6.107}$$

where

$$\begin{aligned} U =& 1 - 4\frac{\bar{C}_{\rm m}}{\Omega^2}\cos\gamma\cos\alpha_s + 2\frac{\bar{C}_{\rm g}}{\Omega^2}(\cos^2\alpha_s - \sin^2\alpha_s\cos^2\beta_s) \\ &+ \left(\frac{\bar{C}_{\rm g}}{\Omega^2}\right)^2(\cos^2\beta_s + \cos^2\alpha_s\sin^2\beta_s)^2. \end{aligned}$$

Expansion to first order in $\bar{C}_{\rm m}/\Omega^2$ and $\bar{C}_{\rm g}/\Omega^2$ gives the solutions

$$\sigma_+^2 = \Omega^2, \tag{6.108}$$

$$\sigma_-^2 = -\bar{C}_\mathrm{m} \sin\gamma \sin\alpha_s(\sin\delta\cos\beta_s + 2\cos\delta\sin\beta_s) \\ + \bar{C}_\mathrm{g}\sin^2\alpha_s(\cos^2\beta_s - \sin^2\beta_s). \qquad (6.109)$$

The nature of the modes can now be investigated. For $\sigma = \Omega$ use of (6.99) in (6.96)–(6.98) yields

$$-I_\perp a + i(I_3 - I_\perp)\sin\alpha_s\cos\alpha_s b + iI_3\sin\alpha_s c = 0,$$

and

$$[(I_\perp - I_3)\sin^2\alpha_s + I_3]b + I_3\cos\alpha_s c = 0,$$

where the torque terms have been dropped, since they are of relative order $\bar{C}_\mathrm{m}/\Omega^2$. Because $\bar{C}_\mathrm{m}/\Omega^2 \sim 10^{-9}$, it follows from (6.83) that $(I_3 - I_\perp)/I_\perp$ must be of this order to allow synchronous solutions. Hence the above equations give

$$a = i\frac{I_3}{I_\perp}\sin\alpha_s c,$$

$$b = -\cos\alpha_s c.$$

The angular velocity for the σ_+ mode is therefore

$$\boldsymbol{\omega}' = i\Omega c\left[i\frac{I_3}{I_\perp}\sin\alpha_s\hat{\boldsymbol{\alpha}}_s + \hat{\boldsymbol{\psi}}_s - (\hat{\boldsymbol{\psi}}_s\cdot\hat{\boldsymbol{\Omega}})\hat{\boldsymbol{\Omega}}\right]\exp(i\Omega t). \qquad (6.110)$$

It follows that $\boldsymbol{\omega}'\cdot\hat{\boldsymbol{\Omega}} = 0$ and , relative to the orbital frame, $\boldsymbol{\omega}'$ rotates with angular frequency Ω. This is essentially the inertial mode corresponding to (6.60), since $I_3/I_\perp$ is close to 1. The Euler angle perturbations vary with a period $2\pi/\Omega$ which is much shorter than the primary's response time to the torque perturbations, so they are ineffective in this mode.

Equation (6.109) shows that the remaining mode frequency has the property $\sigma_-^2 \sim \bar{C}_\mathrm{m} \ll \Omega^2$. Equations (6.96) and (6.97) then yield

$$a = c = 0$$

to high accuracy. Hence this mode has an angular velocity perturbation

$$\boldsymbol{\omega}' = i\sigma_- b\exp(i\sigma_- t)\hat{\boldsymbol{\Omega}}, \qquad (6.111)$$

and is therefore aligned with $\boldsymbol{\Omega}$. Its existence can be seen directly by combining (6.97) and (6.98) to eliminate the $\dot{\alpha}'$ terms. If α' and ψ' are small, this gives a harmonic equation for β' with frequency σ_-. Since any angular velocity perturbation can be expressed as a superposition of an inertial mode and the aligned mode, the sign of σ_-^2 determines the stability of the synchronous state.

6.5.3. DISTRIBUTION OF SYNCHRONOUS STATES

To obtain synchronous orientations the quantity $\bar{C}_\mathrm{m}/\Omega^2$ is taken as 1.5×10^{-9}. Specifying (γ, δ) and using a distortion of $(I_3 - I_\perp)/I_\perp = 10^{-9}$, orientations (α_s, β_s) are found satisfying (6.83). Taking $P = 2.1\,\mathrm{hrs}$, $M_\mathrm{s} = 0.2M_\odot$ and $(B_\mathrm{p})_0 = 2 \times 10^3$ tesla $(= 2 \times 10^7\ \mathrm{G})$, (6.84) and (6.88) determine M_p and then (6.105) fixes $(B_\mathrm{s})_0$. Stability is tested using (6.109). Tables 6.2 and 6.3 show a sample of stable synchronous orientations.

TABLE 6.2. Stable states for $(I_3 - I_\perp)/I_\perp = 10^{-9}$.

γ	δ	α_s	β_s	$M_\mathrm{P}/M_\odot$	$(B_\mathrm{S})_0/10^3$ G
20	125	166.0	135.0	0.41	0.43
20	125	165.0	131.2	0.88	2.68
20	206	164.0	70.7	0.36	0.32
20	206	161.0	73.9	1.13	6.18
40	143	149.0	118.7	0.37	0.33
40	143	145.0	114.8	0.89	2.81
40	216	149.5	61.5	0.32	0.25
40	216	145.5	65.4	0.76	1.77
60	146	130.0	115.9	0.43	0.47
60	146	125.0	112.5	0.93	3.14
60	192	126.0	81.6	0.47	0.58
60	192	120.0	82.9	1.13	6.15
80	118	112.5	146.9	0.37	0.33
80	118	109.5	140.8	0.91	2.96
80	242	112.5	33.1	0.37	0.33
80	242	109.5	39.2	0.91	2.96
100	126	69.0	137.7	0.39	0.37
100	126	73.0	131.2	0.96	3.50
100	242	68.0	34.2	0.43	0.46
100	242	71.0	40.2	1.09	5.22
120	134	48.0	126.3	0.48	0.60
120	134	52.0	122.2	1.05	4.58
120	200	52.0	75.3	0.42	0.43
120	200	57.0	77.2	0.80	2.05
140	156	32.5	108.0	0.36	0.32
140	156	37.5	105.2	0.91	2.91

TABLE 6.3. Stable states for $(I_3 - I_\perp)/I_\perp = 10^{-9}$

γ	δ	α_s	β_s	$M_P/M_\odot$	$(B_S)_0/10^3$ G
140	218	32.0	61.7	0.47	0.57
140	218	34.5	64.1	0.83	2.31
160	120	14.0	138.4	0.62	1.09
160	120	14.5	136.1	1.02	4.15
160	240	13.5	39.2	0.39	0.37
160	240	14.5	43.9	1.02	4.15

TABLE 6.4. Stable states for $(I_3 - I_\perp)/I_\perp = -10^{-9}$

γ	δ	α_s	β_s	$M_P/M_\odot$	$(B_S)_0/10^3$ G
20	125	121.5	106.0	0.42	0.45
20	125	126.5	112.5	0.85	2.47
20	206	117.0	83.1	0.37	0.34
20	206	123.0	80.3	1.09	5.32
40	143	108.0	103.0	0.37	0.35
40	143	110.5	105.8	0.86	2.54
40	216	107.5	77.8	0.33	0.26
40	216	110.0	75.2	0.75	1.72
60	146	100.0	102.9	0.36	0.33
60	146	101.5	105.6	1.07	4.97
60	192	99.5	85.6	0.43	0.47
60	192	100.5	84.9	0.97	3.58
80	118	94.0	118.6	0.33	0.26
80	118	94.5	124.4	0.76	1.80
80	242	94.0	61.4	0.33	0.26
80	242	94.5	55.6	0.76	1.80
100	126	86.0	116.0	0.54	0.79
100	242	86.0	61.4	0.33	0.26
100	242	85.5	55.6	0.76	1.80
120	134	79.0	109.0	0.41	0.43
120	134	77.5	112.9	1.13	6.09
120	200	80.5	82.7	0.38	0.35
120	200	79.5	81.7	0.80	2.08
140	156	73.0	98.1	0.37	0.33
140	156	70.5	99.8	0.92	3.01

TABLE 6.5. Stable states for $(I_3 - I_\perp)/I_\perp = -10^{-9}$

γ	δ	α_s	β_s	$M_P/M_\odot$	$(B_S)_0/10^3$ G
140	218	71.0	75.6	0.50	0.64
140	218	69.5	73.8	0.83	2.26
160	120	54.5	112.6	0.62	1.06
160	120	51.0	117.6	1.03	4.37
160	240	58.0	71.5	0.39	0.38
160	240	51.0	62.4	1.03	4.37

It is seen that the surface magnetic field on the secondary is $\sim 10^3$ G, so justifying neglect of the accretion torque. For given (γ, δ) synchronous states are possible for a wide range of M_P. States in which the accreting pole leads the motion of the line of stellar centres are possible, as well as those in which it lags. It is noted that, for all other parameters fixed, $(B_S)_0$ cannot be chosen independently of (γ, δ) if synchronous states are to result. Equation (6.83) is invariant to the transformation $\beta_s \to -\beta_s$. However, (6.84) does not possess this invariance and hence different values of M_P and $(B_S)_0$ result. The synchronous state distribution generated in this way is similar to that shown in Tables 6.2 and 6.3, but with the south magnetic pole accreting.

Tables 6.4 and 6.5 show synchronous orientations for $(I_3 - I_\perp)/I_\perp = -10^{-9}$. This case could correspond to the distortion caused by an internal toroidal field proposed by Katz (1989), where, from (6.68), $B_0 \sim 10^7$ G.

6.6. Orbital Torques

In the foregoing analysis the magnetic fields of the primary and secondary stars were taken as dipolar, with corresponding moments $\mathbf{m}_P$ and $\mathbf{m}_S$. King, Frank and Whitehurst (1990) pointed out that there is a non-central force between the dipoles, and calculated the resulting magnetic orbital torque.

The primary's magnetic field can be expressed as

$$\mathbf{B}_P = \nabla \wedge \left(\frac{\mu_0 \mathbf{m}_P \wedge \mathbf{r}}{4\pi r^3} \right) = \frac{\mu_0}{4\pi r^3}[3(\hat{\mathbf{r}} \cdot \mathbf{m}_P)\hat{\mathbf{r}} - \mathbf{m}_P]. \tag{6.112}$$

The torque on the secondary is then

$$\mathbf{T}_{mS} = \mathbf{m}_S \wedge \mathbf{B}_P(\mathbf{r}_S) = \frac{\mu_0}{4\pi D^3}[3(\mathbf{n} \cdot \mathbf{m}_P)\mathbf{m}_S \wedge \mathbf{n} - \mathbf{m}_S \wedge \mathbf{m}_P], \tag{6.113}$$

where $\mathbf{n}$ is a unit vector directed from the primary to the centre of the secondary. Similarly the torque on the primary due to interaction with the secondary's magnetic field is

$$\mathbf{T}_{\rm mp} = \frac{\mu_0}{4\pi D^3}[3(\mathbf{n}\cdot\mathbf{m}_{\rm s})\mathbf{m}_{\rm p}\wedge\mathbf{n} - \mathbf{m}_{\rm p}\wedge\mathbf{m}_{\rm s}]. \qquad (6.114)$$

Equations (6.113) and (6.114) show that, in general, $\mathbf{T}_{\rm ms} + \mathbf{T}_{\rm mp} \neq \mathbf{0}$. This result is a consequence of a non-central force between the dipoles, arising from the spatial dependence of their fields. The force on $\mathbf{m}_{\rm s}$ due to the primary's field $\mathbf{B}_{\rm p}$, given by (6.112), is

$$\begin{aligned}\mathbf{F}_{\rm sp} &= [\nabla(\mathbf{m}_{\rm s}\cdot\mathbf{B}_{\rm p})]_{\mathbf{r}=D\mathbf{n}} \\ &= \frac{\mu_0}{4\pi D^4}\left[\{3(\mathbf{m}_{\rm p}\cdot\mathbf{m}_{\rm s}) - 15(\mathbf{n}\cdot\mathbf{m}_{\rm p})(\mathbf{n}\cdot\mathbf{m}_{\rm s})\}\mathbf{n}\right. \\ &\left.+3(\mathbf{n}\cdot\mathbf{m}_{\rm s})\mathbf{m}_{\rm p} + 3(\mathbf{n}\cdot\mathbf{m}_{\rm p})\mathbf{m}_{\rm s}\right]. \qquad (6.115)\end{aligned}$$

The force on the primary due to the secondary's field is $\mathbf{F}_{\rm ps} = -\mathbf{F}_{\rm sp}$. The non-central terms in (6.115) lead to an orbital torque

$$\mathbf{T}_{\rm mo} = D\mathbf{n}\wedge\mathbf{F}_{\rm sp} = \frac{3\mu_0}{4\pi D^3}[(\mathbf{n}\cdot\mathbf{m}_{\rm s})\mathbf{n}\wedge\mathbf{m}_{\rm p} + (\mathbf{n}\cdot\mathbf{m}_{\rm p})\mathbf{n}\wedge\mathbf{m}_{\rm s}]. \qquad (6.116)$$

If follows from (6.113), (6.114) and (6.116) that

$$\mathbf{T}_{\rm ms} + \mathbf{T}_{\rm mp} + \mathbf{T}_{\rm mo} = \mathbf{0}. \qquad (6.117)$$

The angular momentum evolution equations for the primary, secondary and orbit are

$$\frac{d\mathbf{L}_{\rm p}}{dt} = \mathbf{T}_{\rm a} + \mathbf{T}_{\rm mp} + \mathbf{T}_{\rm dp} + \mathbf{T}_{\rm g}, \qquad (6.118)$$

$$\frac{d\mathbf{L}_{\rm s}}{dt} = \mathbf{T}_{\rm br} + \mathbf{T}_{\rm ms} + \mathbf{T}_{\rm ds} + \mathbf{T}_{\rm tid}, \qquad (6.119)$$

$$\frac{d\mathbf{L}_{\rm orb}}{dt} = \mathbf{T}_{\rm gr} - \mathbf{T}_{\rm a} + \mathbf{T}_{\rm mo} + \mathbf{T}_{\rm do} - \mathbf{T}_{\rm g} - \mathbf{T}_{\rm tid}, \qquad (6.120)$$

where the non-dissipative magnetic torques obey (6.117). These evolution equations apply to an asynchronous state, with all torques operable. The dissipative magnetic torques, arising from the process considered in Chapter 4, satisfy

$$\mathbf{T}_{\rm ds} + \mathbf{T}_{\rm dp} + \mathbf{T}_{\rm do} = \mathbf{0}, \qquad (6.121)$$

where $\mathbf{T}_{\rm do}$ is an orbital torque due to a non-central force between the stars. The gravitational torque $\mathbf{T}_{\rm g}$ occurs if the primary is non-spherical, and the tidal torque $\mathbf{T}_{\rm tid}$ results if the secondary has asynchronoous motions with dissipation. A magnetically influenced wind from the secondary generates the braking torque $\mathbf{T}_{\rm br}$, while $\mathbf{T}_{\rm gr}$ is the orbital torque due to gravitational radiation losses. Adding (6.118)–(6.120), then using (6.117) and (6.121), gives

$$\frac{d\mathbf{L}_{\rm orb}}{dt} + \frac{d\mathbf{L}_{\rm s}}{dt} + \frac{d\mathbf{L}_{\rm p}}{dt} = \mathbf{T}_{\rm br} + \mathbf{T}_{\rm gr}. \qquad (6.122)$$

This shows that the total angular momentum of the system evolves under the influence of magnetic wind braking and gravitational waves, both of which remove angular momentum.

The tidal torque is believed to keep the secondary's spin near corotation with the orbit, while the primary may be kept in synchronism by the foregoing locking mechanisms. The stellar spin angular momenta are then given by

$$|\mathbf{L}_{\rm s}| \simeq k_{\rm s}^2 M_{\rm s} R_{\rm s}^2 \Omega, \qquad |\mathbf{L}_{\rm p}| \simeq k_{\rm p}^2 M_{\rm p} R_{\rm p}^2 \Omega, \qquad (6.123\text{a, b})$$

where typically $k_{\rm s}^2 \sim k_{\rm p}^2 \sim 0.2$. The orbital angular momentum can be expressed as

$$|\mathbf{L}_{\rm orb}| = \frac{M_{\rm s} M_{\rm p}}{M} D^2 \Omega. \qquad (6.124)$$

It follows that $|\mathbf{L}_{\rm orb}| \gg |\mathbf{L}_{\rm s}| \gg |\mathbf{L}_{\rm p}|$, while, due to synchronous rotations, the angular momentum evolution time-scales are the same. Hence $|d\mathbf{L}_{\rm orb}/dt| \gg |d\mathbf{L}_{\rm s}/dt| \gg |d\mathbf{L}_{\rm p}/dt|$ and (6.122) can be written

$$\frac{d\mathbf{L}_{\rm orb}}{dt} = \mathbf{T}_{\rm br} + \mathbf{T}_{\rm gr}, \qquad (6.125)$$

to a good approximation. This removal of orbital angular momentum drives mass transfer (see §2.4.2 and §12.3.1) and the binary evolves on the time-scale $\tau_M \sim M_{\rm s}/|\dot{M}_{\rm s}|$. Typically, for an AM Her system, $\tau_M \sim 10^9$ yrs. The characteristic adjustment time of perturbations from synchronism of the primary is given by (6.15), being ~ 40 yrs. Hence the corotating primary has ample time to adjust to the orbital evolution.

6.7. Conclusions

The magneto-gravitational balance, unlike the magneto-accretion balance, can produce the observed states in which the accreting pole leads the motion of the line of stellar centres, as well as other states. Both balances have the property that the secondary's surface magnetic field cannot be chosen independently of its orientation, if synchronous states of the primary are to exist. In both cases these restrictions are lifted if the magnetic torque dominates, so determining (α_s, β_s). This requires $(I_3 - I_\perp)/I_\perp < 10^{-9}$ and $(B_s)_0 \sim 10^3$ G, or larger distortion and $(B_s)_0 > 10^3$ G.

Cases in which the white dwarf has a quadrupolar component to its magnetic field, so far considered, require field-channelling to only be operative very close to the star. The accretion torque then only has a significant vertical component. However, observations suggest that channelling occurs at a distance from the primary which is a significant fraction of its distance from L_1 (e.g. Schwope, Mantel and Horne, 1996).

References

Campbell, C.G., 1985. *Mon. Not. R. Astr. Soc.*, **215**, 509.
Campbell, C.G., 1989. *Mon. Not. R. Astr. Soc.*, **236**, 475.
Campbell, C.G., 1990. *Mon. Not. R. Astr. Soc.*, **244**, 367.
Cropper, M., 1988. *Mon. Not. R. Astr. Soc.*, **231**, 597.
Joss, P.C., Katz, J.I. and Rappaport, S.A., 1979. *Astrophys. J.*, **230**, 176.
Katz, J.I., 1989. *Mon. Not. R. Astr. Soc.* **239**, 751.
King, A.R., Frank, J. and Whitehurst, R., 1990. *Mon. Not. R. Astr. Soc.*, **244**, 731.
Paczyński, B., 1967. *Acta. Astra.*, **17**, 287.
Schwope, A.D., Mantel, K. and Horne, K., 1996. *Astron. Astrophys.*, in press.
Wu, K. and Wickramasinghe, D.T., 1993. *Mon. Not. R. Astr. Soc.*, **260**, 141.

CHAPTER 7

THE ATTAINMENT OF SYNCHRONISM

7.1. The Combined Effect of Torques

Having calculated the torques acting on the primary, their combined effect must be considered. Even if a stable synchronous state exists, it is not clear *a priori* that this state can be reached in the presence of accretion. It will be seen that the situations of initially under and over-synchronous states are not symmetrical as far as attaining corotation is concerned.

In considering the approach to synchronism it is valid to use synodically averaged torques, provided that the instantaneous synchronization time significantly exceeds the synodic period. Hence

$$|\omega| t_s \gg 2\pi \tag{7.1}$$

is necessary. The averaged dissipation torque, given by (4.70), can be written as

$$\mathbf{T}_{\mathrm{D}} = \mp \frac{5\pi (B_{\mathrm{p}})_0^2 R_{\mathrm{s}}^3 R_{\mathrm{p}}^6 \sin^2 \alpha}{\mu_0 D^6} f(|\omega|/\eta)\mathbf{k}, \tag{7.2}$$

where the negative and positive signs apply to over-synchronous and under-synchronous states, respectively. In the absence of other torques this dissipation would synchronize the primary in a characteristic time

$$t_s = \frac{k_{\mathrm{p}}^2 M_{\mathrm{p}} R_{\mathrm{p}}^2 |\omega|}{|T_{\mathrm{D}}|}, \tag{7.3}$$

where k_{p}^2 lies in the range $0.1 < k_{\mathrm{p}}^2 < 0.2$. In general t_s is a function of the degree of asynchronism.

The synodic average of the accretion torque is given by (5.70) as

$$\mathbf{T}_{\mathrm{A}} = A^2 \Omega \dot{M}_{\mathrm{p}} \mathbf{k}. \tag{7.4}$$

If material becomes field-channelled far from L_1 some angular momentum is fed back into the orbit and the magnitude of T_{A} is reduced by $\sim 6\%$.

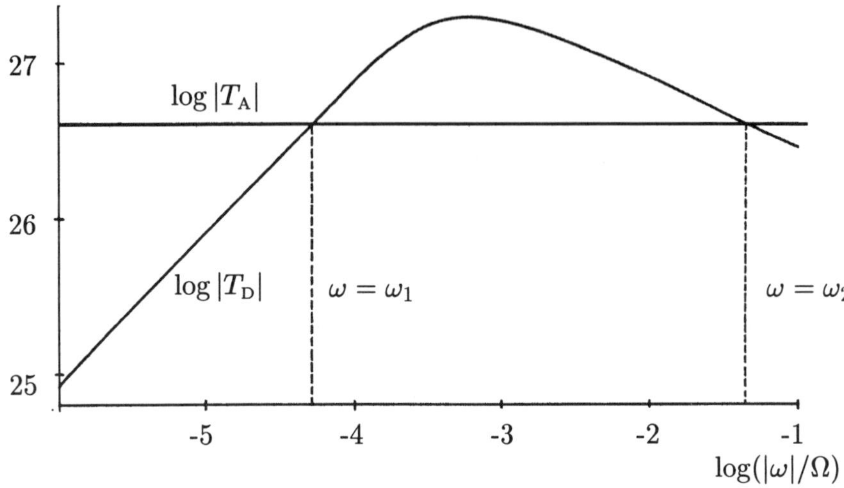

Figure 7.1. The dissipation torque and accretion torque versus degree of asynchronism. Based on Campbell (1986).

For simplicity, the case in which the dipole moment lies in the orbital plane was considered by Campbell (1986). The system DP Leo is believed to have its main accretion column close to the orbital plane (Biermann *et al*, 1985). However, the conclusions reached here regarding the attainment of synchronism should have general validity. When the primary's dipole moment lies in the orbital plane, it follows from (6.21)–(6.23) and (6.69), with $\alpha = \pi/2$ and $\beta = \omega t$, that the non-dissipative magnetic and gravitational torques have zero synodic averages. The spin evolution of the primary is therefore determined by the action of $\mathbf{T}_D$ and $\mathbf{T}_A$.

The magnitudes of T_D and T_A are plotted, on a logarithmic scale, in Figure 7.1 as a function of the asynchronism $|\omega|/\Omega$. Equations (7.2) and (7.4) show that only in the over-synchronous case can the torques have the possibility of cancelling, since in the under-synchronous case T_D and T_A have the same sign. It is noted that T_D vanishes as $|\omega|$ tends to zero, while T_A remains finite. Hence in the absence of a locking mechanism the primary cannot attain exact synchronism. The function $f(|\omega|/\eta)$ has a single maximum so, in general, there will be two finite values of ω at which $T_D + T_A$ vanishes.

The parameters used to plot Figure 7.1 are $P = 1.5\,\text{hr}$, $M_s = 0.14M_\odot$, $M_p = 0.4M_\odot$, $R_s = 10^8\,\text{m}$, $R_p = 1.1 \times 10^7\,\text{m}$, $D = 3.8 \times 10^8\,\text{m}$ and $A = 2.3 \times 10^8\,\text{m}$. These values are consistent with a lobe-filling secondary star, and employ (2.222), (2.225) and (2.226) with $q = 1.1$. The polar magnetic field and accretion rate are taken as $(B_p)_0 = 1.4 \times 10^3$ tesla and $\dot{M}_p =$

$10^{-10}\, M_\odot \mathrm{yr}^{-1}$. It is noted that the qualitative nature of the curves shown in Figure 7.1 is independent of the precise parameters used.

Figure 7.1 illustrates that a critical accretion rate, $\dot{M}_c$, exists at which the accretion line is tangent to the dissipation curve. If $\dot{M}_\mathrm{p}$ exceeds $\dot{M}_c$ then T_A exceeds $|T_\mathrm{D}|$ whatever the degree of asynchronism and the primary is spun up. The function $f(|\omega|/\eta)$ is given by (4.49) and can be written as

$$f(|\omega|/\eta) = f_1(|\omega|/\eta) + \left(\frac{R_\mathrm{s}}{D}\right)^2 f_2(|\omega|/\eta),$$

where

$$f_1 = \frac{3}{4}\left(C_1^2 \delta_1'\right)_s F_1 \quad \text{and} \quad f_2 = \left(C_2^2 \delta_2'\right)_s F_2.$$

It follows from (4.49)–(4.56) that f_1 and f_2 are pure functions of $|\omega|/\eta$, being independent of the orbital parameters. Their maximum values are found to be 8.6×10^{-2} and 0.24, respectively. The factor $(R_\mathrm{s}/D)^2$ is weakly dependent on the stellar mass ratio, for a lobe-filling secondary, so f is essentially independent of the binary parameters. Equating the maximum value of $|T_\mathrm{D}|$ to the accretion torque gives

$$\dot{M}_c = N_1 \frac{\left(\frac{P}{1.5\,\mathrm{hr}}\right)\left(\frac{(B_\mathrm{p})_0}{14\,\mathrm{MG}}\right)^2\left(\frac{R_\mathrm{s}}{10^8\,\mathrm{m}}\right)^3\left(\frac{R_\mathrm{p}}{1.1\times 10^7\,\mathrm{m}}\right)^6}{\left(\frac{A}{2.3\times 10^8\,\mathrm{m}}\right)^2\left(\frac{D}{3.8\times 10^8\,\mathrm{m}}\right)^6}\,\mathrm{M}_\odot \mathrm{yr}^{-1}, \quad (7.5a)$$

where

$$N_1 = 4.4\times 10^{-10}\left[0.86 + 2.42\left(\frac{R_\mathrm{s}}{D}\right)^2\right]\sin^2\alpha. \quad (7.5b)$$

It is noted that $\dot{M}_\mathrm{c}$ does not dependent on η, because f is a function of $|\omega|/\eta$ and changing η simply changes the value of $|\omega|$ at which the maximum in this function occurs.

It follows from Figure 7.1 that, for $\dot{M}_\mathrm{p} < \dot{M}_c$, two synodic frequencies exist at which T_D cancels T_A. It is clear that the frequency ω_2 is unstable, since a small change in ω about this state causes the primary to spin away from it. The smaller frequency ω_1 corresponds to a stable state with zero synodic torque. Hence, in the absence of a locking mechanism, if the initial asynchronism is $< \omega_2/\Omega$, the primary will spin down to the slightly asynchronous state with synodic period $P_1 = 2\pi/\omega_1$.

An expression can be derived for ω_2/Ω, the value of the initial asynchronism above which the primary will be spun away from corotation, since then

T_A exceeds $|T_D|$. This is done by using the fact that, for $\dot{M}_p$ significantly less than $\dot{M}_c$, the spin rate ω_2 lies in the regime of low field penetration in which the dissipation torque is given by the asymptotic form (4.75). Equating this to (7.4) for T_A gives

$$\frac{\omega_2}{\Omega} = N_2 \frac{\left(\frac{\eta}{5 \times 10^8 \, \mathrm{m^2 s^{-1}}}\right)\left(\frac{P}{1.5\,\mathrm{hr}}\right)^3\left(\frac{(B_p)_0}{14\,\mathrm{MG}}\right)^4\left(\frac{R_s}{10^8\,\mathrm{m}}\right)^4\left(\frac{R_p}{1.1 \times 10^7\,\mathrm{m}}\right)^{12}}{\left(\frac{\dot{M}_p}{10^{-10}\,M_\odot \mathrm{yr}^{-1}}\right)^2\left(\frac{A}{2.3 \times 10^8\,\mathrm{m}}\right)^4\left(\frac{D}{3.8 \times 10^8\,\mathrm{m}}\right)^{12}}$$

where

$$N_2 = 2.7 \times 10^{-2}\left[1 + \frac{16}{3}\left(\frac{R_s}{D}\right)^2\right]^2 \sin^4\alpha. \tag{7.6a,b}$$

The synodic period P_1 of the slightly asynchronous state can be found by utilizing the fact that, for $\dot{M}_p$ significantly less than $\dot{M}_c$, the spin rate ω_1 corresponds to the regime of high field penetration in which the dissipation torque is given by (4.77). Equating this to T_A gives

$$P_1 = \frac{N_3\left(\frac{P}{1.5\,\mathrm{hr}}\right)\left(\frac{(B_p)_0}{14\,\mathrm{MG}}\right)^2\left(\frac{R_s}{10^8\,\mathrm{m}}\right)^5\left(\frac{R_p}{1.1 \times 10^7\,\mathrm{m}}\right)^6 \ \mathrm{yr}}{\left(\frac{\eta}{5 \times 10^8\,\mathrm{m^2 s^{-1}}}\right)\left(\frac{\dot{M}_p}{10^{-10}\,M_\odot \mathrm{yr}^{-1}}\right)\left(\frac{A}{2.3 \times 10^8\,\mathrm{m}}\right)^2\left(\frac{D}{3.8 \times 10^8\,\mathrm{m}}\right)^6}$$

where

$$N_3 = 3.0\left[1 + \frac{48}{35}\left(\frac{R_s}{D}\right)^2\right]\sin^2\alpha. \tag{7.7a,b}$$

The synchronization time in the high field penetration regime is given by (4.78), being independent of ω.

Inequality (7.1) gives the condition for the validity of the synodic time average. The synchronization time increases monotonically with ω and only values of $\omega \gtrsim \omega_1$ are of interest here. Condition (7.1) is therefore equivalent to $t_s/P_1 \gg 1$. Using the previously quoted orbital parameters, together with $\alpha = \pi/2$, (4.78) and (7.7) give $t_s/P_1 = 22$, showing that the time averaging is just valid in this case. However, it is noted that t_s/P_1 is proportional to η^2 and the value of η is uncertain by at least an order of magnitude. It follows that, for fixed orbital parameters, there is a minimum value of η below which $t_s/P_1 < 1$ so invalidating the synodic averaging process when

ω is close to ω_1, although it will still be valid for $\omega \gg \omega_1$. Equations (4.78) and (7.7) give the minimum value of η as

$$\eta_0 = \frac{N_4 \left(\frac{k_{\mathrm{p}}^2}{0.18}\right)^{-\frac{1}{2}} \left(\frac{P}{1.5\,\mathrm{hr}}\right)^{\frac{1}{2}} \left(\frac{(B_{\mathrm{p}})_0}{14\,\mathrm{MG}}\right)^2 \left(\frac{R_{\mathrm{s}}}{10^8\,\mathrm{m}}\right)^5 \left(\frac{R_{\mathrm{p}}}{1.1\times 10^7\,\mathrm{m}}\right)^5}{\left(\frac{M_{\mathrm{p}}}{0.4\,M_\odot}\right)^{\frac{1}{2}} \left(\frac{\dot{M}_{\mathrm{p}}}{10^{-10}\,M_\odot \mathrm{yr}^{-1}}\right)^{\frac{1}{2}} \left(\frac{A}{2.3\times 10^8\,\mathrm{m}}\right) \left(\frac{D}{3.8\times 10^8\,\mathrm{m}}\right)^6}\ \mathrm{m\,s^{-1}}$$

where

$$N_4 = 8.6 \times 10^7 \left[1 + \frac{48}{35}\left(\frac{R_{\mathrm{s}}}{D}\right)^2\right] \sin^2\alpha. \qquad (7.8\mathrm{a,b})$$

In the absence of a locking mechanism, the spin rate of the primary will evolve towards ω_1 irrespective of whether η is below or above η_0, or whether its initial spin is under or over-synchronous. For $\eta \gg \eta_0$ the primary will settle at ω_1 in a time-independent state. For $\eta < \eta_0$ the primary will reach a state in which its angular velocity wobbles about ω_1 since $T_{\mathrm{D}} + T_{\mathrm{A}}$ will vary about zero as the dissipation varies, and the star can respond to this. Without a stable locking mechanism the white dwarf can never reach synchronism in the presence of accretion.

7.2. The Effect of a Locking Mechanism

Mechanisms for locking the white dwarf in corotation were investigated in Chapter 6. Assuming that a stable synchronous state exists, it is not obvious that the primary will be able to reach this state while accretion is occurring. The locking mechanism of §6.2 will be considered here and conditions derived under which the primary can attain corotation. These conditions appear to have a general validity, independent of the precise locking mechanism considered.

The sum of the accretion and non-dissipative magnetic torques on the primary is given by (6.8) which, by using the synchronous condition $T = 0$, can be written as

$$\mathbf{T} = A^2 \Omega \dot{M}_{\mathrm{p}} \left[1 - \frac{\sin(\beta - \bar{\delta})}{\sin(\beta_s - \bar{\delta})}\right] \mathbf{k}, \qquad (7.9)$$

where β_s is the stable synchronous orientation, and the angle $\bar{\delta}$ is shown in Figure 6.1.

The work done in rotating the primary from some reference orientation β_0 to a general orientation β is

$$W(\beta) = -\int_{\beta_0}^{\beta} T(\beta) d\beta,$$

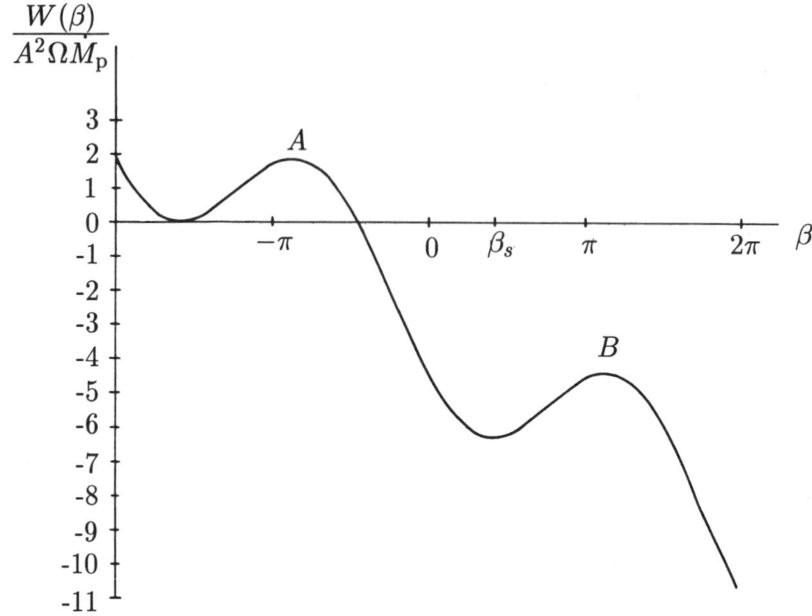

Figure 7.2. The function $W(\beta)/A^2\Omega\dot{M}_p$, showing the synchronous state β_s. Based on Campbell (1986).

so (7.9) gives

$$W(\beta) = -A^2\Omega\dot{M}_p\left[(\beta - \beta_0) + \frac{\cos(\beta - \bar{\delta}) - \cos(\beta_0 - \bar{\delta})}{\sin(\beta_s - \bar{\delta})}\right]. \tag{7.10}$$

It is clear that the form of $W(\beta)$ is independent of the orbital parameters. The first term in the bracket arises from the work done against the accretion torque, while the second term derives from the work done against the magnetic torque. It is seen that for $\beta = \beta_0 \pm 2n\pi$, where n is an integer, the work done against the accretion torque is $\mp 2n\pi A^2\Omega\dot{M}_p$, while no work is done against the magnetic torque for a complete number of rotations of the primary.

Figure 7.2 is a plot of $W(\beta)/A^2\Omega\dot{M}_p$ for $\beta_s = 76°$, $\bar{\delta} = 50°$ and $\beta_0 = \beta_s - 2\pi$. The synchronous orientation occurs at local minima in $W(\beta)$ and adjacent minima differ in height by $2\pi A^2\Omega\dot{M}_p$ due to the work done against T_A. Similarly, adjacent maxima in $W(\beta)$ differ in height by $2\pi T_A = 2\pi A^2\Omega\dot{M}_p$ and as a consequence of this it is necessary to distinguish between the cases of an initially under-synchronous and over-synchronous primary.

An initially under-synchronous primary: In this case $\omega = \dot{\beta}$ is negative and both the dissipation and accretion torques act to reduce ω. Since T_A

and T_D are both positive a state in which they cancel cannot occur as ω tends to zero. The synodic rotational energy of the primary will decrease until it is just sufficient to allow the star to pass through a maximum in $W(\beta)$, such as point B in Figure 7.2. It is then clear that after passing through B the primary cannot gain sufficient rotational energy to allow it to reach the next maximum at A and hence will oscillate in the well, eventually settling at β_s due to the action of the dissipation torque.

An initially over-synchronous primary: For an over-synchronous primary $\omega = \dot{\beta}$ is positive and T_D is negative. Figure 7.1 shows that for $\omega_1 < \omega < \omega_2$, $|T_D|$ exceeds T_A and hence ω decreases. However, synchronism at β_s can only result if a value of ω exists such that when the primary has passed through a maximum in $W(\beta)$ (e.g. point A), at least its existing synodic rotational energy plus $2\pi T_A$ is dissipated before it can reach the next maximum (i.e. point B). The dissipation torque will then vary significantly over a synodic rotation period, but the required condition is approximately given by

$$4\pi(|T_D| - A^2\Omega\dot{M}_p) \geq k_p^2 M_p R_p^2 \omega^2. \tag{7.11}$$

The value of ω satisfying (7.11) will lie in the regime of high field penetration, in which T_D is given by (4.77). The equality in (7.11) then gives a quadratic equation for ω and, for given orbital parameters, the condition for its roots to be real places a maximum value on the magnetic diffusivity of the secondary. To within a factor of $\sqrt{2}$, the expression so obtained for this limiting value of η is the same as that for η_0 given by (7.8). This is a result of the fact that (7.11) is essentially equivalent to the condition $t_s \lesssim 2\pi/\omega_1$.

If $\eta \gg \eta_0$ synchronism cannot be attained in the presence of accretion since the primary will always have too much synodic rotational energy to be dissipated in one synodic cycle. Then, despite the fact that a stable synchronous state exists at β_s, the lowest value attained by ω will be ω_1, as in the case when no locking mechanism is present. If $\eta < \eta_0$ then the primary will ultimately become trapped in the energy well about β_s and settle to the locked state.

7.3. Conclusions

The form of $W(\beta)$, and hence the above conclusions, will be unchanged for any torque $T_B(\beta)$ balancing accretion which has a vanishing integral around one cycle. If the balancing torque has a finite cyclic integral then the height difference of adjacent maxima in $W(\beta)$ would differ from that shown in Figure 7.2. In the special case of the cyclic integral of T_B exactly cancelling that of T_A there would be no height difference in adjacent maxima of W.

Synchronism would then ultimately result whether the primary was initially under or over-synchronous, without a limit on the value of η. However, apart from such a special coincidence, there would still be a limiting value of η necessary to ensure that the primary becomes trapped in the energy well about β_s.

It was shown in §6.3.3 that the characteristic ratio of the accretion torque to the magnetic torque, Q, is given by (6.66). For secondary fields of $(B_s)_0 \gtrsim 6 \times 10^3\,\mathrm{G}$ it follows that, for typical parameters, $Q \lesssim 10^{-2}$. In such cases, η_0, given by (7.8), becomes sufficiently large that the condition $\eta < \eta_0$ for the attainment of synchronism is likely to be satisfied. This corresponds to the heights of the above energy wells becoming nearly equal, so over-shooting from an over-synchronous state would not occur. It is conceivable that some AM Her systems may satisfy the conditions for attaining synchronism, whilst others may not. In the latter cases small degrees of asynchronism would result. Observations indicate that most AM Her systems are close to corotation, but changes in the longitude of the accreting pole are suggested in some systems (e.g. Bailey *et al*, 1993).

Although the two-dimensional case, with accretion in the orbital plane, was considered here, the conditions for attaining synchronism should be similar in the general three-dimensional case. In particular, a limiting value of η will generally result, with η_0 becoming large for $Q \ll 1$.

References

Bailey, J.A., Wickramasinghe, D.T., Ferrario, L., Hough, J.H. and Cropper, M.S., 1993. *Mon. Not. R. Astr. Soc.*, **261**, L31.

Biermann, P., Schmidt, G.D., Liebert, J., Stockman, H.S., Tapia, S., Kuhr, H., Strittmatter, P.A., West, S. and Lamb, D.Q., 1985. *Astrophys. J.*, **293**, 303.

Campbell, C.G., 1986. *Mon. Not. R. Astr. Soc.*, **219**, 589.

CHAPTER 8

BINARY STARS WITH PARTIAL ACCRETION DISCS

8.1. Introduction

The intermediate polars, and some X-ray binary pulsars, are believed to contain magnetic primary stars accreting from partially disrupted discs. The primary magnetic moments are weaker than those occurring in AM Her stars, so partial accretion discs can form. The accreting star has a stronger magnetic interaction with the differentially rotating disc than with the synchronized secondary. As a consequence, the spin behaviour of primary stars in the intermediate polars and X-ray binary pulsars is very different than in the AM Her binaries, where the absence of a disc enables orbital synchronism to be approached.

8.2. The Intermediate Polars

8.2.1. DISCOVERY AND CLASSIFICATION

Charles *et al* (1979) indentified TV Col as an X-ray source with a spectrum like AM Her, but lacking detectable polarization. There were several observed periodicities and some uncertainty in their origin. This was resolved by X-ray observations of Schrijver, Brinkman and van der Woerd (1987) which showed the primary rotation period to be 31.8 mins, and eclipse observations by Hellier, Mason and Mittaz (1991) giving the orbital period as 5 hrs 29.2 mins. Griffiths *et al* (1980) discovered AO Psc. Photometry by Warner (1980) showed a modulation of its spectrum with the orbital period of 3.6 hrs.

TV Col and AO Psc were the first two members of a class of cataclysmic variables now designated the intermediate polars. Photometric and spectroscopic observations led to a model in which a magnetic white dwarf accretes matter transferred from the secondary star via a partially disrupted disc. An X-ray emitting region on the white dwarf is carried around by its rotation, causing periodic illumination of some region fixed in the binary frame (e.g. Hutchings *et al*, 1981; Warner, O'Donoghue and Fairall, 1981; Hassall *et al*, 1981; Patterson and Price, 1981). In contrast to the polars, TV Col

TABLE 8.1. The Intermediate Polars

Name	P (hr)	P_R (min)	$\dot{P}_R/P_R$ (yr^{-1})	$(B_p)_0$ (MG)
GK Per	47.92	5.86	-2.2×10^{-6}	
V1062 Tau	9.95	62.00		
XY Ari	6.06	3.44		
TX Col	5.72	31.84		
TV Col	5.49	31.83		
PQ Gem	5.18	13.89		8
FO Aqr	4.85	20.91	8.6×10^{-7}	
YY Dra	3.91	8.82		
AO Psc	3.59	13.42	-2.6×10^{-6}	
V1223 Sgr	3.37	12.42	9.7×10^{-7}	
BG CMi	3.24	15.22	2.0×10^{-6}	3
EX Hya	1.63	67.03	-3.0×10^{-7}	

TABLE 8.2. The DQ Her Binaries

Name	P (hr)	P_R (min)	$\dot{P}_R/P_R$ (yr^{-1})
AE Aqr	9.88	0.55	5.4×10^{-8}
V533 Her	5.04	1.06	1.5×10^{-7}
DQ Her	4.65	1.18	-3.6×10^{-7}

and AO Psc showed no detectable polarization at optical wavelengths, indicating a weaker magnetic field. A weaker field and the presence of a partial accretion disc are consistent with the observed asynchronous rotation of the primary. Binaries currently classified as definite intermediate polars are hard X-ray emitters and multiperiodic. Table 8.1 lists the main systems in this class.

The DQ Her binaries are a subset of the intermediate polars. They have short primary rotation periods and lack hard X-ray emission. The main systems are shown in Table 8.2. These binaries played an important part in the early work on the nature of cataclysmic variables. AE Aqr was shown by Joy (1954) to be a spectroscopic binary. It was subsequently used by Crawford and Kraft (1956) towards deriving the basic model of a cataclysmic variable. DQ Her is the remnant of Nova Herculis 1934, and was identified by Walker (1954) as an eclipsing binary with an orbital period of 4 hr 39 min.

8.2.2. ROTATION PERIODS AND MAGNETIC FIELDS

Tables 8.1 and 8.2 show the primary star rotation periods, P_R, and their rates of change, expressed as $\dot{P}_R/P_R$. Rotation period changes have been measured in six intermediate polars: GK Per (Patterson, 1991), FO Aqr (e.g. Osborne and Mukai, 1989; Kruszewski and Semeniuk, 1993), AO Psc (Kaluzny and Semeniuk, 1988), V1223 Sgr (van Amerongen, Augusteijn and van Paradijs, 1987), BG CMi (e.g. Patterson and Thomas, 1993), and EX Hya (Vogt, Krzeminski and Sterken, 1980; Bond and Freeth, 1988). The magnitude of typical errors in the quoted values of $\dot{P}_R/P_R$ is $\sim 10\%$. The DQ Her period change observations are: AE Aqr (De Jager *et al*, 1994), V533 Her (Lamb and Patterson, 1983), and DQ Her (Balachandran, Robinson and Kepler, 1983). Both negative and positive values of $\dot{P}_R$ occur.

The white dwarf surface magnetic field values quoted in Table 8.1 are based on circular polarization measurements using broad band polarimetry (Cropper, 1986; Berriman, 1988; Stockman *et al*, 1992). The polarization of BG CMi can be modelled with an accretion arc covering a fractional area of ~ 0.01 of the primary's surface and a polar field of $(B_p)_0 \sim 3\,\text{MG}$ (Wickramasinghe, Wu and Ferrario, 1991). Polarization measurements of PQ Gem (Rosen, Mittaz and Hakala, 1993), and a similar accretion model, give $(B_p)_0 \sim 8\,\text{MG}$. This suggests the intermediate polars have white dwarf surface fields typically an order of magnitude lower than those in the AM Her binaries.

8.3. The X-Ray Binary Pulsars

8.3.1. DISCOVERY AND CLASSIFICATION

X-ray pulsations from an X-ray binary system were first discovered by Giacconi *et al* (1971) during observations of Centaurus X-3 with the Uhuru satellite. The binary nature of Cen X-3 was revealed in a further analysis of the Uhuru data by Schreier *et al* (1972). This showed 2.1 day periodic variations of X-ray intensity with an eclipse, and a sinusoidal variation of the 4.8 s pulsation period, also with a period of 2.1 d. Soon after, Tananbaum *et al* (1972) discovered another X-ray pulsar, Hercules X-1, which showed a periodic intensity variation with a 1.7 d binary period and an associated sinusoidal modulation of the 1.24 s pulse period.

The large luminosities observed can be understood as the gravitational energy released by matter transferred from a normal companion star onto a neutron star. The radius of a neutron star is typically a factor 10^{-3} smaller than that of a white dwarf, so the potential well into which accreting material falls is far deeper in these systems than in standard cataclysmic variables.

Orbital parameters have been determined for several systems from pulse timing analyses, using the Doppler effect due to the binary motion of the compact star. Optical observations of the companion star then lead to accurate values for the stellar parameters, confirming neutron star primaries. The pulse period histories reveal a general tendency for spin-up in the X-ray pulsars, contrary to the general spin-down of radio pulsars. This is consistent with X-ray systems accreting matter through the magnetosphere. However, recent observations have shown a wide variety of pulse period changes in the X-ray pulsars.

X-ray pulsars can be classified into three catagories; (I) binaries with early-type massive companions, (II) binaries with Be star companions, and (III) low mass binaries. Class (I) can be divided into two subclasses; (Ia) short pulse-period systems with very large X-ray luminosity, and (Ib) long pulse-period systems with moderate X-ray luminosity (e.g. Corbet, 1986). Some of the low mass binaries lack pulse periods.

8.3.2. SYSTEMS WITH ACCRETION DISCS

The short pulse-period systems with early-type massive companions and the low mass binaries are thought to have accretion discs around their neutron stars. Observations of Her X-1 in the X-ray and optical bands give direct evidence for an accretion disc (e.g. Middleditch, 1983). Observational evidence of the neutron star magnetic field was reported by Trumper *et al* (1978). They interpreted a prominant emission line feature at 58KeV in the energy spectrum of Her X-1 as cyclotron emission at the magnetic poles of the neutron star. The estimated surface field was 5×10^{12} G. White, Swank and Holt (1983) used cyclotron absorption features of 4U0115+63 to obtain a surface magnetic field of 10^{12} G. Kii *et al* (1986) proposed a magnetic field 8×10^{12} G for 4U1626−67 by simulating the observed energy dependence of the pulse profile with a calculation of anisotropic radiative transfer in a strong magnetic field. The nature of the pulse profiles indicates the presence of accretion columns on the neutron star. Like the intermediate polars, partial discs are believed to be present with material becoming field-channelled after disruption. Consequently, the neutron star cannot become synchronized with the orbital motion, since its magnetic interaction with the secondary star is much weaker than its coupling to the disc.

Table 8.3 shows a representative sample the X-ray binary pulsars believed to contain strongly magnetic neutron stars (i.e. $(B_{\rm p})_0 \sim 10^{12}$ G) accreting from partially disrupted discs. The systems listed are those most extensively observed. Typical rates of change of the neutron star's spin period are given, where available. The associated time-scales are generally shorter than those listed in Tables 8.1 and 8.2, since neutron stars have

TABLE 8.3. Representative Binary X-ray Pulsars believed to have accretion discs

System	P (days)	$P_{\rm R}$ (s)	$\dot{P}_{\rm R}/P_{\rm R}$ (yr^{-1})
GX 1+4	304	114	-2.1×10^{-2}
4U0115+63	24.31	3.61	-3.2×10^{-5}
SMC X-1	3.892	0.72	-6.0×10^{-4}
Cen X-3	2.087	4.84	-2.8×10^{-4}
Her X-1	1.700	1.24	-2.9×10^{-6}
LMC X-4	1.408	13.5	
GX 109.1−1.0	0.080	6.98	
4U1627−67	0.023	7.68	-2.0×10^{-4}

smaller moments of inertia than white dwarfs. It is noted that the unpulsed X-ray binaries are not of interest here, since they are thought to have significantly weaker magnetic fields ($\sim 10^9$ G) than the pulsed sources and consequently lack their magnetically confined hot accretion columns (Verbunt, 1993). Such primary stars would only be weakly magnetically coupled to the disc and secondary star, so the associated torques would have a negligible effect on their spin evolution.

8.3.3. NEUTRON STAR SPIN EVOLUTION

Pulse period monitoring of X-ray binary pulsars has been performed by various satellites, which reveal a wide variety of changes. From 1979 to 1988 the systems SMC X-1 and 4U1626−27 exhibited secular spin-up at almost constant rates of $\dot{P}_{\rm R}/P_{\rm R} = -6.0 \times 10^{-4}\,{\rm yr}^{-1}$ and $\dot{P}_{\rm R}/P_{\rm R} = -2.0 \times 10^{-4}\,{\rm yr}^{-1}$, respectively. Subsequently, Makishima *et al* (1988) used Ginga satellite observations to show that the spin-up rate of SMC X-1 was no longer constant, but decreasing. Between 1971 and 1980 a spin-up rate of $\dot{P}_{\rm R}/P_{\rm R} = -2.6 \times 10^{-2}\,{\rm yr}^{-1}$ was observed for GX 1+4. A short time-scale spin-up episode in this system, superposed on the constant decrease in $P_{\rm R}$, was noted by Doty, Hoffmann and Lewin (1981). Makishima *et al* (1988) found a spin-down in GX 1+4.

Schreier and Fabbiano (1976) noted a secular trend of spin-up for Her X-1 and Cen X-3, from early observations. These systems were subsequently observed to exhibit wavey fluctuations in their spin rates, with time-scales of years. Also, short-term fluctuations of the pulse period on a time-scale of days to months, including evidence of spin-down episodes, were found in Her X-1 (Giacconi, 1974) and Cen X-3 (Fabbiano and Schreier, 1977) from Uhuru satellite observations.

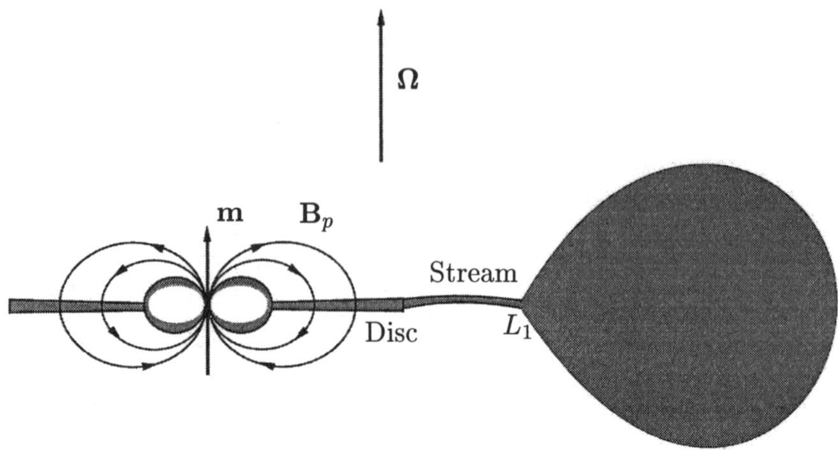

Figure 8.1. Model of an intermediate polar or disc-accreting X-ray binary pulsar. The primary star has a magnetic moment **m** and accretes material along magnetic field lines from the inner edge of the disrupted disc.

8.4. MHD Problems

The main MHD problem in the intermediate polars and the X-ray binary pulsars is the interaction of the primary star's magnetic field with the accretion disc. The stellar magnetic field will penetrate the highly diffusive disc and motions of the plasma across the field will generate electric currents. These currents flow in the disc and in the surrounding magnetosphere. In the outer parts of the disc, where the stellar field is weak, the magnetic force is small and the disc structure is essentially unperturbed. Closer to the primary star the magnetic perturbation becomes significant and ultimately disrupts the disc. The physical mechanism causing disruption is not obvious and needs detailed investigation. The disruption cannot be a simple dynamical effect due to increased magnetic pressure, since the dominant $B_\phi^2/2\mu_0$ term in the vertical equilibrium has a gradient acting to compress the disc. Disruption is found to be due to the effect the magnetic field has on the density variation, resulting from a change in the thermal balance. This causes thermal and viscous instability. Figure 8.1 illustrates the model of these systems.

Chapter 9 investigates the inner disruption of the disc, which requires consideration of a full MHD problem, including the thermal equations. The thin nature of the disc, and its high magnetic diffusivity, enable the equations to be reduced and render the problem tractable. The type of solution depends on the magnetic diffusion mechanism. The disruption radius depends on the stellar spin rate, since this affects the radial distribution of vertical shear in the disc.

Chapter 10 considers the magnetic interaction of the primary star with the undisrupted disc. Provided this part of the disc is weakly perturbed by the magnetic field, the kinematic problem of solving the induction equation using the unperturbed velocity is appropriate. The resulting magnetic stresses transmit a torque to the primary star enabling an exchange of angular momentum. Angular momentum is also transferred due to the magnetic channelling of material accreting onto the star. The interplay of these processes can explain the main spin behaviour observed in the intermediate polars and the X-ray binary pulsars, previously outlined. In particular, a stable equilibrium state can be found in which the disc torque balances the accretion torque. The resulting period expression fits the observations for reasonable parameters.

References

Balachandran, S., Robinson, E.L. and Kepler, S.O., 1983. *Publ. Astr. Soc. Pacific.*, **95**, 653.

Berriman, G., 1988. In *Polarized Radiation of Circumstellar Origin*, eds., Coyne, G.V., *et al.*, Vatican. Obs., Vatican., p281.

Bond, I.A. and Freeth, R.V., 1988. *Mon. Not. R. Astr. Soc.*, **232**, 753.

Charles, P.A., Thorstensen, J., Bowyer, S. and Middleditch, J., 1979. *Astrophys. J.*, **231**, L131.

Corbet, R.H.D., 1986. *Mon. Not. R. Astr. Soc.*, **220**, 1047.

Crawford, J.A. and Kraft, R.P., 1956. *Astrophys. J.*, **123**, 44.

Cropper, M.S., 1986. *Mon. Not. R. Astr. Soc.*, **222**, 225.

De Jager, O.C., Meintjes, P.J., O'Donoghue, D. and Robinson, E.L., 1994. *Mon. Not. R. Astr. Soc.*, **267**, 577.

Doty, J.P., Hoffman, J.A. and Lewin, W.H.G., 1981. *Astrophys. J.*, **243**, 257.

Fabbiano, G. and Schreier, E.J., 1977. *Astrophys. J.*, **214**, 235.

Giacconi, R., 1974. *Astrophysics and Gravitation* (Editions de l'Université de Bruxelles).

Giacconi, R., Gursky, H., Kellogg, E., Schreier, E. and Tananbaum, H., 1971. *Astrophys. J. Lett.*, **167**, L67.

Griffiths, R.E., Lamb, D.Q., Ward, M.J., Wilson, A.J., Charles, P.A., Thorstensen, J., McHardy, I.M. and Lawrence, A., 1980. *Mon. Not. R. Astr. Soc.*, **193**, 25P.

Hassall, B.J.M., Pringle, J.E., Ward, M.J., Whelan, J.A.J., Mayo, S.K., Echevarria, J., Jones, D.H.P., Wallis, R.E., Allen, D.A. and Hyland, A.R., 1981. *Mon. Not. R. Astr. Soc.*, **197**, 275.

Hellier, C., Mason, K.O. and Mittaz, J.P.D., 1991. *Mon. Not. R. Astr. Soc.*, **248**, 5P.

Hutchings, J.B., Crampton, D., Cowley, A.P., Thorstensen, J.R. and Charles, P.A., 1981. *Astrophys. J.*, **249**, 680.

Joy, A.H., 1954. *Astrophys. J.*, **120**, 377.

Kaluzny, J. and Semeniuk, I., 1988. *Info. Bull. Var. Stars.*, No. 3145.

Kii, T., Hayakawa, S., Nagase, F., Ikegami, T. and Kawai, N., 1986. *Publ. Astron. Soc. Japan.*, **38**, 751.

Kruszewski, A. and Semeniuk, I., 1993. *Acta. Astr.*, **43**, 127.

Lamb, D.Q. and Patterson, J., 1983. *Int. Astr. Union. Colloq.*, No. 72., p229.

Makishima, K., *et al*, 1988. *Nature*, **333**, 746.

Middleditch, J., 1983. *Astrophys. J.*, **275**, 278.

Osborne, J.P. and Mukai, K., 1989. *Mon. Not. R. Astr. Soc.*, **238**, 1233.

Patterson, J., 1991. *Publ. Astr. Soc. Pacific.*, **103**, 1149.
Patterson, J. and Price, C., 1981. *Astrophys. J.*, **243**, L83.
Patterson, J. and Thomas, G., 1993. *Publ. Astr. Soc. Pacific.*, **105**, 59.
Rosen, S.R., Mittaz, J.P.D. and Hakala, P.J., 1993. *Mon. Not. R. Astr. Soc.*, **264**, 171.
Schreier, E.J. and Fabbiano, G., 1976. *X-Ray Binaries*, NASA SP-389 (National Technical Information Service, Springfield, Virginia).
Schreier, E., Levinson, R., Gursky, H., Kellogg, E., Tananbaum, H. and Giacconi, R., 1972. *Astrophys. J. Lett.*, **172**, L79.
Schrijver, J., Brinkman, A.C. and van de Woerd, H., 1987. *Astrophys. Sp. Sci.*, **130**, 261.
Stockman, H.S., Schmidt, G.D., Berriman, G., Liebert, J., Moore, R.L. and Wickramasinghe, D.T., 1992. *Astrophys. J.*, **401**, 628.
Tananbaum, H., Gursky, H., Kellogg, E.M., Levinson, R., Schreier, E. and Giacconi, R., 1972. *Astrophys. J. Lett.*, **174**, L143.
Trumper, J., Pietsch, W., Reppin, C., Voges, W., Staubert, R. and Kendziorra, E., 1978. *Astrophys. J. Lett.*, **219**, L105.
van Amerongen, S., Augusteijn, T. and van Paradijs, J., 1987. *Mon. Not. R. Astr. Soc.*, **229**, 245.
Verbunt, F., 1993. *Ann. Rev. Astron. Astrophys.*, **31**, 93.
Vogt, N., Krzeminski, W. and Sterken, C., 1980. *Astron. Astrophys.*, **85**, 106.
Walker, M.F., 1954. *Publ. Astr. Soc. Pacific.*, **66**, 230.
Warner, B., 1980. *Mon. Not. R. Astr. Soc.*, **190**, 69P.
Warner, B., O'Donoghue, D. and Fairall, A.P., 1981. *Mon. Not. R. Astr. Soc.*, **196**, 705.
White, N.E., Swank, J.H. and Holt, S.S., 1983. *Astrophys. J.*, **270**, 711.
Wickramasinghe, D.T., Wu, K. and Ferrario, L., 1991. *Mon. Not. R. Astr. Soc.*, **249**, 460.

CHAPTER 9

BINARIES WITH PARTIAL DISCS: DISC STRUCTURE

9.1. Introduction

The aim of this chapter is to investigate the effect of the stellar magnetic field on the structure of the disc in an intermediate polar or an X-ray binary pulsar, and to understand how disruption occurs. This will also be relevant to the absence of discs in AM Her systems. Early considerations of this problem took the magnetic field to be excluded from the disc, and calculated the resulting external field structure (e.g. Aly, 1980; Kundt and Robnik, 1980; Anzer and Borner, 1983). However, as Ghosh and Lamb (1978) pointed out, turbulent motions and instabilities will allow the stellar field to penetrate the disc, and hence be affected by its motions. Ghosh and Lamb (1979) considered the toroidal field, B_ϕ, produced as a result of the vertical shear between the stellar magnetosphere and the disc. However, Wang (1987) showed that their solution does not satisfy the induction equation; also they do not account for the effect of the magnetic force on the vertical equilibrium.

The effect of the stellar magnetic field on the structure of the accretion disc was considered by Campbell (1992). The unperturbed disc is taken to have the standard structure, with viscosity accounting for angular momentum transport. The disc is surrounded by the primary star's magnetic field and associated magnetosphere. The main uncertainty is the nature of the disc's magnetic diffusivity η. In a turbulent disc η is likely to be at least as large as the coefficient of viscosity, ν. Magnetic buoyancy is also likely to make a major contribution to η. If η had values significantly less than those characteristic of such processes, large values of B_ϕ would result and the ensuing instabilities would diffuse the field, effectively enhancing η. Several forms of η are considered, in order to investigate the sensitivity of the disc's structure to the nature of η.

The stellar magnetic field penetrates the disc and is modified by the vertical shear and inflow. Electric currents flow in the disc and magnetosphere and a $\mathbf{J} \wedge \mathbf{B}$ force is exerted on material. Far from the primary star the magnetic force is weak and the disc will have essentially its unperturbed structure, with viscosity causing the inflow. Closer to the magnetic accretor the toroidal magnetic force, $F_{\mathrm{m}\phi}$, on the disc material will approach that due to viscosity, $F_{\mathrm{v}\phi}$, and ultimately exceeds it. If the disc remains undisrupted when $|F_{\mathrm{m}\phi}| \gg |F_{\mathrm{v}\phi}|$ then the inflow will be caused by the continuous transfer of angular momentum from the disc to the star, via the stressed magnetic field. This transfer requires the angular velocity of disc material to exceed that of the magnetosphere. Vertical equilibrium in the disc must break down at some inner radius, and the flow is subsequently channelled by the magnetic field onto the star. This chapter examines the strongly magnetically affected inner regions of the disc, in order to explain disruption.

Section 9.2 considers the inner part of the disc, where the magnetic perturbations are large. The thin nature of the disc allows some simplification of the equations. A solution can be found for the inner magnetically controlled region, provided the magnetic diffusivity, η, is unspecified. However, when η is specified by a parameterized turbulent form, or a buoyancy form, the disc ends as soon as the magnetic removal of angular momentum becomes comparable to its viscous advection. An expression can be derived for the inner disruption radius of the disc. Section 9.3 considers numerical integration of the disc equations and shows that the effect of the magnetic field on the inner regions of the disc results in viscous instability. The results are discussed in §9.4.

9.2. The Inner Disc Structure

9.2.1. THE MAGNETIC DISC EQUATIONS

The primary star is taken to have a centred dipolar field. The axisymmetric case is considered in which the dipole moment is perpendicular to the orbital plane. Cylindrical polar coordinates (ϖ, ϕ, z) are used, centred on the star.

The steady state equations of momentum, induction and continuity are

$$(\mathbf{v} \cdot \nabla)\mathbf{v} = -\frac{1}{\rho}\nabla P - \nabla\psi + \frac{1}{\mu_0 \rho}(\nabla \wedge \mathbf{B}) \wedge \mathbf{B} + \frac{1}{\rho}\mathbf{F}_{\mathrm{v}}, \tag{9.1}$$

$$\nabla \wedge (\mathbf{v} \wedge \mathbf{B}) - \nabla \wedge (\eta \nabla \wedge \mathbf{B}) = \mathbf{0}, \tag{9.2}$$

$$\nabla \cdot (\rho \mathbf{v}) = 0, \tag{9.3}$$

where ψ is the gravitational potential, $\mathbf{F}_{\mathrm{v}}$ the viscous force per unit volume, and the other symbols have their usual meanings. It is noted that any dynamo generated fields are ignored here. As will be seen in Chapter 11, typical dynamo field strengths are $\sim 10^3$ G. At fields much larger than this the dynamo saturates. In the problem considered here the poloidal field has the primary star as its source. In the inner regions of the disc, where magnetic angular momentum transfer is significant, the field values far exceed those of any dynamo.

The ϖ, ϕ and z-components of (9.1) can be written

$$\frac{v_\phi^2}{\varpi} = \frac{\partial \psi}{\partial \varpi} + \frac{\partial}{\partial \varpi}\left(\frac{v_\varpi^2}{2}\right) + v_z \frac{\partial v_\varpi}{\partial z} + \frac{1}{\rho}\frac{\partial P}{\partial \varpi} + \frac{1}{\varpi^2 \rho}\frac{\partial}{\partial \varpi}\left(\frac{\varpi^2 B_\phi^2}{2\mu_0}\right) - \frac{B_z J_\phi}{\rho}, \tag{9.4}$$

$$\frac{v_\varpi}{\varpi}\frac{\partial}{\partial \varpi}(\varpi^2 \Omega) + \frac{v_z}{\varpi}\frac{\partial}{\partial z}(\varpi^2 \Omega) = \frac{B_\varpi}{\mu_0 \varpi \rho}\frac{\partial}{\partial \varpi}(\varpi B_\phi) + \frac{B_z}{\mu_0 \rho}\frac{\partial B_\phi}{\partial z} + \frac{1}{\varpi^2 \rho}\frac{\partial}{\partial \varpi}\left(\rho \nu \varpi^3 \frac{\partial \Omega}{\partial \varpi}\right), \tag{9.5}$$

$$\frac{\partial \psi}{\partial z} + v_\varpi \frac{\partial v_z}{\partial \varpi} + \frac{\partial}{\partial z}\left(\frac{v_z^2}{2}\right) + \frac{1}{\rho}\frac{\partial}{\partial z}\left(P + \frac{B_\phi^2}{2\mu_0}\right) + \frac{B_\varpi J_\phi}{\rho} = 0, \tag{9.6}$$

where ν is the coefficient of viscosity, $\Omega = v_\phi/\varpi$, and the toroidal current density is

$$J_\phi = \frac{1}{\mu_0}\left(\frac{\partial B_\varpi}{\partial z} - \frac{\partial B_z}{\partial \varpi}\right). \tag{9.7}$$

Since the disc's mass is much less than that of the primary star, ψ can be taken as

$$\psi = -\frac{GM_{\mathrm{p}}}{(\varpi^2 + z^2)^{\frac{1}{2}}}, \tag{9.8}$$

where M_{p} is the stellar mass.

The poloidal and toroidal components of (9.2) yield

$$v_\varpi B_z - v_z B_\varpi + \mu_0 \eta J_\phi = 0, \tag{9.9}$$

$$\frac{\eta}{\varpi}\left(\nabla^2 B_\phi - \frac{B_\phi}{\varpi^2}\right) + \frac{1}{\varpi^2}\frac{d\eta}{d\varpi}\frac{\partial}{\partial \varpi}(\varpi B_\phi) = -\mathbf{B}_p \cdot \nabla \Omega + \nabla \cdot \left(\frac{B_\phi}{\varpi}\mathbf{v}_p\right), \tag{9.10}$$

where the subscript p denotes a poloidal component.

The continuity equation (9.3) is

$$\frac{1}{\varpi}\frac{\partial}{\partial \varpi}(\varpi \rho v_\varpi) + \frac{\partial}{\partial z}(\rho v_z) = 0. \tag{9.11}$$

It is noted that since the disc surfaces are $z = \pm h(\varpi)$, terms involving v_z have been included in the above equations. The significance of these terms will be investigated.

Consider, first, the poloidal magnetic field. Outside the disc, in the primary star's highly conducting magnetosphere, η essentially vanishes and (9.2) becomes

$$\nabla \wedge (\mathbf{v} \wedge \mathbf{B}) = \mathbf{0}.$$

This gives

$$v_\phi - \varpi\alpha = (\hat{\mathbf{B}}_p \cdot \mathbf{v}_p)\frac{B_\phi}{|B_p|}, \tag{9.12}$$

which is a generalization of the isorotation law, where α is constant along the field. Hence $\Omega \simeq \alpha$ holds provided that

$$\frac{|v_p||B_\phi|}{|v_\phi||B_p|} \ll 1. \tag{9.13}$$

For sufficiently strong $|B_p|$ material nearly corotates with the primary star, so $\alpha \simeq \Omega_\mathrm{p}$. Since exact corotation cannot occur, the stress B_pB_ϕ is finite at the stellar surface leading to a torque on the primary star.

The unperturbed stellar poloidal field has the dipole components

$$B_\varpi = \frac{3}{2}(B_\mathrm{p})_0 R_\mathrm{p}^3 \frac{\varpi z}{(\varpi^2 + z^2)^{\frac{5}{2}}}, \tag{9.14}$$

$$B_z = -\frac{1}{2}(B_\mathrm{p})_0 R_\mathrm{p}^3 \frac{(\varpi^2 - 2z^2)}{(\varpi^2 + z^2)^{\frac{5}{2}}}, \tag{9.15}$$

where R_p and $(B_\mathrm{p})_0$ are the primary star's radius and surface polar magnetic field. The stellar field penetrates the disc, in which the density rises to a maximum in the central plane, and is modified by its motions. Equation (9.9) shows that the inflow $\mathbf{v}_p$ across $\mathbf{B}_p$ generates the toroidal current density J_ϕ. If $\mathbf{v}_p$ were effectively vanishingly small then, for finite η, it follows that $J_\phi = 0$ and the stellar poloidal field would penetrate the disc unmodified. The finite inflow through the disc perturbs $\mathbf{B}_p$, (9.9) giving

$$v_\varpi B_z - v_z B_\varpi + v_\varpi B_z' - v_z B_\varpi' = -\mu_0 \eta J_\phi', \tag{9.16}$$

where B_ϖ and B_z are given by (9.14) and (9.15), and primes denote perturbations. It can now be shown that the poloidal field perturbation is small.

Since $\nabla \cdot \mathbf{B}'_p = 0$ for an axisymmetric field, it follows that the components of $\mathbf{B}'_p$ can be expressed in terms of an azimuthal vector potential $A'\hat{\boldsymbol{\phi}}$ as

$$B'_\varpi = -\frac{\partial A'}{\partial z}, \qquad B'_z = \frac{1}{\varpi}\frac{\partial}{\partial \varpi}(\varpi A'). \tag{9.17a,b}$$

The corresponding toroidal current density is

$$J'_\phi = \frac{1}{\mu_0}\left(\frac{\partial B'_\varpi}{\partial z} - \frac{\partial B'_z}{\partial \varpi}\right), \tag{9.18}$$

and hence is related to A' by

$$\nabla^2 A' - \frac{A'}{\varpi^2} = -\mu_0 J'_\phi. \tag{9.19}$$

Large distortions of $\mathbf{B}_p$ require poloidal flows having a kinetic energy density at least comparable to the poloidal magnetic energy density, that is $\rho v_p^2 \gtrsim B_p^2/\mu_0$. In the inner regions, where the stellar magnetic field is strongest, poloidal flows in the low density magnetosphere will not have sufficient ρv_p^2 to cause large $\mathbf{B}'_p$. Hence, in these regions, any large distortion of $\mathbf{B}_p$ would have to be caused by the disc inflow.

For small poloidal flows in the magnetosphere above and below an undisrupted disc, the toroidal current density J'_ϕ will be concentrated in the disc. The magnitude of $|\mathbf{B}'_p|$ in the disc can be investigated using (9.16). Since the unperturbed stellar field is largely vertical for $|z| \leq h$, the relevant quantity is $|B'_z|/|B_z|$. It follows from (9.17) and (9.19) that $\mathbf{B}'_p$ has the vertical length-scale of A' and J'_ϕ, which is h, and hence in the disc (9.18) becomes

$$J'_\phi = \frac{1}{\mu_0}\frac{\partial B'_\varpi}{\partial z}, \tag{9.20}$$

since $h/\varpi \ll 1$. Considering terms on the left hand side of (9.16), equation (9.11) gives

$$|v_z| \sim (h/\varpi)|v_\varpi|. \tag{9.21}$$

Equations (9.14) and (9.15) yield $|B_z| \sim (\varpi/h)|B_\varpi|$ in the disc and hence

$$|v_\varpi B_z| \sim (\varpi/h)^2 |v_z B_\varpi|,$$

so $v_z B_\varpi$ is negligible. Equations (9.17) and (9.21) show that $v_\varpi B'_z$ and $v_z B'_\varpi$ are comparable so (9.16) becomes

$$v_\varpi B_z + v_\varpi B'_z - v_z B'_\varpi = -\mu_0 \eta J'_\phi. \tag{9.22}$$

For $|B_z'| \lesssim |B_z|$, equations (9.17), (9.20) and (9.22) give

$$\frac{|B_z'|}{|B_z|} \sim \frac{|v_\varpi|}{(\eta/\varpi)} \left(\frac{h}{\varpi}\right)^2 . \tag{9.23}$$

In the absence of a magnetic field, viscosity causes the inflow and (9.5) yields

$$|v_{\mathrm{v}\varpi}| \sim \nu/\varpi, \tag{9.24}$$

so, from (9.23), the condition $|B_z'| \ll |B_z|$ can be written

$$|v_\varpi| \ll \frac{\eta}{\nu} \left(\frac{\varpi}{h}\right)^2 |v_{\mathrm{v}\varpi}|. \tag{9.25}$$

For $\eta \gtrsim \nu$, very large magnetically driven inflow speeds are needed to violate this condition, so $|B_z'|$ will be a small perturbation. The consistency of (9.25) will be checked. To first order in z/ϖ, (9.14) and (9.15) therefore give the poloidal field in the disc as

$$B_\varpi = \frac{3}{2}(B_{\mathrm{p}})_0 \left(\frac{R_{\mathrm{p}}}{\varpi}\right)^3 \frac{z}{\varpi}, \tag{9.26}$$

$$B_z = -\frac{1}{2}(B_{\mathrm{p}})_0 \left(\frac{R_{\mathrm{p}}}{\varpi}\right)^3 . \tag{9.27}$$

The result that the poloidal field is predominantly vertical in the disc is also found by Lubow, Papaloizou and Pringle (1994) and Bardou and Heyvaerts (1996).

In order to investigate how disruption occurs, the inner region of the disc will be considered. The question arises as to whether an inner part of the disc can exist in which magnetic extraction of angular momentum dominates that due to viscous advection. This will be defined by

$$|F_{\mathrm{m}\phi}| \geq 10|F_{\mathrm{v}\phi}|. \tag{9.28}$$

Equations are now derived for such an inner region.

The azimuthal magnetic force consists of the last two terms on the right hand side of (9.5). Use of (9.26) and (9.27) shows that the first magnetic term is a factor $\sim (h/\varpi)^2$ smaller than the second, so (9.5) becomes

$$\frac{v_\varpi}{\varpi}\frac{\partial}{\partial \varpi}(\varpi^2\Omega) + \frac{v_z}{\varpi}\frac{\partial}{\partial z}(\varpi^2\Omega) = \frac{B_z}{\mu_0 \rho}\frac{\partial B_\phi}{\partial z}. \tag{9.29}$$

Equation (9.10) describes the diffusion, creation and poloidal advection of B_ϕ. Since B_ϕ has a vertical length-scale of h, the thin disc condition means that the vertical derivative term will be dominant on the left hand side. The ratios of the diffusion to the poloidal advection terms are therefore

$$\frac{|\eta\partial^2 B_\phi/\partial z^2|}{|\partial(v_\varpi B_\phi)/\partial\varpi|} \sim \frac{|\eta\partial^2 B_\phi/\partial z^2|}{|\partial(v_z B_\phi)/\partial z|} \sim \frac{(\eta/\varpi)}{|v_\varpi|}\left(\frac{\varpi}{h}\right)^2 \sim \frac{|B_z|}{|B_z'|},$$

where the last relation follows from (9.23). Since $|B_z'| \ll |B_z|$, vertical diffusion of B_ϕ dominates its poloidal advection. Use of (9.26) and (9.27) in the creation term yields

$$\frac{|B_\varpi\partial\Omega/\partial\varpi|}{|B_z\partial\Omega/\partial z|} \sim \left(\frac{h}{\varpi}\right)^2 \ll 1, \tag{9.30}$$

and hence (9.10) reduces to

$$\eta\frac{\partial^2 B_\phi}{\partial z^2} = -\varpi B_z\frac{\partial\Omega}{\partial z}. \tag{9.31}$$

This equation gives the ratio of the dominant azimuthal magnetic force term to the viscous force term in (9.5) as

$$\frac{|F_{\mathrm{m}\phi}|}{|F_{\mathrm{v}\phi}|} \sim \frac{B_z^2/(\mu_0\rho)}{(\eta/\varpi)(\nu/\varpi)}. \tag{9.32}$$

The magnetically controlled inflow speed can be estimated from (9.29) and (9.31), which give

$$|v_\varpi| \sim \frac{B_z^2/(\mu_0\rho)}{(\eta/\varpi)}. \tag{9.33}$$

Consider, now, the ϖ-momentum given by (9.4). Expanding (9.8) to first order in z/ϖ yields

$$\frac{\partial\psi}{\partial\varpi} = \frac{v_{\mathrm{K}}^2}{\varpi}, \tag{9.34}$$

where the Keplerian speed is

$$v_{\mathrm{K}} = \left(\frac{GM_{\mathrm{p}}}{\varpi}\right)^{\frac{1}{2}}. \tag{9.35}$$

It follows that

$$\frac{|\partial(v_\varpi^2/2)/\partial\varpi|}{|\partial\psi/\partial\varpi|} \sim \frac{|v_z\partial v_\varpi/\partial z|}{|\partial\psi/\partial\varpi|} \sim \left(\frac{v_\varpi}{v_{\mathrm{K}}}\right) \ll 1, \tag{9.36}$$

where the first relation derives from (9.21) and the inequality arises for inflow speeds significantly less than free-fall values. In the central plane $z = 0$, B_ϕ and v_z vanish, so (9.4) becomes

$$\frac{v_\phi^2(\varpi, 0)}{\varpi} = \frac{v_{\rm K}^2}{\varpi} + \frac{1}{\rho_c}\frac{dP_c}{d\varpi} - \frac{B_z J'_\phi}{\rho_c}. \tag{9.37}$$

Use of (9.22) and (9.33) gives

$$\frac{|B_z J'_\phi|}{\rho_c} \sim \frac{v_\varpi^2}{\varpi}, \tag{9.38}$$

so this term can be dropped in (9.37) giving

$$\frac{v_\phi^2(\varpi, 0)}{\varpi} = \frac{v_{\rm K}^2}{\varpi} + \frac{1}{\rho_c}\frac{dP_c}{d\varpi}. \tag{9.39}$$

The perfect gas equation yields

$$\frac{v_\phi^2(\varpi, 0)}{v_{\rm K}^2} \sim 1 - \frac{\Re T_c}{\mu v_{\rm K}^2},$$

where $\Re$ is the gas constant and μ the mean molecular weight. For a central temperature $T_c \lesssim 10^5$ K the last term is $\lesssim 10^{-4}$ and hence

$$v_\phi(\varpi, 0) = v_{\rm K} = \left(\frac{GM_{\rm p}}{\varpi}\right)^{\frac{1}{2}}. \tag{9.40}$$

For small poloidal flows above and below the disc, (9.12) shows that a highly conducting magnetosphere will be essentially corotating with the primary star. The angular velocity of disc material will therefore change from its Keplerian value in the central plane to the stellar value near the surfaces $z = \pm h(\varpi)$. Consequently, there is a large vertical shear $\partial\Omega/\partial z$ in the disc where η is large. The following forms for Ω and η are used to describe this situation;

$$\Omega(\varpi, z) = \begin{cases} \Omega_{\rm K} = (GM_{\rm p}/\varpi^3)^{\frac{1}{2}}, & |z| < h - \Delta, \\ \Omega_{\rm p}, & |z| > h - \Delta, \end{cases} \tag{9.41a}$$

$$\eta = \begin{cases} \eta(\varpi), & |z| < h, \\ \to 0, & |z| > h, \end{cases} \tag{9.41b}$$

where $\Delta/h \ll 1$. These forms are obviously idealizations; the vertical shear is localized just beneath the disc surface, but η is large here so the essential physics is accounted for. The vertical problem is rendered mathematically tractable and physically tenable solutions result. In particular, although (9.41a) concentrates $\partial\Omega/\partial z$ near $z = \pm h$, toroidal magnetic field results throughout the disc when (9.31) is solved. The solution for B_ϕ involves integration over the sources and $B_z(\partial\Omega/\partial z)$ yields the dominant contribution, consistent with (9.30).

Finally, consider the vertical momentum given by (9.6). Use of equations (9.22), (9.24), (9.26), (9.27) and (9.31)–(9.33) yields

$$\frac{B_\varpi J'_\phi}{|\partial(B_\phi^2/2\mu_0)/\partial z)|} \sim \left(\frac{v_\varpi}{v_{\rm K}}\right)^2 \frac{|F_{{\rm v}\phi}|}{|F_{{\rm m}\phi}|} \ll 1. \tag{9.42}$$

Equations (9.8), (9.31) and (9.32) give

$$\frac{\rho|\partial\psi/\partial z|}{|\partial(B_\phi^2/2\mu_0)/\partial z|} \sim \frac{2\mu_0\rho v_{\rm K}^2}{B_\phi^2}\left(\frac{h}{\varpi}\right)^2 \sim \frac{|F_{{\rm v}\phi}|}{|F_{{\rm m}\phi}|} \ll 1. \tag{9.43}$$

It follows from (9.8) and (9.21) that

$$\frac{|v_\varpi(\partial v_z/\partial\varpi)|}{|\partial\psi/\partial z|} \sim \frac{|\partial(v_z^2/2)/\partial z|}{|\partial\psi/\partial z|} \sim \left(\frac{v_\varpi}{v_{\rm K}}\right)^2 \ll 1. \tag{9.44}$$

Equation (9.6) therefore reduces to

$$\frac{\partial}{\partial z}\left(P + \frac{B_\phi^2}{2\mu_0}\right) = 0. \tag{9.45}$$

9.2.2. VERTICAL EQUILIBRIUM AND ANGULAR MOMENTUM ADVECTION

For convenience, the previously derived equations describing the magnetically strong inner region of the disc are gathered together;

$$\Omega = \begin{cases} \Omega_{\rm K} = (GM_{\rm p}/\varpi^3)^{\frac{1}{2}}, & |z| < h - \Delta, \\ \Omega_{\rm p}, & |z| > h - \Delta, \end{cases} \tag{9.46}$$

$$\frac{v_\varpi}{\varpi}\frac{\partial}{\partial\varpi}(\varpi^2\Omega) + \frac{v_z}{\varpi}\frac{\partial}{\partial z}(\varpi^2\Omega) = \frac{B_z}{\mu_0\rho}\frac{\partial B_\phi}{\partial z}, \tag{9.47}$$

$$\frac{\partial}{\partial z}\left(P + \frac{B_\phi^2}{2\mu_0}\right) = 0, \tag{9.48}$$

$$\eta \frac{\partial^2 B_\phi}{\partial z^2} = -\varpi B_z \frac{\partial \Omega}{\partial z}, \tag{9.49}$$

$$\frac{1}{\varpi}\frac{\partial}{\partial \varpi}(\varpi \rho v_\varpi) + \frac{\partial}{\partial z}(\rho v_z) = 0, \tag{9.50}$$

$$B_z = -\frac{B}{2}\left(\frac{R_\mathrm{p}}{\varpi}\right)^3, \tag{9.51}$$

$$\eta = \begin{cases} \eta(\varpi), & |z| < h, \\ \to 0, & |z| > h. \end{cases} \tag{9.52}$$

Equation (9.49) describes the balance between diffusion of the toroidal magnetic field and its creation due to the vertical shearing of poloidal field. Using (9.46) for Ω gives

$$\frac{\partial^2 B_\phi}{\partial z^2} = -\frac{\varpi B_z}{\eta}\left[(\Omega_\mathrm{p} - \Omega_\mathrm{K})\delta(z - h + \Delta) + (\Omega_\mathrm{K} - \Omega_\mathrm{p})\delta(z + h - \Delta)\right], \tag{9.53}$$

where δ is the Dirac delta function. Since the vertical shear is antisymmetric about the $z = 0$ plane, while B_z is symmetric, it follows that B_ϕ is antisymmetric and hence

$$B_\phi(\varpi, 0) = 0. \tag{9.54}$$

The magnetic field in the low density magnetosphere is taken to be essentially force-free (see §2.2.4), so an axisymmetric field obeys the equation

$$\mathbf{B}_p \cdot \nabla(\varpi B_\phi) = 0. \tag{9.55}$$

Close to the disc surface this yields

$$\frac{\partial B_\phi}{\partial z} = 0, \quad h < |z| \ll \varpi. \tag{9.56}$$

The toroidal field distribution satisfying (9.53) and (9.56) in their respective regions, and continuous at h, is

$$B_\phi(\varpi, z) = \frac{\varpi B_z}{\eta}(\Omega_\mathrm{p} - \Omega_\mathrm{K})z, \quad |z| < h, \tag{9.57a}$$

$$B_\phi = B_\phi^+(\varpi) = B_\phi(\varpi, h), \quad h \le |z| \ll \varpi, \tag{9.57b}$$

where the distinction between h and $h-\Delta$ is now dropped, since $\Delta/h \ll 1$.

The vertical equilibrium in the disc is described by (9.48). Integrating this from 0 to z, and using (9.54), gives

$$P + \frac{B_\phi^2}{2\mu_0} = P_c(\varpi),$$

where $P_c(\varpi) = P(\varpi, 0)$. The thermal pressure is relatively small in the magnetosphere, so P is taken to vanish at the disc surface. The vertical equilibrium then becomes

$$P_c = \frac{(B_\phi^+)^2}{2\mu_0}, \tag{9.58}$$

where $B_\phi^+ = B_\phi(\varpi, h)$.

The mass inflow can be related to the magnetic stress by considering (9.47) and (9.50). Combining these equations, using the z-independence of B_z in the disc, yields

$$\frac{\partial}{\partial \varpi}(\varpi \rho v_\varpi \varpi^2 \Omega) + \frac{\partial}{\partial z}(\varpi \rho v_z \varpi^2 \Omega) = \frac{\varpi^2}{\mu_0}\frac{\partial}{\partial z}(B_z B_\phi). \tag{9.59}$$

Integrating this vertically through the disc, using the antisymmetry of B_ϕ and the fact that ρ is essentially zero at $z = \pm h$, gives

$$-\dot{M}\frac{d}{d\varpi}(\varpi^2 \Omega_{\mathrm{K}}) = \frac{4\pi}{\mu_0}\varpi^2 B_z B_\phi^+, \tag{9.60}$$

where

$$\dot{M} = -4\pi \int_0^h \varpi \rho v_\varpi dz \tag{9.61}$$

is the mass inflow rate, obtained by integrating (9.50) using the surface condition on ρ. Equation (9.60) relates the advection of angular momentum to the surface magnetic stress. Angular momentum is continually removed from the disc and transferred to the primary star, via the magnetosphere, allowing matter to spiral inwards. A ring of disc material experiences a magnetic torque, causing a rate of change of its angular momentum, while the primary star feels an opposite torque due to its coupling to the ring. The low density magnetosphere facilitates this coupling by allowing electric currents to flow between the disc and star.

9.2.3. DISSIPATION AND RADIATIVE TRANSFER

The electric currents induced in the disc are dissipated, producing a heat source. The dissipation rate per unit volume is

$$Q_{\mathrm{m}} = \mu_0 \eta \mathbf{J}^2. \tag{9.62}$$

The poloidal components of the current density are

$$J_\varpi = -\frac{1}{\mu_0}\frac{\partial B_\phi}{\partial z}, \qquad J_z = \frac{1}{\mu_0 \varpi}\frac{\partial}{\partial \varpi}(\varpi B_\phi), \tag{9.63}$$

so it follows that in the disc

$$|J_\varpi| \sim \left(\frac{\varpi}{h}\right)|J_z|,$$

and hence J_ϖ dominates J_z. Equations (9.22), (9.49) and (9.63) give

$$\frac{|J_\varpi|}{|J'_\phi|} \sim \frac{\eta |B_\phi|}{h v_\varpi B_z} \sim \frac{|v_\phi|}{|v_\varpi|} \gg 1. \tag{9.64}$$

The heat source, given by (9.62), therefore becomes

$$Q_\mathrm{m} = \frac{\eta}{\mu_0}\left(\frac{\partial B_\phi}{\partial z}\right)^2,$$

and use of (9.57a) yields

$$Q_\mathrm{m} = \frac{\varpi^2 B_z^2 (\Omega_\mathrm{p} - \Omega_\mathrm{K})^2}{\mu_0 \eta}. \tag{9.65}$$

The sum of the heat fluxes through the upper and lower disc faces is

$$2F_+ = \int_{-h}^{h} Q_\mathrm{m} dz,$$

and hence

$$F_+ = \frac{\varpi^2 B_z^2 (\Omega_\mathrm{p} - \Omega_\mathrm{K})^2 h}{\mu_0 \eta}. \tag{9.66}$$

For an optically thick disc radiative diffusion holds, so the heat flux is related to the temperature by

$$F = -\frac{4\sigma_\mathrm{B}}{3\kappa\rho}\frac{\partial}{\partial z}(T^4), \tag{9.67}$$

where σ_B is the Stefan-Boltzmann constant and κ is the opacity. The opacity is taken as that due to free-free and bound-free transitions, in its simplest Kramers form

$$\kappa = K\rho T^{-\frac{7}{2}}, \tag{9.68}$$

where K is a constant (e.g. Cox and Giuli, 1968). Integrating (9.67) vertically, taking

$$\int_0^h F\kappa\rho dz \simeq F_+\kappa_c\rho_c h,$$

and noting that $T_c^4 \gg T_+^4$, gives

$$T_c^4 = \frac{3\tau}{4\sigma_{\rm B}} F_+, \tag{9.69}$$

where the optical depth

$$\tau = \int_0^h \kappa\rho dz \simeq K\rho_c^2 T_c^{-\frac{7}{2}} h. \tag{9.70}$$

Provided the disc is disrupted before radiation pressure becomes important, the equation of state can be taken as

$$P = \frac{\Re}{\mu}\rho T. \tag{9.71}$$

Combining (9.69)–(9.71) yields

$$\rho_c = \frac{(\mu/\Re)^{\frac{15}{19}} P_c^{\frac{15}{19}}}{[(3K/4\sigma_{\rm B})hF_+]^{\frac{2}{19}}}, \tag{9.72}$$

where, by virtue of (9.57b) and (9.66),

$$hF_+ = \frac{(B_\phi^+)^2\eta}{\mu_0}. \tag{9.73}$$

9.2.4. DISC SOLUTION WITH UNSPECIFIED η

If the magnetic diffusivity is taken as an unspecified function, $\eta(\varpi)$, a solution for the inner disc structure can be found. Equations (9.46), (9.51) and (9.60) give the toroidal magnetic field on the disc surface as

$$B_\phi^+ = \frac{\mu_0 (GM_{\rm p})^{\frac{1}{2}} \dot{M}}{4\pi (B_{\rm p})_0 R_{\rm p}^3} \varpi^{\frac{1}{2}}. \tag{9.74}$$

The central thermal pressure then follows from (9.58) as

$$P_c = \frac{\mu_0 G M_{\rm p} \dot{M}^2}{32\pi^2 (B_{\rm p})_0^2 R_{\rm p}^6} \varpi. \tag{9.75}$$

The central density and temperature result from (9.71)–(9.75), which yield

$$\rho_c = \left(\frac{2\sigma_{\rm B}}{3K}\right)^{\frac{2}{19}} \left(\frac{\mu}{\Re}\right)^{\frac{15}{19}} \left(\frac{\mu_0 G M_{\rm p} \dot{M}^2}{32\pi^2 (B_{\rm p})_0^2 R_{\rm p}^6}\right)^{\frac{13}{19}} \frac{\varpi^{\frac{13}{19}}}{\eta^{\frac{2}{19}}}, \tag{9.76}$$

$$T_c = \left(\frac{3K}{2\sigma_{\rm B}}\right)^{\frac{2}{19}} \left(\frac{\mu}{\Re}\right)^{\frac{4}{19}} \left(\frac{\mu_0 G M_{\rm p} \dot{M}^2}{32\pi^2 (B_{\rm p})_0^2 R_{\rm p}^6}\right)^{\frac{6}{19}} \eta^{\frac{2}{19}} \varpi^{\frac{6}{19}}. \tag{9.77}$$

The weak dependence on η in these expressions is noteworthy.

Equations (9.57) and (9.60) give the disc height

$$h = \frac{\mu_0 \dot{M}}{2\pi (B_{\rm p})_0^2 R_{\rm p}^6} \frac{\eta \varpi^4}{(1 - \Omega_{\rm p}/\Omega_{\rm K})}. \tag{9.78}$$

Writing (9.61) as

$$\dot{M} = -4\pi \varpi \rho_c v_\varpi h, \tag{9.79}$$

(9.76) and (9.78) then give the radial inflow speed as

$$v_\varpi = -\left(\frac{3K}{2\sigma_{\rm B}}\right)^{\frac{2}{19}} \left(\frac{\Re}{\mu}\right)^{\frac{15}{19}} \frac{B^2 R_{\rm p}^6}{2\mu_0} \left(\frac{32\pi^2 (B_{\rm p})_0^2 R_{\rm p}^6}{\mu_0 G M_{\rm p} \dot{M}^2}\right)^{\frac{13}{19}} \frac{(1 - \Omega_{\rm p}/\Omega_{\rm K})}{\eta^{\frac{17}{19}} \varpi^{\frac{108}{19}}}. \tag{9.80}$$

It is seen that $\Omega_{\rm K} > \Omega_{\rm p}$ is necessary for $v_\varpi < 0$, so angular momentum is removed from the disc. The magnetically-dominated region must therefore lie inside the radius at which $\Omega_{\rm K} = \Omega_{\rm p}$, for continuous inflow. The consistency of this is shown in the next chapter.

The foregoing disc solution is valid provided $|v_\varpi| \ll v_{\rm K}$. As $|v_\varpi|$ approaches $v_{\rm K}$ then $v_\phi(\varpi, 0)$ will deviate from a Keplerian form and as ρ_c decreases the disc will merge into the magnetosphere. If material is effectively channelled by the stellar magnetic field inside a radius $\varpi_{\rm m}$, then free-fall accretion in an essentially corotating magnetosphere requires $\Omega_{\rm p} < \Omega_{\rm K}(\varpi_{\rm m})$. At higher stellar rotation rates material would experience a centrifugal barrier, since $\varpi_{\rm m}\Omega_{\rm p}^2$ would exceed the ϖ-component of stellar gravity. A fastness parameter can be defined as

$$\xi_{\rm R} = \frac{\Omega_{\rm p}}{\Omega_{\rm K}(\varpi_{\rm m})}. \tag{9.81}$$

The corotation radius at which $\Omega_{\rm K}(\varpi) = \Omega_{\rm p}$ is therefore related to $\varpi_{\rm m}$ by

$$\varpi_{\rm co} = \xi_{\rm R}^{-\frac{2}{3}} \varpi_{\rm m}. \tag{9.82}$$

Taking, as an example, typical X-ray binary parameters of $(B_{\rm p})_0 = 10^8$ tesla, $R = 10^4$ m, $M_{\rm p} = 1.4\,M_\odot$ and $\dot{M} = 10^{-9}\,M_\odot{\rm yr}^{-1}$, together with $\eta = 10^{11}\,{\rm m^2s^{-1}}$, $K = 1.4 \times 10^{19}\,{\rm m^2\,Kg^{-1}}$ and $\mu = 0.62$, gives

$$\frac{|v_\varpi|}{v_{\rm K}} = 6.3 \times 10^{-3} \frac{(1 - \Omega_{\rm p}/\Omega_{\rm K})}{(\eta/10^{11}\,{\rm m^2s^{-1}})^{\frac{17}{19}} (\varpi/4 \times 10^6\,{\rm m})^{\frac{197}{38}}}. \tag{9.83}$$

It is noted that values of η characteristic of turbulence are required to satisfy $|v_\varpi| \ll v_{\rm K}$. The central temperature follows from (9.77) as

$$T_c = 10^5 \left(\frac{\eta}{10^{11}\,{\rm m^2s^{-1}}}\right)^{\frac{2}{19}} \left(\frac{\varpi}{4 \times 10^6\,{\rm m}}\right)^{\frac{6}{19}} {\rm K}, \tag{9.84}$$

while (9.78) gives $h/\varpi \lesssim 10^{-2}$.

For the above parameters, and $0.2 < \xi_{\rm R} < 0.8$, the inner magnetically strong region corresponds to $10^6 \lesssim \varpi \lesssim 4 \times 10^6$ m. Equation (9.60) then gives $|B_\phi^+| < |B_z|$ for this region.

9.2.5. DISC STRUCTURE WITH SPECIFIED η

The foregoing solution was derived by treating η as an unspecified function. It was seen that values of η characteristic of turbulence are required for an inner disc solution. The most likely origins of η are turbulence and magnetic buoyancy. This means that η will depend on disc variables such as B_ϕ and ρ, so affecting the solutions for these and other quantities. In the case of magnetic buoyancy the diffusivity is taken as

$$\eta = \frac{\xi B_\phi^+}{(\mu_0 \rho_c)^{\frac{1}{2}}} h, \tag{9.85}$$

where $\xi < 1$ (see §2.2.10). It is noted that for the inner magnetically strong region B_ϕ^+ is related to P_c by (9.58) and hence (9.85) is equivalent to

$$\eta = \epsilon \left(\frac{P_c}{\rho_c}\right)^{\frac{1}{2}} h, \tag{9.86}$$

where $\epsilon = \sqrt{2}\xi$. This corresponds to a simple turbulent prescription for η in which the rms eddy speed is a fraction ϵ of the isothermal sound speed, and the mixing length is $\sim h$. Equation (9.86) is the form usually adopted for the turbulent viscosity coefficient in standard disc theory.

The surface toroidal magnetic field is given by (9.57b) as

$$B_\phi^+ = \frac{\varpi |B_z|}{\eta} (\Omega_{\rm K} - \Omega_{\rm p}) h. \tag{9.87}$$

Use of (9.85) then gives

$$B_\phi^+ = \left(\frac{2\mu_0}{\epsilon^2}\right)^{\frac{1}{4}} \varpi^{\frac{1}{2}} \rho_c^{\frac{1}{4}} |B_z|^{\frac{1}{2}} (\Omega_{\rm K} - \Omega_{\rm p})^{\frac{1}{2}}, \tag{9.88}$$

so a weak dependence on ρ_c now results. However, the advection of angular momentum due to surface magnetic stress, given by (9.60), yields

$$B_\phi^+ = \frac{\mu_0 \dot{M} d(\varpi^2 \Omega_{\rm K})/d\varpi}{4\pi\varpi^2 |B_z|}. \tag{9.89}$$

The central density in (9.88) must therefore be

$$\rho_c = \frac{\epsilon^2 \mu_0^3 G M_{\rm p} \dot{M}^4}{128\pi^4 (B_{\rm p})_0^6 R_{\rm p}^{18}} \frac{\varpi^9}{(1 - \Omega_{\rm p}/\Omega_{\rm K})}. \tag{9.90}$$

It is seen that ρ_c drops sharply with decreasing ϖ.

The disc height is now determined by vertical equilibrium and the thermal equations. Equations (9.58), (9.72) and (9.73) give

$$h = \frac{4\sqrt{2}\sigma_{\rm B}}{3\epsilon K \mu_0^6} \left(\frac{\mu}{2\Re}\right)^{\frac{15}{2}} \frac{(B_\phi^+)^{12}}{\rho_c^9}. \tag{9.91}$$

Substitution for B_ϕ^+ and ρ_c from (9.88) and (9.90) then yields

$$h \propto \frac{(1 - \Omega_{\rm p}/\Omega_{\rm K})^{18}}{\varpi^{75}}. \tag{9.92}$$

Taking an electron scattering opacity gives $h \propto \varpi^{-38}$. The thin disc condition is therefore strongly violated, the density scale height rapidly diverging with decreasing ϖ.

The foregoing divergence of h is a *reductio ad absurdum* proof that, for turbulent and/or buoyancy forms of η, an inner disc region cannot exist with $|F_{\rm m\phi}|$ significantly larger than $|F_{\rm v\phi}|$. An inner solution with $|F_{\rm m\phi}| \sim |F_{\rm v\phi}|$ is discussed in §9.3, but this is viscously unstable. These results suggest that the disc will end close to the radius at which $F_{\rm m\phi} = F_{\rm v\phi}$. Inside this radius ρ_c rapidly drops, h increases and material merges with the magnetosphere, through a horizontal boundary layer.

The vertical integrals of the viscous and azimuthal force densities are

$$\tilde{F}_{\rm v\phi} = \int_{-h}^{h} F_{\rm v\phi} dz = \frac{1}{\varpi^2} \frac{d}{d\varpi} \left(\varpi^3 \nu \Sigma \frac{d\Omega_{\rm K}}{d\varpi}\right), \tag{9.93}$$

$$\tilde{F}_{\rm m\phi} = \int_{-h}^{h} F_{\rm m\phi} dz = \frac{2}{\mu_0} B_\phi^+ B_z, \tag{9.94}$$

with

$$\Sigma = \int_{-h}^{h} \rho dz. \tag{9.95}$$

The magnetic force term in (9.5) involving B_ϖ has been dropped since, away from $\varpi_{\rm co}$, it is smaller than the above term by a factor $\sim (h/\varpi)^2$. In an unperturbed viscous disc, with $\varpi \gg R_{\rm p}$, (2.277) gives

$$\nu\Sigma = \frac{\dot{M}_{\rm p}}{3\pi}, \tag{9.96}$$

since, $\dot{M} = \dot{M}_{\rm p}$, while, from (2.298),

$$\rho_c = 2.5 \times 10^{-16} \frac{\dot{M}_{\rm p}^{\frac{11}{20}} M_{\rm p}^{\frac{5}{8}}}{\epsilon^{\frac{7}{10}} \varpi^{\frac{15}{8}}}. \tag{9.97}$$

In the main body of the disc (9.93)–(9.96) give

$$\left| \frac{\tilde{F}_{{\rm m}\phi}}{\tilde{F}_{{\rm v}\phi}} \right| = \frac{8\pi\varpi B_\phi^+ B_z}{\mu_0 \dot{M}_{\rm p} \Omega_{\rm K}}, \tag{9.98}$$

which is valid provided this ratio is significantly < 1. When the ratio approaches unity disruption will occur over a short length-scale. The inner radius of the disc can be estimated by taking $\tilde{F}_{{\rm m}\phi} = 2\tilde{F}_{{\rm v}\phi}$ in (9.98). Use of (9.46), (9.51), (9.81), (9.88) and (9.97) then gives

$$\varpi_{\rm m} = 10^7 \frac{\left(\frac{(B_{\rm p})_0}{10^8\,{\rm T}}\right)^{\frac{16}{29}} \left(\frac{R_{\rm p}}{10^4\,{\rm m}}\right)^{\frac{48}{29}} (1-\xi_{\rm R})^{\frac{16}{87}}}{\left(\frac{\epsilon}{0.1}\right)^{\frac{36}{145}} \left(\frac{M_{\rm p}}{1.4\,M_\odot}\right)^{\frac{1}{29}} \left(\frac{\dot{M}_{\rm p}}{10^{-9}\,M_\odot{\rm yr}^{-1}}\right)^{\frac{46}{145}}}\ {\rm m}. \tag{9.99}$$

9.2.6. DISCUSSION

The foregoing analysis illustrates the important role played by the magnetic diffusivity η. In §9.2.4, η was taken as an unspecified function of ϖ and a self-consistent thin disc model could be constructed for the magnetically strong inner region. The inflow is caused by the transfer of angular momentum from the disc to the star, via magnetic stresses. The disc ends when the inflow speed becomes a significant fraction of the Keplerian speed, this occurring in the region of the free-fall Alfvén radius. In §9.2.5 a definite form was adopted for η, corresponding to magnetic buoyancy or a simple representation of turbulence. The disc then ends as the magnetic azimuthal force starts to exceed the viscous force. The resulting disruption radius is significantly larger than the free-fall Alfvén radius, given by (2.142).

The foregoing difference in the behaviour of h arises from the contribution made by the thermal problem. In the first case, in which η is taken as an abstract function of ϖ, the local disc hight h is determined from the azimuthal components of the induction and momentum equations, being decoupled from the thermal problem. In the second case, in which η depends on ρ, the azimuthal induction and momentum equations lead to a sharp drop in ρ and the coupling to the thermal problem then results in the divergence of h. Whilst the former case is of interest, the latter case is the most physically plausible.

9.3. Numerical Solutions

9.3.1. THE MAGNETO-VISCOUS EQUATIONS

In order to investigate further the effect of the stellar magnetic field on the disc structure, Campbell and Heptinstall (1997) performed a numerical integration of the equations. The viscous force term was included, since the whole disc structure was considered in addition to investigating its inner disruption.

The disc rotation remains Keplerian in its central plane, but the azimuthal and vertical momentum equations now become

$$v_\varpi \frac{\partial}{\partial \varpi}(\varpi^2 \Omega) + v_z \frac{\partial}{\partial z}(\varpi^2 \Omega) = \frac{1}{\varpi \rho} \frac{\partial}{\partial \varpi}\left(\rho \nu \varpi^3 \frac{\partial \Omega}{\partial \varpi}\right) + \frac{\varpi}{\mu_0 \rho} \frac{\partial}{\partial z}(B_\phi B_z), \tag{9.100}$$

$$\frac{z}{\varpi}\frac{v_{\mathrm{K}}^2}{\varpi} + \frac{1}{\rho}\frac{\partial}{\partial z}\left(P + \frac{B_\phi^2}{2\mu_0}\right) = 0. \tag{9.101}$$

The azimuthal induction equation is still given by (9.49). Combining (9.100) with the continuity equation (9.50) yields

$$\frac{\partial}{\partial \varpi}(\varpi \rho v_\varpi \varpi^2 \Omega) + \frac{\partial}{\partial z}(\varpi \rho v_z \varpi^2 \Omega) = \frac{\partial}{\partial \varpi}\left(\rho \nu \varpi^3 \frac{\partial \Omega}{\partial \varpi}\right) + \frac{1}{\mu_0}\frac{\partial}{\partial z}(\varpi^2 B_\phi B_z).$$

Vertical integration through the disc, using $\rho(\varpi, \pm h) = 0$, and the symmetry of $B_\phi B_z$, gives

$$-\dot{M}\frac{d}{d\varpi}(\varpi^2 \Omega_{\mathrm{K}}) = \frac{d}{d\varpi}(2\pi \varpi^3 \Omega' \nu \Sigma) + \frac{4\pi}{\mu_0}\varpi^2 B_\phi^+ B_z. \tag{9.102}$$

This represents the advection of angular momentum due to the viscous and magnetic torques. Using $P(\varpi, h) = 0$ and the vertical mean $\langle z\rho \rangle \simeq \rho_c h/2$, vertical integration of (9.101) yields

$$P_c = \frac{(B_\phi^+)^2}{2\mu_0} + \frac{1}{2}\Omega_{\mathrm{K}}^2 h^2 \rho_c. \tag{9.103}$$

Inclusion of the viscous dissipation gives a total dissipation per unit volume of

$$Q = \rho\nu(\varpi\Omega_{\mathrm{K}}')^2 + \frac{\eta}{\mu_0}\left(\frac{\partial B_\phi}{\partial z}\right)^2. \tag{9.104}$$

The surface heat flux is

$$F_+ = \int_0^h Q dz.$$

Using (9.57a) in (9.104), and then integrating vertically through the disc, leads to

$$F_+ = \frac{1}{2}\nu\Sigma(\varpi\Omega_{\mathrm{K}}')^2 + \frac{\varpi^2 B_z^2}{\mu_0\eta}(\Omega_{\mathrm{p}} - \Omega_{\mathrm{K}})^2 h. \tag{9.105}$$

Equation (9.72) relating ρ_c to P_c remains valid.

The magnetic diffusivity is taken to be due to buoyancy and turbulence, with

$$\eta = \frac{\xi B_\phi^+}{(\mu_0\rho_c)^{\frac{1}{2}}} h + \epsilon c_{\mathrm{s}} h, \tag{9.106}$$

while the viscous coefficient used is

$$\nu = \epsilon c_{\mathrm{s}} h, \tag{9.107}$$

where the dimensionless parameters ξ, and ϵ are < 1. The viscosity contains a simple modification to allow for the effect of rotation on the turbulence (see §11.3.1).

9.3.2. SOLUTIONS AND VISCOUS INSTABILITY

Equations (9.57), (9.71), (9.72), (9.102), (9.103) and (9.105), with (9.51), (9.106) and (9.107), were integrated numerically from the outer part of the disc inwards. An iterative procedure was used, starting with the unperturbed viscous disc solution where the stellar field is weak. Outside the corotation radius, ϖ_{co}, the magnetic perturbation to the disc is small.

Some distance inside $\varpi_{\rm co}$ a steady disc solution appears to be possible with $|F_{{\rm v}\phi}| \sim |F_{{\rm m}\phi}|$. This can only be achieved by the gradient $d\Sigma/d\varpi$ steepening to reverse the sign of the viscous force in (9.102) and apparently prevent the asymptotic limit $|F_{{\rm m}\phi}| \gg |F_{{\rm v}\phi}|$ being approached. However, as explained below, the solutions with $F_{{\rm v}\phi} > 0$ are viscously unstable.

The time-dependent, vertically integrated equations of continuity and angular momentum advection are

$$-\frac{\partial \Sigma}{\partial t} = \frac{1}{\varpi}\frac{\partial}{\partial \varpi}(\varpi \Sigma v_\varpi),$$

$$\varpi \Sigma v_\varpi \frac{d}{d\varpi}(\varpi^2 \Omega_{\rm K}) = \frac{\partial}{\partial \varpi}\left(\varpi^3 \nu \Sigma \frac{d\Omega_{\rm K}}{d\varpi}\right) + \frac{2\varpi^2}{\mu_0} B_\phi^+ B_z,$$

noting $\partial\Omega/\partial t = 0$. Elimination of $\varpi\Sigma v_\varpi$ between these equations leads to the diffusion equation

$$\frac{\partial \Sigma}{\partial t} = \frac{3}{\varpi}\frac{\partial}{\partial \varpi}\left(\varpi^{\frac{1}{2}}\frac{\partial}{\partial \varpi}\left(\varpi^{\frac{1}{2}}\nu\Sigma\right)\right) - \frac{4}{\varpi}\frac{\partial}{\partial \varpi}\left(\frac{\varpi B_\phi^+ B_z}{\mu_0 \Omega_{\rm K}}\right), \tag{9.108}$$

where B_z is the stellar field (9.51), but B_ϕ^+ is now a solution of the time-dependent azimuthal induction equation.

Equation (9.108) describes the diffusion of the surface density function Σ. It can be used to investigate the linear diffusive stability of the foregoing steady disc solution. Defining

$$\mu(\varpi, \Sigma) = \nu\Sigma, \tag{9.109}$$

it follows that for Eulerian perturbations (i.e. small variations at fixed ϖ),

$$\delta\mu = \left(\frac{\partial \mu}{\partial \Sigma}\right)_0 \delta\Sigma, \tag{9.110}$$

where the subscript zero refers to the steady solution. Since B_ϕ^+ can also be expressed as $B_\phi^+(\varpi, \Sigma)$, then

$$\delta B_\phi^+ = \left(\frac{\partial B_\phi^+}{\partial \Sigma}\right)_0 \delta\Sigma. \tag{9.111}$$

Taking a first order perturbation of (9.108), and using (9.110) and (9.111) to eliminate $\delta\Sigma$ and δB_ϕ^+ in terms of $\delta\mu$, yields

$$\frac{\partial}{\partial t}(\delta\mu) = \frac{3}{\varpi}\left(\frac{\partial \mu}{\partial \Sigma}\right)_0 \frac{\partial}{\partial \varpi}\left(\varpi^{\frac{1}{2}}\frac{\partial}{\partial \varpi}\left(\varpi^{\frac{1}{2}}\delta\mu\right)\right) - \frac{4}{\varpi}\left(\frac{\partial \mu}{\partial \Sigma}\right)_0 \frac{\partial}{\partial \varpi}\left(f(\varpi)\delta\mu\right), \tag{9.112a}$$

where

$$f(\varpi) = \frac{\varpi B_z}{\mu_0 \Omega_{\rm K}} \left(\frac{\partial B_\phi^+}{\partial \Sigma} \right)_0 \left(\frac{\partial \mu}{\partial \Sigma} \right)_0^{-1} . \tag{9.112b}$$

The magnitude of the dimensionless function $f(\varpi)$ is given by

$$f(\varpi) \sim \frac{\varpi B_z B_\phi^+}{\mu_0 \Omega_{\rm K} \nu h \rho_c} \sim \frac{\eta}{\nu} \frac{(B_\phi^+)^2}{\mu_0 \rho_c h^2 \Omega_{\rm K}^2},$$

using (9.57) to eliminate B_z. It follows from (9.103) that $(B_\phi^+)^2/\mu_0\rho_c \sim h^2\Omega_{\rm K}^2$ and hence

$$f(\varpi) \sim \frac{\eta}{\nu}.$$

Equations (9.106) and (9.107) give $\eta \sim \nu$, and hence $f(\varpi) \sim 1$.

Quantities in the steady disc have a radial length-scale of $\sim \varpi$. Local, short wavelength perturbations can be Fourier analysed with the form

$$\delta\mu = A \exp[i(k\varpi - \sigma t)], \tag{9.113}$$

with A constant, provided $k\varpi \gg 1$. Variations in the radial disc structure can then be ignored over a wavelength $2\pi/k$. Substitution of (9.113) in (9.112a) shows that the second radial derivative in the viscous diffusion term, containing $(k\varpi)^2$, dominates in determining σ. Since the largest part of the magnetic term contains only $k\varpi$, the magnetic diffusion effect is small for short wavelength modes and can be neglected. The sign of σ therefore depends on the sign of $(\partial\mu/\partial\Sigma)_0$. For $(\partial\mu/\partial\Sigma)_0 < 0$, the time-dependence of $\delta\mu$ becomes a growing exponential corresponding to viscous instability. The foregoing steady solutions with $F_{\mathrm{v}\phi} > 0$, due to a magnetically steepened radial density gradient within $\varpi_{\rm co}$, are found to be viscously unstable, leading to break up of the disc. This result reinforces the conclusion that the disc cannot exist with $|F_{\mathrm{m}\phi}| \gg |F_{\mathrm{v}\phi}|$, and ends when these forces become comparable.

9.4. Conclusions

It is noted from expression (9.99) for the disruption radius that typically $\varpi_{\rm m} \gtrsim 2r_{\rm A}$, where $r_{\rm A}$ is the spherical Alfvén radius given by (2.142). It is shown in the next chapter that reducing the vertical shear by factors of 2 to 3, corresponding to some break down of corotation in the magnetosphere, does not significantly affect this result. It is also noted that, unlike $r_{\rm A}$, the radius $\varpi_{\rm m}$ depends on the stellar rotation rate. Hence $r_{\rm A}$ should not be used to calculate disc disruption radii.

The difference between r_{A} and ϖ_{m} arises because r_{A} is related to the dynamical effect the magnetic field has on a free-fall, highly conducting flow, whilst, as shown in this chapter, the disruption of a diffusive disc by the stellar field has magnetically induced thermal and viscous causes.

The analysis of this chapter justifies the assumption that such discs will end when $|F_{\mathrm{m}\phi}| \gtrsim |F_{\mathrm{v}\phi}|$, made by Bath, Evans and Pringle (1974) and Wang (1987).

References

Aly, J.J., 1980. *Astron. Astrophys.*, **86**, 192.
Anzer, U. and Borner, G., 1983. *Astron. Astrophys.*, **122**, 73.
Bardou, A. and Heyvaerts, J., 1996. *Astron. Astrophys.*, in press.
Bath, G.T., Evans, W.D. and Pringle, J.E., 1974. *Mon. Not. R. Astr. Soc.*, **166**, 113.
Campbell, C.G., 1992. *Geophys. Astrophys. Fluid. Dynam.*, **63**, 179.
Campbell, C.G. and Heptinstall, P., 1997. *Mon. Not. R. Astr. Soc.*, submitted.
Cox, J.P. and Giuli, R.T., 1968. *Principles of Stellar Structure*, Vol I, Gordon and Breach.
Ghosh, P. and Lamb, F.K., 1978. *Astrophys. J.*, **223**, 296.
Ghosh, P. and Lamb, F.K., 1979. *Astrophys. J.*, **234**, 296.
Kundt, W. and Robnik, M., 1980. *Astron. Astrophys.*, **91**, 305.
Lubow, S.H., Papaloizou, J.C.B. and Pringle, J.E., 1994. *Mon. Not. R. Astr. Soc.*, **267**, 235.
Wang, Y.M., 1987. *Astron. Astrophys.*, **183**, 257.

CHAPTER 10

BINARIES WITH PARTIAL DISCS: SPIN EVOLUTION

10.1. Introduction

In this chapter the angular momentum transfer to the primary star will be considered. This occurs through magnetic interaction with the disc and via magnetically channelled accretion. The magnetic field exerts a torque on the disc material and an equal and opposite torque is felt by the primary star, as a result of currents flowing in the low density magnetosphere. If a solution can be found for the magnetic field in the disc, then integration of the associated surface stresses yields the torque. The accretion torque arises from the flow of material along stellar field lines in the inner disrupted region. The stellar field exerts a torque on the stream material and the primary star feels the reaction to this. The torques depend on the stellar spin rate $\Omega_{\rm p}$, and an equilibrium period can be found at which the total torque vanishes.

The most physically plausible cases of a turbulent or buoyant η are considered here, following Campbell (1992). The magnetic torque in the case with a buoyant diffusivity was also considered by Wang (1987). As shown in Chapter 9, the disc then ends where the azimuthal magnetic force significantly exceeds the viscous force. In §10.2 expressions are found for the surface azimuthal magnetic field and the inner edge of the disc, for the two forms of η. The remainder of the disc will be unperturbed provided the ratio $|\tilde{F}_{\rm m\phi}/\tilde{F}_{\rm v\phi}|$, given by equation (9.98), is significantly less than unity. The problem is then kinematic, requiring solution of the induction equation to obtain the disc magnetic field, followed by integration of the surface stresses to give the stellar magnetic torque. In §10.3 some exact solutions of the induction equation are presented. In §10.4 an expression is derived for the stellar magnetic torque as an integral over the disc surface stresses. The accretion torque is found by using angular momentum conservation at the disc's inner edge. The spin evolution of the star is considered in §10.5 and expressions are derived for its rate of change of period, and for the equilibrium period at which the accretion and disc torques cancel. The results are discussed in §10.6.

10.2. The Effect of Azimuthal Magnetic Force

10.2.1. THE KINEMATIC CONDITION

The azimuthal magnetic field at the disc's surfaces is given by (9.57). Allowing for some uncertainty in the vertical shear, this can be written as

$$B_\phi^+ = \frac{\varpi |B_z|}{\eta} \gamma (\Omega_\mathrm{K} - \Omega_\mathrm{p}) h. \tag{10.1}$$

The magnetic buoyancy and turbulent forms of η are taken as

$$\eta_\mathrm{B} = \frac{|B_\phi^+|}{(\mu_0 \rho_c)^{\frac{1}{2}}} h, \tag{10.2}$$

$$\eta_\mathrm{T} = \epsilon \left(\frac{P_c}{\rho_c} \right)^{\frac{1}{2}} h. \tag{10.3}$$

It follows that for $\gamma \sim 1$ these forms of η result in $|B_\phi^+|$ significantly exceeding $|B_z|$. However, the situation $|B_\phi^+| \gg |B_z|$ would result in dynamical instabilities with associated dissipation. This would cause a break down of near-corotation in the magnetosphere and a reduction in the vertical shear, giving $\gamma < 1$ in (10.1). An enhancement of η in the disc, due to such instabilities, would also result in $\gamma < 1$. It will be seen that most of the results of this chapter are only weakly dependent on γ.

The magnetic transfer of angular momentum between the disc and the primary star is via the $B_\phi^+ B_z$ stress, a torque being transmitted through the magnetosphere to the star whose spin acts as a sink or source of angular momentum. A corotation radius, ϖ_co, exists at which $\Omega_\mathrm{K} = \Omega_\mathrm{p}$. Inside this radius the magnetic stress extracts angular momentum from the disc, while the reverse is true for $\varpi > \varpi_\mathrm{co}$. The force ratio $|\tilde{F}_{\mathrm{m}\phi}/\tilde{F}_{\mathrm{v}\phi}|$ will become small as ϖ_co is approached since at this radius the vertical shear, which is the dominant source of B_ϕ^+, vanishes. Inside ϖ_co the force ratio $|\tilde{F}_{\mathrm{m}\phi}/\tilde{F}_{\mathrm{v}\phi}|$ increases monotonically until disruption occurs. Outside ϖ_co the ratio will rise to a maximum and then fall towards zero with increasing ϖ. The main body of the disc will be magnetically kinematic provided the maximum value of $|\tilde{F}_{\mathrm{m}\phi}/\tilde{F}_{\mathrm{v}\phi}|$ is significantly less than unity. This can be tested for the two forms of η.

The following results are supported by the numerical work of Campbell and Heptinstall (1997), described in §9.3. This showed that, for $\gamma < 1$, the main body of the disc is only weakly perturbed by the stellar magnetic field, so its structure is close to that of the viscous disc.

10.2.2. MAGNETIC BUOYANCY DIFFUSIVITY

In this case (10.1) and (10.2) give the surface azimuthal field as

$$B_\phi^+ = \pm(2\mu_0)^{\frac{1}{4}}\gamma^{\frac{1}{2}}\varpi^{\frac{1}{2}}\rho_c^{\frac{1}{4}}|B_z|^{\frac{1}{2}}|\Omega_{\mathrm{K}} - \Omega_{\mathrm{p}}|^{\frac{1}{2}}, \tag{10.4}$$

where the negative sign applies for $\varpi > \varpi_{\mathrm{co}}$. Equation (9.98) then gives

$$\left|\frac{\tilde{F}_{\mathrm{m}\phi}}{\tilde{F}_{\mathrm{v}\phi}}\right| = \frac{8\pi(2\mu_0)^{\frac{1}{4}}\gamma^{\frac{1}{2}}\varpi^{\frac{3}{2}}\rho_c^{\frac{1}{4}}|B_z|^{\frac{3}{2}}(\Omega_{\mathrm{p}} - \Omega_{\mathrm{K}})^{\frac{1}{2}}}{\mu_0\dot{M}\Omega_{\mathrm{K}}}. \tag{10.5}$$

Using (9.46), (9.51) and (9.97), differentiation shows that this force ratio has a maximum where $\Omega_{\mathrm{K}} = (21/29)\Omega_{\mathrm{p}}$, which occurs at $\varpi = 1.24\varpi_{\mathrm{co}}$. It then follows that

$$\left|\frac{\tilde{F}_{\mathrm{m}\phi}}{\tilde{F}_{\mathrm{v}\phi}}\right|_{\max} = 0.35\frac{\gamma^{\frac{1}{2}}\left(\frac{(B_{\mathrm{p}})_0}{10^8\,\mathrm{T}}\right)^{\frac{3}{2}}\left(\frac{R_{\mathrm{p}}}{10^4\,\mathrm{m}}\right)^{\frac{9}{2}}}{\left(\frac{M_{\mathrm{p}}}{1.4\,M_\odot}\right)\left(\frac{\dot{M}_{\mathrm{p}}}{10^{-9}\,M_\odot\mathrm{yr}^{-1}}\right)^{\frac{69}{80}}\left(\frac{P_{\mathrm{R}}}{20\,\mathrm{s}}\right)^{\frac{29}{16}}}, \tag{10.6}$$

where $P_{\mathrm{R}} = 2\pi/\Omega_{\mathrm{p}}$. It is seen that for typical low mass X-ray binary parameters, $|\tilde{F}_{\mathrm{m}\phi}/\tilde{F}_{\mathrm{v}\phi}|_{\max}$ is significantly less than unity since $\gamma < 1$. Using Table 8.1, with other typical parameters for the intermediate polars, yields $|\tilde{F}_{\mathrm{m}\phi}/\tilde{F}_{\mathrm{v}\phi}|_{\max} \lesssim 1$. The values of this maximum force ratio nearer to unity occur for systems with the shorter primary spin periods. In these systems some disruption of the disc may occur where $\tilde{F}_{\mathrm{m}\phi}$ approaches $\tilde{F}_{\mathrm{v}\phi}$. Inside the corotation radius $|\tilde{F}_{\mathrm{m}\phi}/\tilde{F}_{\mathrm{v}\phi}|$ increases monotonically with decreasing ϖ. The very rapid divergence of $h(\varpi)$, given by (9.92), indicates that the disc will have ended when $|\tilde{F}_{\mathrm{m}\phi}/\tilde{F}_{\mathrm{v}\phi}| \gtrsim 2$. The disc ends in a boundary layer in which $\Omega(\varpi, 0)$ changes from Ω_{K} to Ω_{p} as the highly conducting magnetosphere is entered. The disc's inner edge ϖ_{m} can be estimated by taking $\tilde{F}_{\mathrm{m}\phi} = 2\tilde{F}_{\mathrm{v}\phi}$, so (9.46), (9.51), (9.97) and (10.5) then give

$$\varpi_{\mathrm{m}} = 10^7\frac{\gamma^{\frac{16}{87}}\left(\frac{(B_{\mathrm{p}})_0}{10^8\,\mathrm{T}}\right)^{\frac{16}{29}}\left(\frac{R_{\mathrm{p}}}{10^4\,\mathrm{m}}\right)^{\frac{48}{29}}(1-\xi_{\mathrm{R}})^{\frac{16}{87}}}{\left(\frac{M_{\mathrm{p}}}{1.4\,M_\odot}\right)^{\frac{1}{29}}\left(\frac{\dot{M}_{\mathrm{p}}}{10^{-9}\,M_\odot\mathrm{yr}^{-1}}\right)^{\frac{46}{145}}}\,\mathrm{m}, \tag{10.7}$$

where the fastness parmeter ξ_{R} is defined in (9.81). For $\gamma = 0.1$ and $\xi_{\mathrm{R}} = 0.6$, in an X-ray binary pulsar (10.7) typically yields $\varpi_{\mathrm{m}} \sim 6 \times 10^6\,\mathrm{m} \sim 2r_{\mathrm{A}}$, where r_{A} is the spherical Alfvén radius given by (2.142). The longer orbital periods in these systems result in $\varpi_{\mathrm{m}} \ll D$.

Intermediate polar parameters also typically give $\varpi_{\rm m} \sim 2r_{\rm A}$. The shorter orbital period systems should have more extensively disrupted discs, since $\varpi_{\rm m}$ is a larger fraction of the stellar separation D. This is consistent with the observational values of P and $P_{\rm R}$ shown in Table 8.1, which illustrate that the shorter period systems are nearer to synchronism and hence more like the discless AM Her binaries.

10.2.3. TURBULENT DIFFUSIVITY

Equations (10.1) and (10.3) give the surface azimuthal field as

$$B_\phi^+ = \gamma \frac{\varpi\,|B_z|}{\epsilon c_{\rm s}} (\Omega_{\rm K} - \Omega_{\rm p}), \tag{10.8}$$

where the isothermal sound speed for a viscous disc is

$$c_{\rm s} = 0.2 \left(\frac{\Re}{\mu}\right)^{\frac{1}{2}} \frac{M_{\rm p}^{\frac{1}{8}} \dot{M}_{\rm p}^{\frac{3}{20}}}{\epsilon^{\frac{1}{10}} \varpi^{\frac{3}{8}}}. \tag{10.9}$$

The ratio of the vertically integrated azimuthal forces is then

$$\left|\frac{\tilde{F}_{{\rm m}\phi}}{\tilde{F}_{{\rm v}\phi}}\right| = \frac{8\pi\gamma\varpi^2 B_z^2 |\Omega_{\rm K} - \Omega_{\rm p}|}{\mu_0 \epsilon \dot{M}_{\rm p} c_{\rm s} \Omega_{\rm K}}, \tag{10.10}$$

which has a maximum where $\Omega_{\rm K} = (17/29)\Omega_{\rm p}$, occurring at $\varpi = 1.43\varpi_{\rm co}$. It follows that

$$\left|\frac{\tilde{F}_{{\rm m}\phi}}{\tilde{F}_{{\rm v}\phi}}\right|_{\max} = \frac{0.46\gamma \left(\frac{(B_{\rm p})_0}{10^8\,{\rm T}}\right)^2 \left(\frac{R_{\rm p}}{10^4\,{\rm m}}\right)^6}{\left(\frac{\epsilon}{0.1}\right)^{\frac{9}{10}} \left(\frac{M_{\rm p}}{1.4\,M_\odot}\right)^{\frac{4}{3}} \left(\frac{\dot{M}_{\rm p}}{10^{-9}\,M_\odot{\rm yr}^{-1}}\right)^{\frac{23}{20}} \left(\frac{P_{\rm R}}{30\,{\rm s}}\right)^{\frac{29}{12}}}. \tag{10.11}$$

Remembering that $\gamma < 1$, typical parameters give $|\tilde{F}_{{\rm m}\phi}/\tilde{F}_{{\rm v}\phi}|_{\max}$ significantly less than unity. Taking the disruption radius at $\tilde{F}_{{\rm m}\phi} = 2\tilde{F}_{{\rm v}\phi}$, equations (9.46), (9.51), (9.81), (10.9) and (10.10) yield

$$\varpi_{\rm m} = 1.7 \times 10^7 \frac{\gamma^{\frac{8}{29}} \left(\frac{(B_{\rm p})_0}{10^8\,{\rm T}}\right)^{\frac{16}{29}} \left(\frac{R_{\rm p}}{10^4\,{\rm m}}\right)^{\frac{48}{29}} (1-\xi_{\rm R})^{\frac{8}{29}}}{\left(\frac{\epsilon}{0.1}\right)^{\frac{36}{145}} \left(\frac{M_{\rm p}}{1.4\,M_\odot}\right)^{\frac{1}{29}} \left(\frac{\dot{M}_{\rm p}}{10^{-9}\,M_\odot{\rm yr}^{-1}}\right)^{\frac{46}{145}}}\ {\rm m}. \tag{10.12}$$

Provided the stellar spin period, $P_{\rm R}$, does not become too short, (10.6) and (10.11) show that the main body of the disc is only weakly magnetically perturbed. The self-consistency of the period assumption will be checked when the stellar spin evolution is considered. For a kinematic disc, solutions of the toroidal induction equation for B_ϕ can be found using the unperturbed disc structure. The stellar torques and spin evolution can then be considered.

10.3. Exact Solutions of the Induction Equation

The foregoing toroidal fields, given by (10.4) and (10.8), are solutions of (9.49). This was an approximation of (9.10), the toroidal part of the induction equation, employing the thin nature of the disc. Solutions of the induction equation, with its radial derivatives retained, were found by Campbell (1987). Ignoring the advection term, (9.10) becomes

$$\nabla^2 B_\phi - \frac{B_\phi}{\varpi^2} + \frac{1}{\varpi\eta}\frac{d\eta}{d\varpi}\frac{\partial}{\partial\varpi}(\varpi B_\phi) = -\frac{\varpi}{\eta}\mathbf{B}_p \cdot \nabla\Omega. \tag{10.13}$$

The angular velocity given by (9.46) yields the components of $\nabla\Omega$ as

$$\frac{\partial\Omega}{\partial\varpi} = \frac{\partial\Omega_{\rm K}}{\partial\varpi} + [\Omega_{\rm K}(\varpi_{\rm m}) - \Omega_{\rm p}]\delta(\varpi - \varpi_{\rm m}), \tag{10.14}$$

$$\frac{\partial\Omega}{\partial z} = (\Omega_{\rm p} - \Omega_{\rm K})\delta(z - h) + (\Omega_{\rm K} - \Omega_{\rm p})\delta(z + h). \tag{10.15}$$

Use of (10.4), together with the viscous disc solution, in (10.2) and (10.3) gives

$$\eta_{\rm B} \propto \varpi^{\frac{19}{32}}|\Omega_{\rm K} - \Omega_{\rm p}|^{\frac{1}{2}} \quad \text{and} \quad \eta_{\rm T} \propto \varpi^{\frac{3}{4}}.$$

It is seen that in both cases η is slowly varying. An exact solution of (10.13) is possible if η is taken as constant.

10.3.1. CONSTANT DIFFUSIVITY

For constant η, (10.13) becomes

$$\nabla^2 B_\phi - \frac{B_\phi}{\varpi^2} = -\frac{\varpi}{\eta}\mathbf{B}_p \cdot \nabla\Omega, \tag{10.16}$$

where it is remembered that $B_\phi = B_\phi(\varpi, z)$. This equation can be transformed to a Poisson form, through multiplication by a factor $\exp(i\phi)$. This gives

$$\nabla^2(B_\phi e^{i\phi}) = -\frac{\varpi}{\eta}\mathbf{B}_p \cdot \nabla\Omega e^{i\phi}. \tag{10.17}$$

The Poisson integral yields

$$B_\phi = \frac{1}{4\pi\eta}\int_{-h}^{h}\int_{\varpi_m}^{\varpi_D}\int_0^{2\pi}\frac{\varpi_0^2(\mathbf{B}_p\cdot\nabla\Omega)_0}{|\mathbf{r}-\mathbf{r}_0|}e^{i(\phi_0-\phi)}d\phi_0 d\varpi_0 dz_0, \qquad (10.18)$$

where

$$|\mathbf{r}-\mathbf{r}_0| = [\varpi^2+\varpi_0^2+(z-z_0)^2-2\varpi\varpi_0\cos(\phi-\phi_0)]^{\frac{1}{2}},$$

and the integral is over the volume of the disc. The periodic nature of the azimuthal integral facilities its reduction to the form

$$G(\varpi,z|\varpi_0,z_0) = \int_0^{2\pi}\frac{\cos\alpha d\alpha}{[\varpi^2+\varpi_0^2+(z-z_0)^2-2\varpi\varpi_0\cos\alpha]^{\frac{1}{2}}}. \qquad (10.19)$$

This can be expressed as

$$G(\varpi,z|\varpi_0,z_0) = \frac{2}{b(a+b)^{\frac{1}{2}}}\left[2aF\left(\frac{\pi}{2},k\right)-(a+b)E\left(\frac{\pi}{2},k\right)\right.$$
$$\left.-(a-b)\Pi\left(\frac{\pi}{2},k^2,k\right)\right], \qquad (10.20)$$

where

$$a=\varpi^2+\varpi_0^2+(z-z_0)^2, \qquad b=2\varpi\varpi_0, \qquad (10.21a)$$

and

$$k=\left[\frac{4\varpi\varpi_0}{(\varpi+\varpi_0)^2+(z-z_0)^2}\right]^{\frac{1}{2}}, \qquad (10.21b)$$

with F, E and Π being elliptic integrals of the first, second and third kinds (see Appendix). The azimuthal field solution for constant η is therefore

$$B_\phi = \frac{1}{4\pi\eta}\int_{-h}^{h}\int_{\varpi_m}^{\varpi_D}\varpi_0^2(\mathbf{B}_p\cdot\nabla\Omega)_0 G(\varpi,z|\varpi_0,z_0)d\varpi_0 dz_0, \qquad (10.22)$$

where the components of $\mathbf{B}_p$ is given by (9.14) and (9.15).

The azimuthal field is found from (10.22) using (10.14), (10.15) and (10.20), together with the dipolar field $\mathbf{B}_p$. It is convenient to measure lengths in units of the inner disc radius ϖ_m, so defining the following dimensionless quantities;

$$s=\frac{\varpi}{\varpi_m}, \qquad \bar{h}=\frac{z}{\varpi_m}, \qquad (10.23)$$

$$\bar{h}_+ = \frac{h}{\varpi_\mathrm{m}}, \qquad \bar{h}_- = -\frac{h}{\varpi_\mathrm{m}}. \tag{10.24}$$

The toroidal magnetic field can then be expressed as

$$B_\phi = \frac{3(B_\mathrm{p})_0 R_\mathrm{p}^3}{8\pi\varpi_\mathrm{m}\eta}\left(\frac{GM_\mathrm{p}}{\varpi_\mathrm{m}^3}\right)^{\frac{1}{2}} J(s,h), \tag{10.25}$$

where

$$J(s,\bar{h}) = \left(\frac{GM_\mathrm{p}}{\varpi_\mathrm{m}^3}\right)^{-\frac{1}{2}}(I_1 + I_2 + I_3), \tag{10.26}$$

with

$$I_1 = \int_{\bar{h}_-}^{\bar{h}_+}\int_1^{s_\mathrm{D}} \frac{s_0^3\bar{h}_0 K(s,\bar{h}|s_0,\bar{h}_0)}{(s_0^2+\bar{h}_0^2)^{\frac{5}{2}}}\frac{d\Omega_\mathrm{K}}{ds_0} ds_0 d\bar{h}_0, \tag{10.27}$$

$$I_2 = -\frac{1}{3}\int_1^{s_\mathrm{D}} \frac{s_0^2(s_0^2 - 2\bar{h}_+^2)}{(s_0^2+\bar{h}_+^2)^{\frac{5}{2}}}\Delta K[\Omega_\mathrm{p} - \Omega_\mathrm{K}(s_0)]ds_0, \tag{10.28}$$

$$I_3 = [\Omega_\mathrm{K}(\varpi_\mathrm{m}) - \Omega_\mathrm{p}]\int_{\bar{h}_-}^{\bar{h}_+}\frac{\bar{h}_0 K(s,\bar{h}|1,\bar{h}_0)}{(1+\bar{h}_0^2)^{\frac{5}{2}}} d\bar{h}_0, \tag{10.29}$$

and

$$K(s,\bar{h}|s_0,\bar{h}_0) = \frac{2}{\beta(\alpha-\beta)^{\frac{1}{2}}}\left[2\alpha F\left(\frac{\pi}{2},k\right) - (\alpha+\beta)E\left(\frac{\pi}{2},k\right) - (\alpha-\beta)\Pi\left(\frac{\pi}{2},k^2,k\right)\right], \tag{10.30}$$

$$\Delta K = K(s,\bar{h}|s_0,\bar{h}_+) - K(s,\bar{h}|s_0,\bar{h}_-),$$

$$\alpha = s^2 + s_0^2 + (\bar{h}-\bar{h}_0)^2, \qquad \beta = 2ss_0, \tag{10.30a}$$

$$k = \left[\frac{4ss_0}{(s+s_0)^2 + (\bar{h}-\bar{h}_0)^2}\right]^{\frac{1}{2}}. \tag{10.30b}$$

The integrals I_i were evaluated numerically, using the surface for a standard viscous disc. The dimensionless disc radius was taken as $s_\mathrm{D} = 15$; beyond

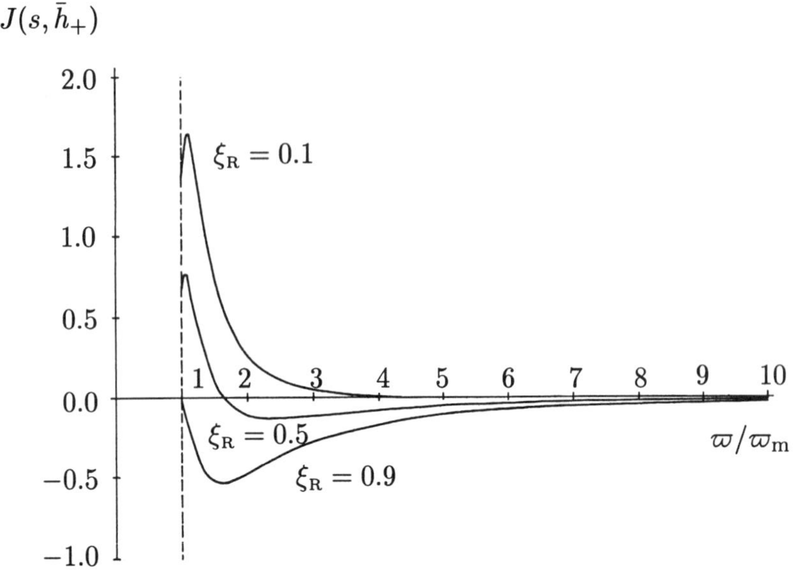

Figure 10.1. The surface toroidal field for constant diffusivity. Based on Campbell (1987).

this radius the source in the integrand of (10.22) makes no significant contribution to B_ϕ. It is noted that $\varpi = 15\varpi_{\rm m}$ is well within a tidal radius, so justifying treating the disc as axisymmetric.

Figure 10.1 is a plot of $B_\phi(\varpi, h)$, represented by its dimensionless form $J(s, \bar{h}_+)$, for three values of the fastness parameter $\xi_{\rm R}$. It is noted that for intermediate values of $\xi_{\rm R}$ the sign of B_ϕ^+ changes. For all $\xi_{\rm R}$ values B_ϕ^+ has essentially vanished for $\varpi \gtrsim 10\varpi_{\rm m}$.

10.3.2. VARIABLE DIFFUSIVITY

It is of interest that another exact solution of the induction equation can be found if η is taken as

$$\eta = K\varpi^2, \tag{10.31}$$

where K is a constant. Substituting this in (10.13) gives

$$\frac{\partial^2 B_\phi}{\partial \varpi^2} + \frac{3}{\varpi}\frac{\partial B_\phi}{\partial \varpi} + \frac{B_\phi}{\varpi^2} + \frac{\partial^2 B_\phi}{\partial z^2} = -\frac{\varpi}{\eta}\mathbf{B}_p \cdot \nabla\Omega. \tag{10.32}$$

This equation can be written in the Poisson form

$$\nabla^2(\varpi B_\phi) = -\frac{\varpi^2}{\eta}\mathbf{B}_p \cdot \nabla\Omega. \tag{10.33}$$

The integral solution yields

$$B_\phi = \frac{1}{4\pi\varpi}\int_{-h}^{h}\int_{\varpi_\mathrm{m}}^{\varpi_\mathrm{D}}\int_0^{2\pi}\frac{\varpi_0^3(\mathbf{B}_p\cdot\nabla\Omega)_0}{\eta(\varpi_0)|\mathbf{r}-\mathbf{r}_0|}d\phi_0 d\varpi_0 dz_0. \tag{10.34}$$

The azimuthal integral can be expressed as

$$W(\varpi, z|\varpi_0, z_0) = \frac{F(\pi/2, k)}{[(\varpi+\varpi_0)^2+(z-z_0)^2]^{\frac{1}{2}}}, \tag{10.35}$$

where F is an elliptic integral of the first kind and k is defined by (10.21b). The toroidal field solution for quadratic η is therefore

$$B_\phi = \frac{1}{4\pi\varpi}\int_{-h}^{h}\int_{\varpi_\mathrm{m}}^{\omega_\mathrm{D}}\frac{\varpi_0^3(\mathbf{B}_p\cdot\nabla\Omega)_0 F(\pi/2,k)}{\eta(\varpi_0)[(\varpi+\varpi_0)^2+(z-z_0)^2]^{\frac{1}{2}}}d\varpi_0 dz_0. \tag{10.36}$$

Equations (9.14), (9.15), (10.14), (10.15), (10.31) and (10.36) yield

$$B_\phi = \frac{3(B_\mathrm{p})_0 R_\mathrm{p}^3}{2\pi\varpi_\mathrm{m}\eta(\varpi_\mathrm{m})}\left(\frac{GM_\mathrm{p}}{\varpi_\mathrm{m}^3}\right)^{\frac{1}{2}} U(s,\bar{h}), \tag{10.37}$$

where

$$U(s,\bar{h}) = \left(\frac{GM_\mathrm{p}}{\varpi_\mathrm{m}^3}\right)^{-\frac{1}{2}}\frac{1}{s}(K_1+K_2+K_3), \tag{10.38}$$

with

$$K_1 = \int_{\bar{h}_-}^{\bar{h}_+}\int_1^{s_\mathrm{D}}\frac{s_0^2\bar{h}_0}{(s_0^2+\bar{h}_0^2)^{\frac{5}{2}}}\frac{d\Omega_\mathrm{K}}{ds_0}Q(s,\bar{h}|s_0,\bar{h}_0)ds_0 d\bar{h}_0, \tag{10.39}$$

$$K_2 = -\frac{1}{3}\int_1^{s_\mathrm{D}}\frac{s_0(s_0^2-2\bar{h}_+^2)}{(s_0^2+\bar{h}_+^2)^{\frac{5}{2}}}\Delta Q[\Omega_\mathrm{p}-\Omega_\mathrm{K}(s_0)]ds_0, \tag{10.40}$$

$$K_3 = [\Omega_\mathrm{K}(\varpi_\mathrm{m})-\Omega_\mathrm{p}]\int_{\bar{h}_-}^{\bar{h}_+}\frac{\bar{h}_0 Q(s,\bar{h}|1,\bar{h}_0)}{(1-\bar{h}_0^2)^{\frac{5}{2}}}d\bar{h}_0, \tag{10.41}$$

and

$$Q(s,\bar{h}|s_0,\bar{h}_0) = \frac{F\left(\frac{\pi}{2},k\right)}{[(s+s_0)^2+(\bar{h}-\bar{h}_0)^2]^{\frac{1}{2}}}, \tag{10.42a}$$

$$\Delta Q = Q(s,\bar{h}|s_0,\bar{h}_+) - Q(s,\bar{h}|s_0,\bar{h}_-), \tag{10.42b}$$

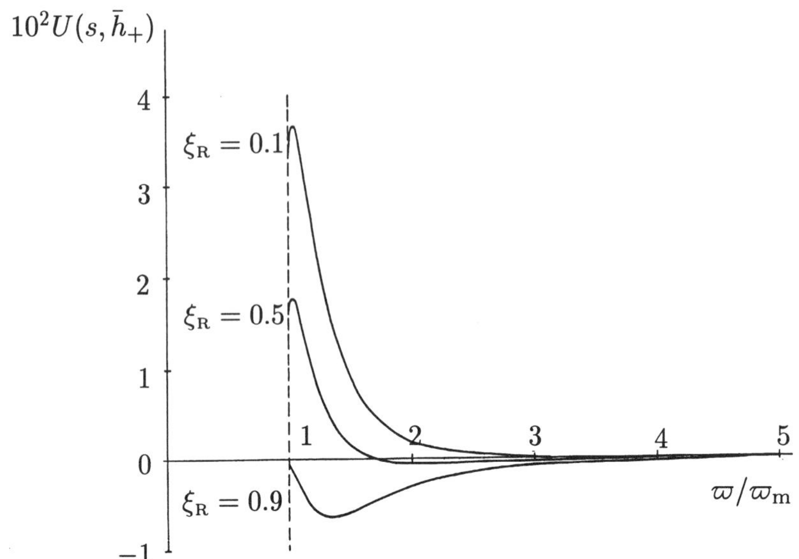

Figure 10.2. The toroidal field for quadratic diffusivity. Based on Campbell (1987).

where s, $\bar{h}$ and k are defined by (10.23) and (10.30b). Figure 10.2 is a plot of the dimensionless toroidal field on the disc surface, for three values of ξ_R. The similar qualitative form to the solutions in Figure 10.1 is noted. However, in the present case B_ϕ^+ falls more rapidly with increasing ϖ, being reduced by the increasing diffusivity.

The approximate expressions (10.4) and (10.8) for B_ϕ^+, found by dropping the radial derivatives in the induction equation, have similar radial variations to the solutions shown in Figures 10.1 and 10.2.

10.4. The Stellar Torques

10.4.1. MAGNETIC AND ACCRETION TORQUES

A magnetic torque is exerted on the disc material and an equal and opposite torque is transmitted to the primary star through electric currents in the low density magnetosphere. It follows that the primary star experiences a torque

$$\mathbf{T}_m = -\frac{1}{\mu_0} \int_{V_D} \mathbf{r} \wedge [(\nabla \wedge \mathbf{B}) \wedge \mathbf{B}] dV. \tag{10.43}$$

It is straightforward to show that this can be expressed in terms of surface integrals as

$$\mathbf{T}_{\mathrm{m}} = -\frac{1}{\mu_0}\int_{S_{\mathrm{D}}}(\mathbf{r}\wedge\mathbf{B})\mathbf{B}\cdot d\mathbf{S} + \frac{1}{2\mu_0}\int_{S_{\mathrm{D}}} B^2\mathbf{r}\wedge d\mathbf{S}. \tag{10.44}$$

From the symmetry of B^2 about $z = 0$, it follows that the second surface integral vanishes. A thin disc has $dh/d\varpi \ll 1$, so $d\mathbf{S}$ is nearly parallel to $\hat{\mathbf{z}}$ on the upper and lower surfaces and $\hat{\mathbf{r}}$ is nearly coincident with $\hat{\boldsymbol{\varpi}}$. The magnetic torque then becomes

$$\mathbf{T}_{\mathrm{m}} = -\frac{4\pi}{\mu_0}\left[\int_{S_+}\varpi^2 B_\phi B_z d\varpi - \frac{\varpi_{\mathrm{m}}^2}{2}\int_{S_{\mathrm{m}}} B_\varpi B_\phi dz\right]\hat{\mathbf{z}}, \tag{10.45}$$

where S_+ and S_{m} are the upper and inner disc faces.

Material leaving the inner edge of the disc is channelled by the primary star's magnetic field. The accretion torque can be found by multiplying the specific angular momentum of material at ϖ_{m} by the mass transfer rate $\dot{M}_{\mathrm{p}}$, giving

$$\mathbf{T}_{\mathrm{a}} = \dot{M}_{\mathrm{p}}\varpi_{\mathrm{m}}^2\Omega_{\mathrm{K}}(\varpi_{\mathrm{m}})\hat{\mathbf{z}}. \tag{10.46}$$

10.4.2. MAGNETIC BUOYANCY

The magnetic buoyancy solution for B_ϕ^+, given by (10.4), was derived using the thin disc condition $dh/d\varpi \ll 1$. It is therefore appropriate when using this solution to calculate the magnetic torque to drop the second integral in (10.45), since it is a factor $\sim (h/\varpi)^2$ smaller than the first. Use of (10.4), together with (9.51), (9.81) and (9.97), in (10.45) yields

$$T_{\mathrm{m}} = \frac{4\pi}{\mu_0}\left|\varpi^3 B_\phi^+ B_z\right|_{\varpi_{\mathrm{m}}}\frac{I_{\mathrm{B}}(\xi_{\mathrm{R}})}{(1-\xi_{\mathrm{R}})^{\frac{1}{2}}}, \tag{10.47}$$

where

$$I_{\mathrm{B}}(\xi_{\mathrm{R}}) = \int_1^{x_{\mathrm{co}}}\frac{(1-\xi_{\mathrm{R}}x^{\frac{3}{2}})^{\frac{1}{2}}}{x^{\frac{103}{32}}}dx + \int_{x_{\mathrm{co}}}^{x_{\mathrm{D}}}\frac{(\xi_{\mathrm{R}}x^{\frac{3}{2}}-1)^{\frac{1}{2}}}{x^{\frac{103}{32}}}dx. \tag{10.48}$$

The dimensionless variable $x = \varpi/\varpi_{\mathrm{m}}$, and the integral is split about the corotation point. Using the transformation $u = \xi_{\mathrm{R}}x^{3/2}$ gives

$$I_{\mathrm{B}}(\xi_{\mathrm{R}}) = -\frac{4}{9}\xi_{\mathrm{R}}^{\frac{3}{2}}\left[\left(1-\frac{1}{\xi_{\mathrm{R}}x_{\mathrm{D}}^{\frac{3}{2}}}\right)^{\frac{3}{2}} - \left(\frac{1}{\xi_{\mathrm{R}}}-1\right)^{\frac{3}{2}}\right].$$

The dimensionless disc radius $x_{\rm D} = \varpi_{\rm D}/\varpi_{\rm m} \gg 1$ and so $1/\xi_{\rm R} x_{\rm D}^{3/2} \ll 1$ for the values of $\xi_{\rm R}$ considered here. It follows that, to a good approximation,

$$I_{\rm B}\left(\xi_{\rm R}\right) = -\frac{4}{9}\left[\xi_{\rm R}^{\frac{3}{2}} - \left(1-\xi_{\rm R}\right)^{\frac{3}{2}}\right]. \tag{10.49}$$

The magnetic torque can be related to the accretion torque by noting that $\varpi_{\rm m}$ was found from $\tilde{F}_{{\rm m}\phi} = 2\tilde{F}_{{\rm v}\phi}$, so (9.98) yields

$$\frac{4\pi}{\mu_0}\left|\varpi^3 B_\phi^+ B_z\right|_{\varpi_{\rm m}} = \dot{M}_{\rm p}\varpi_{\rm m}^2\Omega_{\rm K}(\varpi_{\rm m}). \tag{10.50}$$

Equations (10.46), (10.47) and (10.50) give the total torque acting on the primary star as

$$T = \dot{M}_{\rm p}\varpi_{\rm m}^2\Omega_{\rm K}(\varpi_{\rm m})\left(1 - \frac{4}{9}\frac{\left[\xi_{\rm R}^{\frac{3}{2}} - \left(1-\xi_{\rm R}\right)^{\frac{3}{2}}\right]}{\left(1-\xi_{\rm R}\right)^{\frac{1}{2}}}\right), \tag{10.51}$$

where $\varpi_{\rm m}$ is given by (10.7) as a function of the binary parameters and $\xi_{\rm R}$.

10.4.3. TURBULENT DIFFUSION

In this case the magnetic torque is given by (10.8), (10.9) and the first integral in (10.45) as

$$T_{\rm m} = \frac{4\pi}{\mu_0}\left|\varpi^3 B_\phi^+ B_z\right|_{\varpi_{\rm m}}\frac{I_{\rm T}\left(\xi_{\rm R}\right)}{\left(1-\xi_{\rm R}\right)}, \tag{10.52}$$

where

$$I_{\rm T}\left(\xi_{\rm R}\right) = \int_1^{x_{\rm D}}\frac{\left(1-\xi_{\rm R}x^{\frac{3}{2}}\right)}{x^{\frac{33}{8}}}dx. \tag{10.53}$$

Since $x_{\rm D} \gg 1$ this integral gives

$$I_{\rm T}\left(\xi_{\rm R}\right) = \frac{8}{25}\left(1 - \frac{25}{13}\xi_{\rm R}\right). \tag{10.54}$$

The sum of the magnetic and accretion torques is

$$T = \dot{M}_{\rm p}\varpi_{\rm m}^2\Omega_{\rm K}(\varpi_{\rm m})\left(1 + \frac{8}{25}\frac{\left(1-\frac{25}{13}\xi_{\rm R}\right)}{\left(1-\xi_{\rm R}\right)}\right), \tag{10.55}$$

with $\varpi_{\rm m}$ given by (10.12).

10.5. Stellar Spin Evolution

The spin evolution of the primary star is determined by

$$I\dot{\Omega}_{\mathrm{p}} = T, \tag{10.56}$$

where the moment of the inertia is taken as $I = 0.4 M_{\mathrm{p}} R_{\mathrm{p}}^2$. The quantity $\dot{P}_{\mathrm{R}}/P_{\mathrm{R}}$, where $P_{\mathrm{R}} = 2\pi/\Omega_{\mathrm{p}}$, can then be expressed as

$$\frac{\dot{P}_{\mathrm{R}}}{P_{\mathrm{R}}} = -\frac{\dot{\Omega}_{\mathrm{p}}}{\Omega_{\mathrm{p}}} = -\frac{5}{2}\frac{T}{M_{\mathrm{p}} R_{\mathrm{p}}^2 \Omega_{\mathrm{p}}}, \tag{10.57}$$

and so can be related to the fastness parameter ξ_{R}, for the two diffusivity cases.

10.5.1. MAGNETIC BUOYANCY

Use of (10.7) and (10.51) in (10.57) gives

$$\frac{\dot{P}_{\mathrm{R}}}{P_{\mathrm{R}}} = -\frac{1.8}{10^3} \frac{\gamma^{\frac{32}{87}} \left(\frac{(B_{\mathrm{p}})_0}{10^8\,\mathrm{T}}\right)^{\frac{32}{29}} \left(\frac{R_{\mathrm{p}}}{10^4\,\mathrm{m}}\right)^{\frac{38}{29}} \left(\frac{\dot{M}_{\mathrm{p}}}{10^{-9}\,M_{\odot}\mathrm{yr}^{-1}}\right)^{\frac{53}{145}}}{\left(\frac{M_{\mathrm{p}}}{1.4\,M_{\odot}}\right)^{\frac{31}{29}}} f_{\mathrm{B}}(\xi_{\mathrm{R}})\,\mathrm{yr}^{-1}, \tag{10.58}$$

where

$$f_{\mathrm{B}}(\xi_{\mathrm{R}}) = \frac{(1-\xi_{\mathrm{R}})^{\frac{32}{87}}}{\xi_{\mathrm{R}}} \left(1 - \frac{4}{9}\frac{\left[\xi_{\mathrm{R}}^{\frac{3}{2}} - (1-\xi_{\mathrm{R}})^{\frac{3}{2}}\right]}{(1-\xi_{\mathrm{R}})^{\frac{1}{2}}}\right). \tag{10.59}$$

The parameters in (10.58) are normalized with respect to values typical of X-ray binary pulsars. For given system parameters, a value of ξ_{R} can be found which corresponds to an observed value of $\dot{P}_{\mathrm{R}}/P_{\mathrm{R}}$. Equation (10.7) then yields the disc's inner radius ϖ_{m}.

An equilibrium period exists at which $\dot{P}_{\mathrm{R}}$ vanishes. It follows from (10.58) that $f_{\mathrm{B}}(\xi_{\mathrm{R}}) = 0$ at this period and (10.59) gives

$$\frac{9}{4}(1-\xi_{\mathrm{R}})^{\frac{1}{2}} = \xi_{\mathrm{R}}^{\frac{3}{2}} - (1-\xi_{\mathrm{R}})^{\frac{3}{2}},$$

which has the solution $\xi_{\mathrm{R}}^{\mathrm{eq}} = 0.88$. The equilibrium period can then be found from

$$P_{\mathrm{R}}^{\mathrm{eq}} = \frac{2\pi}{\xi_{\mathrm{R}}^{\mathrm{eq}}} \frac{\varpi_{\mathrm{m}}^{\frac{3}{2}}}{(GM_{\mathrm{p}})^{\frac{1}{2}}}. \tag{10.60}$$

TABLE 10.1. The spin evolution function $f_{\rm B}(\xi_{\rm R})$

$\xi_{\rm R}$	$f_{\rm B}(\xi_{\rm R})$	$\xi_{\rm R}$	$f_{\rm B}(\xi_{\rm R})$
0.100	13.33	0.525	1.26
0.125	10.42	0.550	1.12
0.150	8.48	0.575	1.13
0.175	7.09	0.600	1.01
0.200	6.04	0.625	0.90
0.225	5.22	0.650	0.80
0.250	4.57	0.675	0.70
0.275	4.03	0.700	0.60
0.300	3.58	0.725	0.51
0.325	3.19	0.750	0.43
0.350	2.86	0.775	0.34
0.375	2.58	0.800	0.26
0.400	2.32	0.825	0.18
0.425	2.10	0.850	0.10
0.450	1.90	0.875	0.01
0.475	1.72	0.900	-0.07
0.500	1.55	0.925	-0.17

Equations (10.7) and (10.60) yield this period as

$$P_{\rm R}^{\rm eq} = 9.2 \frac{\gamma^{\frac{8}{29}} \left(\frac{(B_{\rm p})_0}{10^8\,{\rm T}}\right)^{\frac{24}{29}} \left(\frac{R_{\rm p}}{10^4\,{\rm m}}\right)^{\frac{72}{29}}}{\left(\frac{M_{\rm p}}{1.4\,M_\odot}\right)^{\frac{16}{29}} \left(\frac{\dot{M}_{\rm p}}{10^{-9}\,M_\odot{\rm yr}^{-1}}\right)^{\frac{69}{145}}}\ {\rm s}. \tag{10.61}$$

The weak dependencies of (10.58) and (10.61) on γ are noted. Table 10.1 shows the dimensionless function $f_{\rm B}(\xi_{\rm R})$, for $0.1 \leq \xi_{\rm R} < 1$.

10.5.2. TURBULENT DIFFUSION

The relation between $P_{\rm R}/\dot{P}_{\rm R}$ and ξ follows from (10.12), (10.55) and (10.57) as

$$\frac{\dot{P}_{\rm R}}{P_{\rm R}} = -\frac{5.2}{10^3} \frac{\gamma^{\frac{16}{29}} \left(\frac{(B_{\rm p})_0}{10^8\,{\rm T}}\right)^{\frac{32}{29}} \left(\frac{R_{\rm p}}{10^4\,{\rm m}}\right)^{\frac{38}{29}} \left(\frac{\dot{M}_{\rm p}}{10^{-9}\,M_\odot{\rm yr}^{-1}}\right)^{\frac{53}{145}}}{\left(\frac{\epsilon}{0.1}\right)^{\frac{72}{145}} \left(\frac{M_{\rm p}}{1.4\,M_\odot}\right)^{\frac{31}{29}}} f_{\rm T}(\xi_{\rm R})\,{\rm yr}^{-1}, \tag{10.62}$$

TABLE 10.2. The spin evolution function $f_{\mathrm{T}}(\xi_{\mathrm{R}})$

ξ_{R}	$f_{\mathrm{T}}(\xi_{\mathrm{R}})$	ξ_{R}	$f_{\mathrm{T}}(\xi_{\mathrm{R}})$
0.100	12.14	0.525	1.26
0.125	9.50	0.550	1.12
0.150	7.73	0.575	1.00
0.175	6.46	0.600	0.88
0.200	5.51	0.625	0.77
0.225	4.77	0.650	0.67
0.250	4.17	0.675	0.56
0.275	3.68	0.700	0.46
0.300	3.27	0.725	0.37
0.325	2.92	0.750	0.27
0.350	2.62	0.775	0.17
0.375	2.35	0.800	0.07
0.400	2.12	0.825	-0.03
0.425	1.91	0.850	-0.15
0.450	1.72	0.875	-0.27
0.475	1.55	0.900	-0.42
0.500	1.40	0.925	-0.60

where

$$f_{\mathrm{T}}(\xi_{\mathrm{R}}) = \frac{(1-\xi_{\mathrm{R}})^{\frac{16}{29}}}{\xi_{\mathrm{R}}} \left(1 + \frac{8}{25} \frac{(1-\frac{25}{13}\xi_{\mathrm{R}})}{(1-\xi_{\mathrm{R}})}\right). \tag{10.63}$$

At equilibrium $f_{\mathrm{T}}(\xi_{\mathrm{R}}) = 0$ and $\xi_{\mathrm{R}}^{\mathrm{eq}} = 0.82$, so (10.12) and (10.60) give

$$P_{\mathrm{R}}^{\mathrm{eq}} = 19.5 \frac{\gamma^{\frac{12}{29}} \left(\frac{(B_{\mathrm{p}})_0}{10^8\,\mathrm{T}}\right)^{\frac{24}{29}} \left(\frac{R_{\mathrm{p}}}{10^4\,\mathrm{m}}\right)^{\frac{72}{29}}}{\left(\frac{\epsilon}{0.1}\right)^{\frac{54}{145}} \left(\frac{M_{\mathrm{p}}}{1.4\,M_{\odot}}\right)^{\frac{16}{29}} \left(\frac{\dot{M}_{\mathrm{p}}}{10^{-9}\,M_{\odot}\mathrm{yr}^{-1}}\right)^{\frac{69}{145}}}\ \mathrm{s}. \tag{10.64}$$

Table 10.2 shows the dimensionless function $f_{\mathrm{T}}(\xi_{\mathrm{R}})$. Tables 10.1 and 10.2 illustrate that the equilibrium periods are stable in both cases, since the net torques act to return the primary to equilibrium.

Since the primary star's spin change time-scale of $P_{\mathrm{R}}/|\dot{P}_{\mathrm{R}}|$ is much longer than the Keplerian flow time-scale, the disc will evolve through quasi-steady states.

10.6. Conclusions

The results for the buoyant and turbulent forms of η can be compared. The disruption radii, given by (10.7) and (10.12), are of similar magnitude but with the former having a weaker dependence on γ, the uncertainty in the vertical shear. This also applies to the equilibrium periods, given by (10.61) and (10.64). If the binary parameters and surface field $(B_{\rm p})_0$ were known for a given system, then the observed value of $\dot{P}_{\rm R}/P_{\rm R}$ would yield $\xi_{\rm R}$ from (10.58) or (10.62). The inner radius could then be found from (10.7) or (10.12). The definition of $\xi_{\rm R}$ given by (9.81) would give values for $\Omega_{\rm p}$ and hence $P_{\rm R}$ in each diffusivity case, and these could be compared with the observed values listed in Tables 8.1 and 8.3. However, the values of $(B_{\rm p})_0$ are not accurately known for the intermediate polars or the X-ray binary pulsars, at present. Nevertheless, the foregoing formulae can reproduce the observed values of $\dot{P}_{\rm R}/P_{\rm R}$ and $P_{\rm R}$ for plausible values of $(B_{\rm p})_0$, (i.e. $\sim 10^6$ G for intermediate polars and $\sim 10^{12}$ G for X-ray binary pulsars). The cases in which the primary star is observed to be spinning up correspond to positive values of $f_{\rm B}(\xi_{\rm R})$ and $f_{\rm T}(\xi_{\rm R})$ in Tables 10.1 and 10.2, since then $\dot{P}_{\rm R} < 0$ in (10.59) and (10.62).

The observed erratic period behaviour of some X-ray binary pulsars may be related to a local break down of the steady disc structure beyond $\varpi_{\rm co}$, when $P_{\rm R}$ is near $P_{\rm R}^{\rm eq}$. In the case of $\eta_{\rm B}$, (10.6) and (10.61) show that the ratio $|\tilde{F}_{{\rm m}\phi}/\tilde{F}_{{\rm v}\phi}|_{\rm max}$ is independent of γ at $P_{\rm R} = P_{\rm R}^{\rm eq}$ and, for the parameters shown, has a value of 1.4. The same property occurs in the case of $\eta_{\rm T}$, with $|\tilde{F}_{{\rm m}\phi}/\tilde{F}_{{\rm v}\phi}|_{\rm max} = 1.3$ at $P_{\rm R} = P_{\rm R}^{\rm eq}$. Since the radius at which $|\tilde{F}_{{\rm m}\phi}/\tilde{F}_{{\rm v}\phi}|_{\rm max}$ occurs exceeds the corotation radius $\varpi_{\rm co}$, the magnetic stress $B_\phi^+ B_z$ feeds angular momentum into the disc in the region of the maximum. The vertical integral of the azimuthal force equation (9.5) then shows that steady inflow will not be possible since $|\tilde{F}_{{\rm m}\phi}|$ exceeds $|\tilde{F}_{{\rm v}\phi}|$ locally. The results of Chapter 9 indicate that local thermal equilibrium will break down in such a region. This perturbation to the steady state would make a fluctuating contribution to the torque integral, which might account for the observed erratic spin period variations occurring in some systems. Assuming the neutron stars have similar values of $(B_{\rm p})_0$, $R_{\rm p}$ and $M_{\rm p}$, the strength of this effect, operable as $P_{\rm R}$ approaches $P_{\rm R}^{\rm eq}$, would depend on the value of $\dot{M}_{\rm p}$.

This chapter has estimated the behaviour of $|\tilde{F}_{{\rm m}\phi}/\tilde{F}_{{\rm v}\phi}|$ outside $\varpi_{\rm co}$. The present results are supported by the numerical work described in §9.3. It is noted that if the disc turbulence is quenched inside $\varpi_{\rm co}$, due to $|B_z|$ becoming larger, then the disruption radius $\varpi_{\rm m}$ is increased. Such quenching might occur when the magnetic energy density exceeds the turbulent energy density. Papaloizou and Szuszkiewicz (1992) showed that the Balbus-Hawley instability, believed to initiate turbulence in accretion discs, is suppressed when a sufficiently large vertical magnetic field is applied to the disc. The condition for such quenching is

$$\frac{|B_z|}{(\mu_0 \rho)^{\frac{1}{2}}} \geq \sqrt{3} \left(\frac{h}{R_{\rm D}}\right)^{\frac{1}{2}} \left(\frac{R_{\rm D}}{\varpi}\right)^2 v_{\rm K}, \tag{10.65}$$

where $R_{\rm D}$ is the disc radius. In the absence of quenching the disc still ends, due to magnetically induced viscous instability (see §9.3).

Bardou and Heyvaerts (1996) argue that large values of $B_\phi^+/|B_z|$ lead to inflation of the poloidal field. In the disc $\mathbf{B}_p$ is still predominantly vertical, but has a different radial dependence than a dipole field. This assumes that the magnetosphere remains highly conducting in the regions of the disc surfaces, giving large vertical shear (i.e. $\gamma \sim 1$). Assuming this were the case, inflation is largest for $\varpi > \varpi_{\rm co}$. However, as illustrated in Figures 10.1 and 10.2, the toroidal field B_ϕ^+ falls rapidly with increasing ϖ beyond $\varpi_{\rm co}$ and the contribution of $B_\phi^+ B_z$ to the magnetic torque is small in most of this outer region. As discussed in §10.2.1, large values of $B_\phi^+/|B_z|$ are dynamically unstable and this is liable to enhance η in the magnetosphere near the disc surfaces. The consequent break down of corotation would reduce $\partial\Omega/\partial z$ and hence lower $B_\phi^+/|B_z|$. These effects need further investigation.

Lovelace, Romanova and Bisnovatyi-Kogan (1995) suggest that if conditions allow $|B_\phi^+|/|B_z| \gg 1$ the stellar field becomes open, leading to the possibility of a disc wind. Such a picture requires the rate of opening of the field to balance its rate of reconnection, in a stable way. It is shown in Chapter 12 that wind flows from binary star accretion discs have tight restrictions on the poloidal field inclination for self-consistency. Blandford and Payne (1982) found that a critical angle, $i_c = 60°$, exists for the inclination of the poloidal field to the central plane of the disc. If this angle is exceeded by more than a few degrees the wind mass flux becomes small, since material must surmount a high effective potential barrier in order to pass through the sonic point and be accelerated away from the disc by centrifugal force. A small mass flux would be unlikely to keep the stellar field lines open in the systems considered here. Conversely, for $i < i_c$, there is no effective potential barrier to the flow and the resulting large mass flux would evaporate the disc on the Keplerian time-scale of $\sim 1/\Omega_{\rm K}$. The angle range above i_c for a tenable wind is $\lesssim 5°$. In order for i to lie close to i_c the disc must have a self-regulating mechanism, related to its magnetic diffusivity and radial inflow. This might be possible with a dynamo-generated field, but it seems unlikely with the severe constraint of an imposed stellar field occurring here.

It is likely that there is some poloidal flow between the surfaces of the undisrupted disc and the primary star. Self-consistency would require the associated mass flux to be small. The effect of such a flow would be to reduce values of $|B_\phi^+|/|B_z|$ due to advection of magnetic stress.

References

Bardou, A. and Heyvaerts, J., 1996. *Astron. Astrophys.*, in press.
Blandford, R.D. and Payne, D.G., 1982. *Mon. Not. R. Astr. Soc.*, **199**, 183.
Campbell, C.G., 1987. *Mon. Not. R. Astr. Soc.*, **229**, 405.
Campbell, C.G., 1992. *Geophys. Astrophys. Fluid. Dynam.*, **66**, 243.
Campbell, C.G. and Heptinstall, P., 1997. *Mon. Not. R. Astr. Soc.*, submitted.
Lovelace, R.V.E., Romanova, M.M. and Bisnovatyi-Kogan, G.S., 1995. *Mon. Not. R. Astr. Soc.*, **275**, 244.
Papaloizou, J.C.B. and Szuszkiewicz, E., 1992. *Geophys. Astrophys. Fluid. Dynam.*, **66**, 223.
Wang, Y.M., 1987. *Astron. Astrophys.*, **183**, 257.

CHAPTER 11

INTRINSIC MAGNETISM IN ACCRETION DISCS

11.1. Introduction

In a standard cataclysmic binary an accretion disc forms around a non-magnetic white dwarf, and extends down to the stellar surface. The disc is fed by the stream originating from the L_1 region of the secondary star. In a steady state, matter will be transported through the disc at the rate it is supplied by the stream and angular momentum will be advected outwards. Angular momentum advection requires coupling between rings of material and, as shown in §2.4.3, ordinary molecular viscosity is far too weak to provide this. Hence some form of anomalous viscosity must be invoked to explain mass flow through the disc. The standard accretion disc model of Shakura and Sunyaev (1973) employs a parameterized form of turbulent viscosity to supply the necessary coupling. However, no instability has yet been found which leads to turbulence in a non-magnetic Keplerian disc.

The foregoing situation led to a search for an alternative coupling mechanism to generate the anomalous viscosity in accretion discs. Magnetic fields present the most plausible means of angular momentum advection, through the stresses they exert on material. The possibility of magnetic viscosity was suggested early in the development of accretion disc theory (e.g. Lynden-Bell, 1969; Shakura and Sunyaev, 1973; Eardley and Lightman, 1975; Ichimaru, 1977; Coroniti, 1981). These models are based on the notion that the disc contains small-scale magnetic cells which are sheared by the differential rotation, resulting in reconnection of oppositely directed fields and hence a magnetic torque. Such topological dissipation produced by reconnection was discussed by Parker (1972). The associated magnetic field has a turbulent configuration. However, Galeev, Rosner and Vaiana (1979) and Stella and Rosner (1984) showed that magnetic buoyancy effects should transport the strong toroidal magnetic field vertically through the disc in less than the radial inflow time of material. Hence a dynamo mechanism is required to regenerate the magnetic field. As seen in §2.3.1, differential rotation creates toroidal magnetic field from poloidal field (the ω-effect), but an α-effect is required to generate the poloidal field from the toroidal and give a self-sustaining dynamo. The α-effect might be generated from the magnetic turbulence, but at the time of these early models no mechanism had been found to create such turbulence. Hence such models were subject to the same fundamental dilemma that existed in the non-magnetic turbulent disc model.

Blandford and Payne (1982) developed a model of a hydromagnetic wind from an accretion disc, as a means of removing angular momentum and driving the inflow. This idea has been investigated by subsequent authors and is discussed in Chapter 12. However, the mechanism relies on a source of poloidal field and hence, for binary star discs, requires a dynamo.

A viscosity mechanism based on the inward propagation of an incoherent field of slightly non-axisymmetric internal waves was considered in models by Vishniac and Diamond (1989, 1992) and Vishniac, Jin and Diamond (1990). The asymmetry in the wave distribution creates a non-zero mean helicity and hence supplies the α-effect for an $\alpha\omega$-dynamo. The generation of such waves requires a continuous energy source, and the accretion stream impact and tidal forcing in the outer parts of the disc were suggested for this.

A major step towards explaining angular momentum transport in accretion discs was taken by Balbus and Hawley (1991). They applied a linear, hydromagnetic instability, first discovered by Velikhov (1959) and Chandrasekhar (1960, 1961), to show that a Keplerian disc containing a weak poloidal magnetic field is dynamically unstable to axisymmetric disturbances. The instability derives its energy from the strong radial shear in the disc. The authors suggested that this may be a major source of turbulence in accretion discs. Subsequent work confirmed such turbulent generation (e.g. Hawley and Balbus, 1991, 1992; Brandenburg *et al*, 1995; Hawley, Gammie and Balbus, 1995).

The advent of the Balbus-Hawley instability makes the role of the foregoing internal wave mechanism a more complex issue. In particular, the turbulence generated by magnetic shear instabilities may damp internal waves, rendering them ineffective in producing an α-effect. However, Zhang, Diamond and Vishniac (1994) argue that low frequency magnetic shearing instability cells can coexist with high frequency radial internal waves, at least in the case when the mean magnetic field is vertical. Nevertheless, magnetic shear instabilities alone appear to generate the turbulence necessary for radial advection of angular momentum in Keplerian discs.

Tout and Pringle (1992) developed an accretion disc dynamo model incorporating the Balbus-Hawley instability, together with the Parker instability and reconnection. The model gives rise to finite, but non-stationary, magnetic field configurations.

Having established a large-scale, mean magnetic field in the disc by one, or several, of the foregoing mechanisms, its effect on angular momentum advection and the large-scale structure of the disc must be investigated. Campbell (1992) constructed a dynamical model of a magnetic accretion disc, incorporating a turbulent $\alpha\omega$-dynamo. For a buoyancy diffusivity, the magnetic stress dominates the viscous stress in the advection of angular momentum, and only weak turbulence is required for the necessary dynamo α-effect. More general solutions were found by Campbell and Caunt (1997), allowing for a sum of buoyancy and turbulent diffusivities.

It should be noted that convection, and finite instabilities developing from a defect in the shear flow, have been considered as a means of angular momentum transport in discs. However, since these were investigated in a non-magnetic context, they are beyond the scope of this book. The reader is referred to the review by Papaloizou and Lin (1995) for a discussion of such processes.

Section 11.2.1 considers the properties of $\alpha\omega$-dynamos operating in thin discs. In §11.2.2 the possible generation of an α-effect by internal waves is outlined. Section 11.2.3 discusses the magnetic shear instability, while in §11.2.4 dynamo models with hydromagnetic turbulence are reviewed. Section 11.3 presents a dynamical disc model, incorporating an $\alpha\omega$-dynamo, with a self-consistent solution for the disc structure and a magnetically-driven inflow. Section 11.4 considers magneto-viscous discs, and future areas of research are discussed in §11.5.

11.2. Accretion Disc Dynamos

11.2.1. TURBULENT $\alpha\omega$-DYNAMOS IN THIN DISCS

This section discusses the properties of thin disc, mean-field $\alpha\omega$-dynamos. It is assumed that some form of turbulence generates the α-effect by producing a finite mean helicity (see §2.3.1). Possible mechanisms for generating the poloidal field are subsequently considered.

The equations appropriate to an $\alpha\omega$-dynamo follow from (2.192) and (2.193) as

$$\frac{\partial B_\phi}{\partial t} = \varpi \mathbf{B}_p \cdot \nabla\Omega + \eta\left(\nabla^2 B_\phi - \frac{B_\phi}{\varpi^2}\right) - \varpi\nabla\cdot\left(\frac{B_\phi}{\varpi}\mathbf{v}_p\right), \qquad (11.1)$$

$$\frac{\partial A}{\partial t} = \alpha B_\phi + \eta\left(\nabla^2 A - \frac{A}{\varpi^2}\right) - \frac{1}{\varpi}\mathbf{v}_p\cdot\nabla(\varpi A), \qquad (11.2)$$

where

$$\mathbf{B}_p = \nabla\wedge(A\hat{\boldsymbol{\phi}}), \qquad (11.3)$$

and the α terms have been dropped in (11.1) since they are negligble in this regime. In thin disc dynamo models the poloidal advection terms are usually small, so the field equations to be solved in the disc are

$$\frac{\partial B_\phi}{\partial t} = \varpi \mathbf{B}_p \cdot \nabla\Omega + \eta \left(\nabla^2 B_\phi - \frac{B_\phi}{\varpi^2} \right), \tag{11.4}$$

$$\frac{\partial A}{\partial t} = \alpha B_\phi + \eta \left(\nabla^2 A - \frac{A}{\varpi^2} \right). \tag{11.5}$$

It is often reasonable to treat the surrounding medium as a vacuum, giving

$$B_\phi = 0, \tag{11.6}$$

$$\nabla^2 A - \frac{A}{\varpi^2} = 0, \tag{11.7}$$

where (11.6) follows from axisymmetry Equation (11.3) gives the poloidal field components as

$$B_\varpi = -\frac{\partial A}{\partial z}, \qquad B_z = \frac{1}{\varpi}\frac{\partial}{\partial \varpi}(\varpi A). \tag{11.8a,b}$$

These are continuous across the disc surfaces $z = \pm h(\varpi)$, while continuity of B_ϕ is satisfied by its vanishing on the boundaries.

The basic properties of dynamos in thin galactic discs, formulated by Parker (1971), are relevant to binary star accretion discs. The key idea is that in a thin disc, for which $h(\varpi)/\varpi \ll 1$, solutions of (11.4) and (11.5) should exist which have a radial length-scale long compared to the characteristic disc height. The vertical derivatives therefore dominate in the diffusion terms, and local magnetic Reynolds numbers can be defined as

$$R_\alpha = \frac{h\tilde{\alpha}}{\eta}, \qquad R_\omega = \frac{h^2}{\eta}\varpi\frac{d\Omega}{d\varpi}, \tag{11.9a,b}$$

where Ω is taken as independent of z and the α-effect is described by the separable form

$$\alpha = \tilde{\alpha}(\varpi) f(\zeta), \tag{11.10}$$

with

$$\zeta = \frac{z}{h}, \tag{11.11}$$

and $f(\zeta)$ antisymmetric. The local dynamo number is given by

$$D = R_\alpha R_\omega, \tag{11.12}$$

which principally determines the field generation at a given radius.

Parker (1971) considered a thin disc of constant height h and sought local solutions of the form

$$B_\phi = B_0 R_\omega \bar{B}(\zeta) \exp[i(k\varpi - \omega t)], \tag{11.13}$$

$$A = B_0 h \bar{A}(\zeta) \exp[i(k\varpi - \omega t)], \tag{11.14}$$

where B_0 is a typical magnetic field, and ζ is the scaled z-coordinate defined by (11.11). Substitution in (11.4) and (11.5) shows that the dimensonless functions $\bar{B}$ and $\bar{A}$ must satisfy

$$\bar{B}'' - (\bar{k}^2 - i\omega^*)\bar{B} = \bar{A}', \tag{11.15}$$

$$\bar{A}'' - (\bar{k}^2 - i\omega^*)\bar{A} = -Df\bar{B}, \tag{11.16}$$

where the dimensionless radial wavenumber and frequency are

$$\bar{k} = hk, \qquad \omega^* = \tau_{\mathrm{d}}\omega, \tag{11.17a,b}$$

with the vertical diffusion time being

$$\tau_{\mathrm{d}} = \frac{h^2}{\eta}, \tag{11.17c}$$

and primes denote differentiation with respect ζ. The function $f(\zeta)$ occurring in (11.16) is the vertical dependence of α, in accordance with (11.10). Matching the solutions (11.13) and (11.14) to an exterior vacuum field leads to the boundary conditions

$$\bar{B} = 0, \qquad \bar{A}' \pm |\bar{k}|\bar{A} = 0, \quad \text{on } \zeta = \pm 1. \tag{11.18a,b}$$

The foregoing derivations assume that the modes have radial wavelengths satisfying $h \ll \lambda < \varpi$. Such modes will therefore have $\bar{k} \ll 1$. Radial derivatives of R_ω and k are dropped in obtaining (11.15) and (11.16), the radial diffusion being represented by the $\bar{k}^2$ terms. Since $f(\zeta)$ is antisymmetric the modes are of two types. The even modes of quadrupole parity have

$$\bar{B}(-\zeta) = \bar{B}(\zeta), \qquad \bar{A}(-\zeta) = -\bar{A}(\zeta), \tag{11.19a,b}$$

which imply that

$$\bar{B}'(0) = \bar{A}(0) = 0. \tag{11.20a,b}$$

The odd modes of dipole parity have

$$\bar{B}(-\zeta) = -\bar{B}(\zeta), \qquad \bar{A}(-\zeta) = \bar{A}(\zeta), \tag{11.21a,b}$$

giving

$$\bar{B}(0) = \bar{A}'(0) = 0. \tag{11.22a,b}$$

The most readily excited modes of a given class are usually of long wavelength, satisfying $\bar{k} \ll 1$. For this reason, Moffatt (1978) proposed the approximation of setting $\bar{k} = 0$ in (11.15), (11.16) and (11.18). For steady modes, these equations then give

$$\bar{B}''' + Df\bar{B} = 0, \tag{11.23}$$

$$\bar{A} = K + \bar{B}', \tag{11.24}$$

where K is a constant. Since (11.23) is a third order equation, the possibility arises that the four boundary conditions on the original fourth order system may be inconsistent. The boundary conditions on the steady quadrupole modes reduce consistently to

$$\bar{B}'(0) = 0, \qquad \bar{B}(1) = \bar{B}''(1) = 0, \tag{11.25}$$

with $K = 0$. An inconsistency arises for steady dipole modes because four independent boundary conditions emerge for $\bar{B}$. This is overcome by considering small but finite $\bar{k}$ and using the expansions

$$\bar{A} = \bar{k}^{-1}\bar{A}_{-1} + \bar{A}_0 + O(\bar{k}), \tag{11.26}$$

and

$$\bar{B} = \bar{B}_0 + O(\bar{k}), \tag{11.27}$$

where $\bar{A}_{-1}$ is independent of ζ. At zeroth order $\bar{B}$ is again governed by (11.23). The symmetry conditions (11.22a,b) on $\zeta = 0$ and the vanishing of $\bar{B}$ on $\zeta = 1$ imply that

$$\bar{B}(0) = \bar{B}''(0) = 0, \qquad \bar{B}(1) = 0, \tag{11.28a}$$

while the remaining boundary condition at $\zeta = 1$, given by (11.18b), yields

$$\bar{A}_{-1} = -\bar{B}_0''(1). \tag{11.28b}$$

The dynamo numbers for steady modes with $\bar{k} \ll 1$ can be expanded in the form

$$D = D_0 + D_1\bar{k} + D_2\bar{k}^2 + O(\bar{k}^3), \tag{11.29}$$

where D_0, D_1 and D_2 all have the same sign. The quadrupole and dipole modes have eigenvalues of oposite signs, being negative and positive respectively. At lowest order (11.23) is solved subject to either (11.25) or (11.28a), and the eigenvalue problem yields D_0. At first order the radial structure influences the solution via (11.18b), which expresses the match of the poloidal disc field to an external potential field. At this order the external field links different parts of the disc. Radial diffusion only becomes effective at second order through the $\bar{k}^2$ terms in (11.15) and (11.16).

Account can be taken of a slowly varying disc height $h(\varpi)$ and WKBJ approximations have been used to allow for the effect of radial disc structure (see Soward, 1992a, and references therein). For most dynamo modes in a thin disc the foregoing asymptotic analysis gives reasonable results. It should be noted that the most readily excited oscillatory dipole mode has a radial length-scale comparable to h, so $\bar{k}$ is of order unity for this mode and a different analysis is required (see Soward, 1992b).

Simple forms for a turbulent η and the separable radial dependence of α can be defined as

$$\eta = \epsilon c_{\mathrm{S}} h, \tag{11.30}$$

$$\tilde{\alpha} = \epsilon c_{\mathrm{S}}, \tag{11.31}$$

with $\epsilon = v_{\mathrm{T}}/c_{\mathrm{S}}$, where v_{T} is the rms turbulent speed, so $\epsilon < 1$ for subsonic turbulence. For a Keplerian disc, with a vertical equilibrium weakly affected by $\mathbf{B}$, $c_{\mathrm{S}} = h\Omega_{\mathrm{K}}$ so substitution of (11.30) and (11.31) in (11.9) and (11.12) yields

$$\left|\frac{R_\alpha}{R_\omega}\right| = \frac{2}{3}\frac{\tilde{\alpha}}{h\Omega_{\mathrm{K}}} = \frac{2}{3}\epsilon, \tag{11.32}$$

$$D = -\frac{3h^3\tilde{\alpha}\Omega_{\mathrm{K}}}{2\eta^2} = -\frac{3}{2\epsilon}, \tag{11.33}$$

Taking the turbulent Mach number ϵ as independent of ϖ therefore gives a spatially uniform dynamo number in this case. If the turbulent viscosity is $\nu \sim \eta$, then a viscously driven inflow has $|v_\varpi| \sim \eta/\varpi$. Then the poloidal advection terms in (11.1) and (11.2) are of the same order as the radial diffusion terms, and hence are ignorable to a good approximation. This is a consequence of the vertical diffusion time being much shorter than the radial inflow time.

Pudritz (1981) considered dynamo action in a turbulent Keplerian disc. Forms for η and $\tilde{\alpha}$ similiar to (11.30) and (11.31) were used, so yielding a uniform D. The total α function was taken as

$$\alpha = \tilde{\alpha}(\varpi)\frac{z}{h}, \tag{11.34}$$

with a constant height h. Separable, steady state solutions of (11.4) and (11.5) were found and matched to the solutions of the vacuum equations, using the properties $\epsilon < 1$ and $\varpi/h \ll 1$. The quadrupole solution, for $z > 0$, is

$$B_\varpi = \beta B(\kappa)\lambda^{(1-\kappa^2)/4}e^{-3\lambda/8}\sin\left(\frac{3\sqrt{3}}{8}(\lambda_0 - \lambda) + \frac{2\pi}{3}\right) J_1(\kappa^{\frac{1}{2}}s), \tag{11.35}$$

$$B_\phi = B(\kappa)\lambda^{-(1+\kappa^2)/4}e^{-3\lambda/8}\sin\left(\frac{3\sqrt{3}}{8}(\lambda_0 - \lambda)\right) J_1(\kappa^{\frac{1}{2}}s), \tag{11.36}$$

$$B_z = -\beta B(\kappa)\lambda^{-\kappa^2/4}e^{-3\lambda/8}\cos\left(\frac{3\sqrt{3}}{8}(\lambda_0 - \lambda) - \frac{2\pi}{3}\right) J_0(\kappa^{\frac{1}{2}}s), \tag{11.37}$$

where

$$\beta = \left|\frac{R_\alpha}{R_\omega}\right|^{\frac{1}{2}}, \qquad s = |D|^{\frac{1}{4}}\frac{\varpi}{h}, \qquad \lambda = |D|^{\frac{1}{3}}\left|\frac{z}{h}\right|^{\frac{4}{3}}, \tag{11.38a,b,c}$$

$\kappa^{1/2}$ is a dimensionless radial wavenumber, being equivalent to $\bar{k}$ defined by (11.17a), $B(\kappa)$ is the field amplitude and J_0 and J_1 are Bessel functions. These are the most readily excited modes, the critical dynamo number being $D_c = 13.8$. Consistency of the model requires $\kappa^{1/2} < \epsilon$.

Stepinski and Levy (1990) consider a Keplerian disc in which

$$\alpha = \tilde{K}\Omega_K z, \tag{11.39}$$

where $\tilde{K}$ is a constant. The magnetic diffusivity is taken as uniform, giving a local dynamo number

$$D = \tilde{D}\left(\frac{h}{\varpi}\right)^3, \tag{11.40}$$

with $\tilde{D}$ constant. The fields are expressed in the forms

$$B_\phi = e^{\gamma t} \sum_{k=1}^{\infty} B_k(z) J_1 \left(\alpha_k \frac{\varpi}{R_{\rm D}} \right), \tag{11.41}$$

$$A = e^{\gamma t} \sum_{k=1}^{\infty} A_k(z) J_1 \left(\alpha_k \frac{\varpi}{R_{\rm D}} \right), \tag{11.42}$$

where $R_{\rm D}$ is the disc radius, J_1 is a Bessel function and, in general, γ is complex. Solutions are sought for which $Re(\gamma) = 0$, resulting in oscillatory or steady behaviour. Substitution of (11.39)–(11.42) in (11.4) and (11.5), radial integration over the disc using the orthogonality property of J_1 given by (A23), and truncation of the summations at N gives a set of 2N ordinary differential equations containing 2N z-dependent functions. The problem can then be expressed in matrix form with eigenvalues γ, and eigenvectors containing A_n and B_n. The critical values of the global dynamo number, corresponding to $Re(\gamma) = 0$, yield quadrupole and dipole solutions when the appropriate boundary conditions are applied.

Two types of disc surface boundary conditions were used, by considering the surrounding medium to have high or low conductivity. In the very high conductivity case the field is unable to diffuse beyond the disc, so $\mathbf{B}$ is contained in $|z| < h$. The steady dipole modes are the most easily excited. With a contained field the dipole and quadrupole modes both have a strong component of $\mathbf{B}_p$ across the Keplerian shear surfaces, leading to the generation of B_ϕ in (11.4). However, the dipole mode has its poloidal field lines crossing the central plane, while the quadrupole mode does not. Hence the vertical spatial scale of $\mathbf{B}_p$ in the basic dipole mode is larger than the corresponding scale in the quadrupole mode. The dipole mode is therefore dissipated less rapidly and is the more easily generated mode in this case. The fields are localized to the inner parts of the disc, this being a consequence of D, given by (11.40), having its largest values in this region. The lowest order mode consists of a single structure centred in the innermost part of the disc. The higher order modes have several structures with the field spreading out further radially.

In the case of a vacuum exterior the steady quadrupole modes are the most easily excited. The field structures are radially localized in a similar way to the dipole modes generated with highly conducting surroundings. The result that the quadrupole modes are the most easily excited when the exterior has low conductivity agrees with all the foregoing calculations, which made a variety of assumptions about the spatial variation of the dynamo number. The reason for this is that with a vacuum exterior the poloidal field lines extend beyond the disc. In a dipole mode $\mathbf{B}_p$ is therefore nearly vertical through the disc, and so nearly parallel to the surfaces of Keplerian shear. The shear term $\varpi B_\varpi \Omega'_K$, which creates toroidal field in (11.4), is consequently small and a relatively high dynamo number is required to sustain the field. Conversely, in a quadrupole mode the lines of $\mathbf{B}_p$ do not cross the central plane and hence there is a relatively large B_ϖ component crossing the Keplerian shear surfaces, leading to generation of B_ϕ via radial shear at lower dynamo numbers.

The magnetic fields generated in this model differ in radial structure to those found in Pudritz (1981), given by (11.35)–(11.37). The difference arises from assumptions made about the spatial variation of the local dynamo number $D(\varpi)$. Pudritz used forms for η and $\tilde{\alpha}$ similar to (11.30) and (11.31) and hence found a uniform D. Stepinski and Levy (1990) assumed the form (11.39) for α (i.e. $\tilde{\alpha} = \tilde{K} h \Omega_K$) together with constant η, so obtaining $D \propto \varpi^{-3}$. A uniform D produces a field of a global nature, extending radially through the disc. A form $D(\varpi)$ with maximum values produces a field with localized structures. Stepinski and Levy took the inner five percent by radius of their disc to have uniform rotation. $D(\varpi)$ therefore has one maximum near the disc centre, and most of the field is generated in this region. They found that for higher global dynamo numbers, characterized by $G_m = (h^3 \tilde{D})^{1/2}$ in (11.40), field generation occurred somewhat further from the disc centre. However, the value of the local dynamo number $D(\varpi)$ evaluated at the centre of the field structures only increased slightly at higher values of G_m. This indicates a self-similar structure of the steady state magnetic field generated in discs with variable D.

Dynamo action in Keplerian viscous discs was also considered by Rudiger *et al*, (1995). They took a uniform D and solved (11.4) and (11.5) numerically. For a vacuum exterior their results agree with those of Pudritz (1981), the steady quadrupole being the most easily excited mode and the field structure extending radially. The case in which the exterior has a magnetic diffusivity satisfying $0 < \eta^{\text{ext}} < \eta^{\text{disc}}$ is also considered. For $\eta^{\text{ext}} = 0.01\eta^{\text{disc}}$ the quadrupole modes are still the most readily generated, in fact occurring at lower dynamo numbers than in the vacuum case. The condition $\eta^{\text{ext}} \ll 0.01\eta^{\text{disc}}$ is required to produce the dipole modes found by Stepinski and Levy (1990) to be the most easily excited in the case $\eta^{\text{ext}} \to 0$.

Rudiger *et al* (1995) also solved the dynamo equations, for the case with a vacuum exterior, but using an α-quenching term

$$\alpha = \frac{\alpha_0}{1 + (B/B_{\text{eq}})^2}, \tag{11.43}$$

where $B_{\text{eq}} = (\mu_0 \rho v_T^2)^{1/2}$ is an equilibrium magnetic field obtained by equating the turbulent kinetic energy density to the magnetic energy density. The field solution has approximately quadrupolar symmetry, but with a radial structure differing from that of the corresponding normal mode. The magnetic field strength has a radial profile similar to that of B_{eq}, having a maximum in the innermost part of the disc. Typically $B \sim 9B_{\mathrm{eq}}$ through the disc.

11.2.2. DYNAMO ACTION BY INTERNAL WAVES

The foregoing mean-field $\alpha\omega$-dynamo models assume some mechanism generates the required turbulence for the α-effect. Vishniac and Diamond (1989) and Vishniac, Jin and Diamond (1990) proposed that a spectrum of low frequency, slightly non-axisymmetric internal waves could provide a turbulent background having a finite mean helicity, $\langle \mathbf{v}_{\mathrm{T}} \cdot \nabla \wedge \mathbf{v}_{\mathrm{T}} \rangle$. A seed magnetic field is then amplified by $\alpha\omega$-dynamo action to produce a large-scale field. Tidal forcing in the outer parts of the disc provides a mechanism for exciting non-axisymmetric disturbances (e.g. Papaloizou and Pringle, 1977; Paczyński, 1977; Lin and Papaloizou, 1979).

Vishniac and Diamond (1989) and Vishniac, Jin and Diamond (1990) consider linear perturbations of the shear flow in a thin Keplerian disc. The dispersion relation for a local analysis has an internal wave branch of low frequency, approximately incompressible modes. To linear order, the total energy flux and the radial flux of angular momentum carried by a wave packet are conserved. Perturbations are expressed in the form $Af(z)\exp[i(\omega t + k_{\varpi}\varpi + m\phi)]$, where A is an amplitude and $f(z)$ a vertical eigenfunction. The nonlinear dissipation is estimated using the theory of resonant interactions, which assumes that the fluid motion is dominated by linear modes with independent phases. Wave interaction processes lead to a turbulent background which is then used to estimate an α-effect, assuming a seed magnetic field is present.

The authors employ mean-field theory, using first order smoothing (see §2.3.1), to estimate the elements of an α-tensor. The element $\alpha_{\phi\phi}$, which converts B_ϕ to B_ϖ, is found to be

$$\alpha_{\phi\phi} = 2\left(\frac{h}{\varpi}\right)^3 v_{\mathrm{K}}.$$

The main uncertainty lies in the role played by small-scale waves. The authors assume that the local wave energy per logarithmic interval is transferred locally in wavenumber space.

The internal wave mechanism assumes the seed magnetic field does not affect the stability of the perturbations. However, it was subsequently discovered that the presence of a weak magnetic field in a Keplerian disc leads to a magnetic shear instability. This is discussed below.

11.2.3. SHEAR INSTABILITIES IN WEAKLY MAGNETIZED DISCS

A very promising mechanism for generating turbulence in accretion discs was presented by Balbus and Hawley (1991). They applied a linear instability, originally discovered by Velikhov (1959) and Chandrasekhar (1961, 1962), to show that an accretion disc is dynamically unstable to axisymmetric disturbances in the presence of a weak poloidal magnetic field. The decrease of the angular velocity, Ω, with distance from the accretor plays a vital role in the instability. The basic nature of the instability can be understood by considering a differentially rotating disc containing a vertical magnetic field. In a highly conducting disc a fluid element will be essentially tethered by the magnetic field. If the element is displaced radially outwards the magnetic field will try to keep it rotating with the angular velocity of its ring of origin. Consequently, in its new position the element will not be in centrifugal balance with the radial component of the accretor's gravity. For outwardly decreasing Ω the force imbalance will accelerate the element away from its equilibrium position. There is a restoring force due to the elastic nature of the magnetic field, but at wavelengths longer than a critical value this is weak and destabilization wins.

The disc is taken to contain an initially weak poloidal magnetic field, so its dynamical effect on the unperturbed state is negligible. Axisymmetric, large wavenumber linear perturbations are considered which take the form $\exp[i(k_\varpi \varpi + k_z z - \omega t)]$, where the local spatial variations of the components of $\mathbf{k}$ can be ignored since the ϖ and z length-scales of the unperturbed disc structure are long relative to the wavelengths in those directions. Adiabatic perturbations are considered and the Boussinesq approximation is used, which consists of ignoring pressure perturbations in all equations except the momentum equation. The special case of vanishing radial field, B_ϖ, is first addressed. The linearized equations of momentum, continuity, induction and heat lead to the dispersion relation

$$\frac{k^2}{k_z^2}\sigma^4 - \left[K^2 + \left(\frac{k_\varpi}{k_z}N_z - N_\varpi\right)^2\right]\sigma^2 - 4\Omega^2 k_z^2 v_{\mathrm{A}z}^2 = 0, \tag{11.44}$$

where

$$\sigma^2 = \omega^2 - k_z^2 v_{\mathrm{A}z}^2, \tag{11.45a}$$

$$v_{\mathrm{A}z}^2 = \frac{B_z^2}{\mu_0 \rho}, \tag{11.45b}$$

$$K^2 = \frac{2\Omega}{\varpi}\frac{d}{d\varpi}(\varpi^2\Omega), \tag{11.45c}$$

$$N_{\varpi}^2 + N_z^2 = -\frac{3}{5\rho}(\nabla P) \cdot [\nabla \ln(P\rho^{-\frac{5}{3}})], \tag{11.45d}$$

with B_z, P and ρ denoting unperturbed disc quantities. In a Keplerian accretion disc,

$$\Omega = \Omega_{\mathrm{K}} = \left(\frac{GM_{\mathrm{p}}}{\varpi^3}\right)^{\frac{1}{2}}. \tag{11.46}$$

Then K, given by (11.45c), is real. A non-magnetized disc is stable to inviscid adiabatic perturbations if N_{ϖ} and N_z, given by (11.45d), are also real. By solving (11.44) for σ^2, it follows that σ^2 (and hence ω^2) is real. Disc stability can therefore be investigated by considering conditions in the neighbourhood of $\omega^2 = 0$, at which $\sigma^2 = -k_z^2 v_{\mathrm{A}z}^2$. In this limit (11.44) becomes

$$(k_z^2 v_{\mathrm{A}z}^2 + N_z^2) k_{\varpi}^2 - 2N_{\varpi} N_z k_z k_{\varpi} + k_z^2 \left(\frac{d\Omega^2}{d\ln\varpi} + N_{\varpi}^2 + k_z^2 v_{\mathrm{A}z}^2\right) = 0. \tag{11.47}$$

Regarded as a quadratic for k_{ϖ}, this equation does not allow real solutions, so assuring stability since ω^2 could then not pass through zero, provided the discriminant is negative. This stability condition can be expressed as

$$k_z^4 v_{\mathrm{A}z}^4 + k_z^2 v_{\mathrm{A}z}^2 \left(N^2 + \frac{d\Omega^2}{d\ln\varpi}\right) + N_z^2 \frac{d\Omega^2}{d\ln\varpi} \geq 0. \tag{11.48}$$

Since $N_z^2 > 0$, the inequality can only be satisfied for all non-vanishing k_z if

$$\frac{d\Omega^2}{d\varpi} \geq 0. \tag{11.49}$$

This is clearly violated in a Keplerian disc, so giving instability for values of k_z less than the critical value yielded from the equality in (11.48) as

$$(k_z)_{\mathrm{crit}} = \frac{1}{v_{\mathrm{A}z}} \left|\frac{d\Omega_{\mathrm{K}}^2}{d\ln\varpi}\right|^{\frac{1}{2}} = \sqrt{3}\frac{\Omega_{\mathrm{K}}}{v_{\mathrm{A}z}}, \tag{11.50}$$

using $N_z^2 \gg N_{\varpi}^2$ for a thin disc. The inclusion of a radial field, B_{ϖ}, in the unperturbed disc does not affect the stability criterion, given by (11.49).

In a thin Keplerian disc $K^2 = \Omega_{\mathrm{K}}^2$ and this term dominates in the coefficient of σ^2 in (11.44). The unstable root of this equation, together with (11.45a), leads to a time dependence $\exp(\gamma t)$ in which the growth rate is

$$\gamma = \left[-\frac{\Omega_{\mathrm{K}}^2}{2}\frac{k_z^2}{k^2} - k_z^2 v_{\mathrm{A}z}^2 + \frac{1}{2}\left(\Omega_{\mathrm{K}}^4 \frac{k_z^4}{k^4} + 16\Omega_{\mathrm{K}}^2 \frac{k_z^4}{k^2} v_{\mathrm{A}z}^2\right)^{\frac{1}{2}}\right]^{\frac{1}{2}}. \tag{11.51}$$

For $k_z v_{Az} \ll \Omega_K$ first order expansion yields $\gamma \simeq \sqrt{3} k_z v_{Az}$. Differentiation of (11.51) with respect to k_z, using $k_z \gg k_\varpi$, gives a maximum growth rate

$$\gamma_{max} \simeq \frac{\Omega_K}{\sqrt{2}} \text{ at } k_z = \frac{\Omega_K}{v_{Az}}, \tag{11.52}$$

so γ_{max} is independent of the magnetic field.

The foregoing analysis ignored dissipative effects. However, for sufficiently small $|B_z|$ the value of $(k_z)_{crit}$ given by (11.50) will become so large that there will be wavenumbers at which dissipation is important. This value of $|B_z|$ can be estimated for ordinary ohmic dissipation, with the associated diffusivity

$$\eta_{ohm} = 5.2 \times 10^7 \ln \Lambda \, T^{-\frac{3}{2}} \, \mathrm{m^2 s^{-1}}, \tag{11.53}$$

discussed in §2.2.10. The dissipation and growth time-scales are

$$\tau_d = \frac{4\pi^2}{\eta_{ohm} k_z^2}, \qquad \tau_g = \frac{2\pi}{\gamma}.$$

Using typical values of $\gamma \sim \Omega_K$ and $k_z \sim \Omega_K / v_{Az}$, the condition $\tau_d \gg \tau_g$ for negligible dissipation becomes

$$v_{Az}^2 \gg \eta_{ohm} \frac{\Omega_K}{2\pi},$$

which can be expressed as

$$\frac{B_z^2}{2\mu_0 P} \gg \frac{\eta_{ohm} \Omega_K}{4\pi c_s^2}, \tag{11.54}$$

where $c_s^2 = (\Re/\mu)T$. Taking $\ln \Lambda = 10$, (11.53) and (11.54) give

$$\frac{B_z^2}{2\mu_0 P} \gg 3.5 \times 10^{-9} \left(\frac{M_p}{M_\odot}\right)^{\frac{1}{2}} \left(\frac{T}{10^4\,\mathrm{K}}\right)^{-\frac{5}{2}} \left(\frac{\varpi}{10^8\,\mathrm{m}}\right)^{-\frac{3}{2}}, \tag{11.55}$$

where M_p is the mass of the accreting primary star. This indicates that quite weak initial fields can be considered for which the dissipation associated with the linear instability is ignorable. For $\rho \sim 10^{-6}\,\mathrm{Kg\,m^{-3}}$, (11.55) yields $|B_z| \sim 10^{-6}$ tesla for the field below which ohmic dissipation becomes effective. Similar field values apply to the case of thermal conductivity.

The maximum value of k_z for instability is given by (11.50). A minimum value for k_z is set by the wavelength condition $\lambda_z \lesssim 2h$, where h is the disc scale height, so $(k_z)_{min} \sim \pi/h$. The validity of the local analysis requires $\lambda_z \ll 2h$ and hence $k_z \gg \pi/h$. For a thin Keplerian disc with a weak magnetic field, $h \sim c_s/\Omega_K$, so the range of k_z for which the local instability is operable is

$$\pi \frac{\Omega_K}{c_s} \ll k_z < \sqrt{3}\frac{\Omega_K}{v_{Az}}. \tag{11.56}$$

This leads to $B_z^2/2\mu_0 P \ll 1$ which can be consistent with the negligble dissipation condition (11.55).

The foregoing derivation of the magnetic shear instability is a local analysis, and used the Boussinesq approximation. Papaloizou and Szuszkiewicz (1992) considered an axisymmetric, compressible, differentially rotating, non self-gravitating fluid containing a poloidal magnetic field. They performed a global stability analysis of axisymmetric, adiabatic modes by formulating a variational principle. The modes have a time dependence $\exp(\sigma t)$ and the self-adjoint nature of the operators leads to real values for σ^2. Stability therefore requires $\sigma^2 < 0$ and, for a weak magnetic field, this gives the condition

$$\varpi(\mathbf{e}\cdot\nabla\varpi)(\mathbf{e}\cdot\nabla\Omega^2) - \left(\mathbf{e}\cdot\frac{\nabla P}{\rho}\right)\left[\mathbf{e}\cdot\left(\frac{\nabla P}{\Gamma_1 P} - \frac{\nabla\rho}{\rho}\right)\right] \geq 0, \tag{11.57}$$

where $\mathbf{e}$ is an arbitrary vector and the adiabatic exponent Γ_1 is defined by (2.87). It follows that for $\mathbf{e}$ tangent to surfaces of constant pressure, or of constant entropy, stability requires

$$(\mathbf{e}\cdot\nabla\varpi)(\mathbf{e}\cdot\nabla\Omega^2) \geq 0. \tag{11.58}$$

For a thin disc this condition is essentially the same as (11.49), confirming the Balbus-Hawley result that a Keplerian disc is unstable. Gammie and Balbus (1994) showed that the field magnitude required for stabilization in a global problem may depend on the detailed boundary conditions.

Dubrulle and Knobloch (1993) consider incompressible motions in a thin disc, but allow the basic state to have azimuthal as well as vertical magnetic field. With unperturbed velocities $v_\varpi = v_z = 0$, together with $B_\varpi = 0$, the unperturbed Alfvén speeds $v_{A\phi}$ and v_{Az} are necessarily independent of z. Axisymmetric perturbations are considered of the separable form $f(\varpi)\exp[i(k_z z + \sigma t)]$, leading to a second order radial eigenvalue problem. When v_{Az} is constant the sufficient condition for stability becomes

$$\varpi^2 \frac{d}{d\varpi}(\Omega^2) - \frac{1}{\varpi^2}\frac{d}{d\varpi}\left(\frac{\varpi^2 B_\phi^2}{\mu_0 \rho}\right) \geq 0, \tag{11.59}$$

and is found not to be sensitive to the boundary conditions imposed. In the weak field limit, this result reduces to (11.49).

Balbus and Hawley (1991) pointed out that there are two possible outcomes of the magnetic shearing instability. Firstly, in the absence of sufficient dissipation, the magnetic field will grow to the point where the minimum critical wavelength exceeds the disc scale height. This occurs when the Alfvén and sound speeds are comparable, the magnetic field being left in a dynamically important state but no longer prone to shearing instability. The second, and more likely, possibility is that reconnection dissipates the growing field and a state is reached in which the growth rate of the instability is counterbalanced by dissipation at the smallest scales. Hence classical turbulence results.

Subsequent work strongly supported a turbulent outcome for the hydromagnetic instability. Hawley and Balbus (1991, 1992) performed nonlinear simulations in both two and three dimensions. Their local analysis considers unstratified fluid in a box with periodic shearing boundary conditions. For initial conditions appropriate to axisymmetric modes, the three-dimensional case has a period of exponential growth after which the solutions break up into a turbulent state.

Hawley, Gammie and Balbus (1995) showed that the magnetic shearing instability also operates in the presence of an irregular magnetic field. This field can sustain self-excited turbulence, even in the absence of stratification. In the fully turbulent state, they find that the $B_{\varpi}B_{\phi}$ magnetic stress dominates the viscous stress so angular momentum transport is largely magnetic. Stone *et al* (1996) included density stratification and found similar turbulence. They showed that the saturated state is essentially independent of the initial magnetic field geometry.

Numerical simulations of magnetic shear instabilities by Brandenburg *et al* (1995) give rise to turbulence with a finite mean helicity. A self-sustaining $\alpha\omega$-dynamo therefore generates mean magnetic fields, leading to angular momentum advection. This work is discussed in more detail in the next section.

The advent of the Balbus-Hawley instability makes the internal wave mechanism, discussed in §11.2.2, a more complex issue. Vishniac and Diamond (1989) assumed that weak magnetic fields could be ignored in the early development of internal waves. However, as seen above, the magnetic shear instability operates for weak magnetic fields. Balbus and Hawley (1991) therefore suggested that this instability would rapidly quench internal waves, so rendering them ineffective in generating an α-effect. However, Zhang, Vishniac and Diamond (1994) argue that high frequency radial internal waves can coexist with low frequency magnetic shear instability cells, at least when the magnetic field is vertical. Nevertheless, magnetic shear instabilities now appear to be the most viable origin of angular momentum advection in Keplerian discs.

11.2.4. DISC DYNAMOS WITH HYDROMAGNETIC TURBULENCE

Tout and Pringle (1992) developed a model for a magnetic accretion disc dynamo. The dynamo mechanism depends on the Balbus-Hawley instability, together with the Parker buoyancy instability and reconnection. The authors concentrate on a local description of the physical mechanisms involved, rather than detailed spatial dependencies. Their azimuthal induction equation is

$$\frac{dB_\phi}{dt} = \frac{3}{2}\Omega_{\rm K}B_\varpi - \frac{B_\phi}{\tau_{\rm P}}. \tag{11.60}$$

This is the standard form for this equation, the azimuthal field being created by shearing of poloidal field and diffused, in this case by the Parker instability. The origin of this instability was discussed in §2.2.10, and the corresponding magnetic diffusivity is given by (2.177). The Parker diffusion time is therefore

$$\tau_{\rm P} = \frac{h}{\xi v_{\rm A\phi}}, \tag{11.61}$$

where h is the vertical scale height, $v_{\rm A\phi} = |B_\phi|/(\mu_0\rho)^{1/2}$, and $\xi < 1$.

The Balbus-Hawley instability creates radial field from vertical and Parker buoyancy diffuses the radial field. The radial equation, averaged over wavenumbers, is taken to be

$$\frac{dB_\varpi}{dt} = \begin{cases} \bar{\gamma}_{\rm max}\Omega_{\rm K}B_z - B_\varpi/\tau_{\rm P}, & v_{\rm Az}/c_{\rm s} \le \sqrt{2}/\pi, \quad (11.62a)\\ \bar{\gamma}_{\rm BH}\Omega_{\rm K}B_z - B_\varpi/\tau_{\rm P}, & \sqrt{2}/\pi < v_{\rm Az}/c_{\rm s} \le \sqrt{6}/\pi, \quad (11.62b)\\ -B_\varpi/\tau_{\rm P}, & v_{\rm Az}/c_{\rm s} > \sqrt{6}/\pi, \quad (11.62c) \end{cases}$$

where

$$\bar{\gamma}_{\rm BH} = \bar{\gamma}_{\rm max}\left[1 - \frac{(1 - \pi v_{\rm Az}/\sqrt{2}c_{\rm s})^2}{(1-\sqrt{3})^2}\right]^{\frac{1}{2}}, \tag{11.62d}$$

and $\bar{\gamma}_{\rm max} \sim 0.71$. Equation (11.62a) arises when the wavelength of the fastest growing mode of the Balbus-Hawley instability is $\lesssim 2h$. From (11.52) this mode has a vertical wavelength

$$\lambda_z^{\rm max} = \frac{2\pi v_{\rm Az}}{\Omega_{\rm K}}, \tag{11.63}$$

where $v_{\rm Az} = |B_z|/(\mu_0\rho)^{1/2}$. It then follows from (11.50) that

$$\lambda_z^{\rm max} = \sqrt{3}\lambda_z^{\rm crit}, \tag{11.64}$$

where the instability is cut off when $\lambda_z < \lambda_z^{\rm crit}$. Hence in this first case the range of unstable wavelengths is $\lambda_z^{\rm max}/\sqrt{3} \le \lambda_z \le \lambda_z^{\rm max}$, with $\lambda_z^{\rm max} \lesssim 2h$. The unstable modes therefore all have growth rates near the maximum value of $\sim \bar{\gamma}_{\rm max}\Omega_{\rm K}$, this being taken as the average value in equation (11.62a). The associated inequality $v_{\rm Az}/c_{\rm S} \lesssim \sqrt{2}/\pi$ arises from $\lambda_z^{\rm max} \lesssim 2h$, together with the assumption of an isothermal vertical structure which yields $h\Omega_{\rm K} = \sqrt{2}c_{\rm S}$. Equation (11.62b) gives the intermediate regime, while (11.62c) arises when $\lambda_z^{\rm crit} \lesssim 2h$, so there are no unstable wavelengths in the disc and the generation of B_ϖ ceases. Equation (11.62d) is an analytic fit to the results of Balbus and Hawley (1991).

The vertical equation is taken as

$$\frac{dB_z}{dt} = \frac{B_\phi}{\tau_{\rm P}} - \frac{B_z}{\tau_{\rm rec}}. \tag{11.65}$$

The Parker instability creates B_z from B_ϕ, while reconnection dissipates B_z. The reconnection time-scale is

$$\tau_{\rm rec} = \frac{\lambda_{\rm rec}}{\Gamma v_{\rm Az}}, \tag{11.66}$$

where $\lambda_{\rm rec}$ is the mean distance between patches of B_z of opposite sign and $\Gamma^{-1} \sim \ln R_{\rm m}$, with $R_{\rm m}$ the magnetic Reynolds number.

The authors first investigate the linear stability of the trivial solution $\mathbf{B} = \mathbf{0}$. They find this solution to be unstable with a growth time of order $(2/3)\Omega_{\rm K}^{-1}$. An equilibrium solution with finite $\mathbf{B}$ is then sought. The growth terms for B_ϕ and B_ϖ involve the time-scale $\Omega_{\rm K}^{-1}$, whereas the loss terms involve the larger time-scales $\tau_{\rm P}$ and $\tau_{\rm rec}$. Equilibrium can therefore only occur when the growth of B_ϖ via the Balbus-Hawley instability is inhibited by the presence of strong B_z. The equilibrium solution has $B_\phi^2/2\mu_0 \sim P$ and is shown to be overstable.

The nonlinear evolution of the dynamo is investigated using numerical integration. In all cases the magnetic field remains finite, but oscillates about the equilibrium state. The cycle is largely controlled by how far B_z is from equilibrium. For weaker B_z the Balbus-Hawley instability leads to a rapid growth in B_ϖ, and more slowly in B_ϕ. These lead to growth in B_z towards equilibrium and then B_ϖ and B_ϕ decay, followed by B_z. Although B_ϖ and B_ϕ escape from the disc on the time-scale $\tau_{\rm P} \sim 4\Omega_{\rm K}^{-1}$, B_ϖ is converted to B_ϕ on the shear time-scale of $\sim 0.7\Omega_{\rm K}^{-1}$. Hence in equilibrium $B_\phi \sim 6B_\varpi$.

Brandenburg *et al* (1995) used numerical methods to simulate the nonlinear evolution of magnetized Keplerian shear flows in a local, three-dimensional model, including compressibility and stratification. The Balbus-Hawley instability was found to generate motions which regenerate a turbulent magnetic field which, in turn, reinforces the turbulence.

Local Cartesian coordinates (x, y, z) are used, with origin at a cylindrical radius ϖ_0 and unit vectors $\hat{\mathbf{x}} = \hat{\boldsymbol{\varpi}}$, $\hat{\mathbf{y}} = \hat{\boldsymbol{\phi}}$. The origin is taken to have the angular velocity $\Omega_0 = \Omega_{\rm K}(\varpi_0)$ so, to linear order, the Keplerian shear flow in this frame is

$$v_y^0(x) = -\frac{3}{2}\Omega_0 x. \tag{11.67}$$

The equations of momentum, induction, continuity and heat are solved in a local box to obtain the deviations, $\mathbf{v}$, from this flow. Equation (11.67) is consistent with no systematic variation of quantities with x, since only $dv_y^0(x)/dx$ occurs in the governing equations. A standard form is adopted for the viscous force, and the perfect gas equation is used. A simple form for the cooling rate is taken as

$$Q = -\Omega_0(E - E_0), \tag{11.68}$$

where

$$E = \frac{P}{(\gamma - 1)\rho}$$

is the thermal energy per unit mass, with E_0 its initial value. An initial isothermal stratification is assumed and the insulating boundary condition $\partial E/\partial z = 0$ at $z = \pm h$ is employed.

The velocity is taken to obey stress-free conditions at the upper and lower boundaries, so

$$\frac{\partial v_x}{\partial z} = \frac{\partial v_y}{\partial z} = v_z = 0 \quad \text{at} \quad z = \pm h. \tag{11.69}$$

The magnetic field is taken to be purely vertical at these boundaries. Using $\mathbf{B} = \nabla \wedge \mathbf{A}$, then gives the surface conditions $B_x = B_y = 0$ in terms of the vector potential as

$$\frac{\partial A_x}{\partial z} = \frac{\partial A_y}{\partial z} = A_z = 0 \quad \text{at} \quad z = \pm h. \tag{11.70}$$

Periodic boundary conditions are adopted in the y-direction. The quasi-periodic condition

$$F(\Delta x, y, z) = F(0, y + \frac{3}{2}\Omega_0 t \Delta x, z) \tag{11.71}$$

is used for any quantity on the radial boundaries, where Δx is the radial extent of the box. This sliding condition accounts for the effect of the Keplerian shear in the y-direction. The horizontal boundary conditions lead to vanishing vertical flux, since

$$\int\!\!\int B_z dx dy = \int_0^{\Delta y} [A_y]_0^{\Delta x}\, dy - \int_0^{\Delta x} [A_x]_0^{\Delta y}\, dx = 0. \tag{11.72}$$

This enables the field to decay to zero if the motions become too weak to sustain it.

The governing equations are solved using sixth-order compact derivatives, and a third-order Hyman scheme for the time-stepping (Nordlund and Stein, 1990). Random velocity perturbations are taken, with an initial Mach number of $v_{\rm rms}/c_{\rm s} = 0.002$. The initial field is $\mathbf{B} = B_0 \sin(2\pi x/\Delta x)\hat{\mathbf{z}}$ with B_0 satisfying $2\mu_0 P/B_0^2 = 100$.

The Balbus-Hawley magnetic shear instability generates turbulence. Energy flows from the Keplerian motion into both magnetic and turbulent kinetic energies in the ratio of ~ 6 to 1. However, the energy transfer rates (=energy flux/energy content) into these two reservoirs are approximately equal at $\sim 0.3\Omega_{\rm K}$. The Lorentz force pumps half this magnetic energy into turbulent kinetic energy. Hence three-quarters of the energy going from the Keplerian motion into turbulence first passes through a phase of magnetic energy. The magnetic and turbulent energies are subsequently dissipated and heat the disc. Poloidal magnetic field is regenerated at a rate of $\sim 0.6\Omega_0$, comparable to the growth rate of the magnetic shear instability.

The large-scale toroidal magnetic field is mainly of quadrupolar parity, and exhibits cyclic behaviour. The ratio of the mean magnetic energy density to the thermal energy density is $B^2/2\mu_0 P \lesssim 0.1$. The dynamo α-function is found to be negative for $0 < z < h$, contrary to standard theory which uses the Parker instability and Coriolis force to generate α. As discussed in Brandenburg and Donner (1997), the reason for this sign difference appears to be the strong effect of the magnetic shear instability. Oscillatory quadrupole modes are then favoured, but the average value of $B_\varpi B_\phi$ is still negative, as required for outward radial advection of angular momentum.

The standard inner boundary condition of vanishing radial shear, described in §2.4.3, leads to the stress equation

$$\left\langle \rho v'_\varpi v'_\phi - \frac{1}{\mu_0} B'_\varpi B'_\phi \right\rangle = -\rho\nu\varpi \frac{d\Omega_{\rm K}}{d\varpi}, \tag{11.73}$$

where primes denote turbulent components and the radial integration constant can be ignored for $\varpi \gg R_{\rm p}$, with $R_{\rm p}$ the radius of the accretor. In the Shakura-Sunyaev model the turbulent viscosity coefficient is expressed as

$$\nu = \epsilon c_{\rm s} h, \tag{11.74}$$

with $\epsilon < 1$. The value of ϵ found here fluctuates in time, with a mean of $\epsilon \sim 10^{-2}$. Brandenburg *et al* (1996) derive a parabolic fit for ϵ of the form

$$\epsilon \simeq \epsilon_0 + \epsilon_B \frac{B_\phi^2}{B_{\rm eq}^2}, \tag{11.75}$$

where $B_{\mathrm{eq}} = (\mu_0 \rho c_{\mathrm{s}}^2)^{1/2}$ is the equipartition field with respect to the thermal energy density and $\epsilon_B \simeq 0.5$.

The numerical simulations are consistent with a dynamo α-function above the central plane of

$$\alpha \simeq -10^{-3} \Omega_{\mathrm{K}} h. \tag{11.76}$$

The turbulent magnetic diffusivity is

$$\eta_{\mathrm{T}} \simeq 8 \times 10^{-3} \Omega_{\mathrm{K}} h^2. \tag{11.77}$$

The ratio of field components is typically $|B_{\varpi}/B_{\phi}| \sim 10^{-2}$, characteristic of an $\alpha\omega$-type dynamo.

11.3. A Magnetically-Driven Disc

11.3.1. MAGNETIC DISC EQUATIONS

The foregoing dynamo models address the problems associated with magnetic field generation in accretion discs. The presence of a large-scale magnetic field then leads to the possibility of angular momentum advection via magnetic stresses. Campbell (1992) constructed a magnetohydrodynamic model of an accretion disc as an alternative to the standard viscous disc. It is assumed that the Balbus-Hawley instability generates some turbulence and this leads to an α-effect. Buoyancy and turbulent reconnection are taken to cause the flux loss. Both viscous and magnetic stresses will be present and their relative strengths are of central importance to angular momentum advection through the disc. The following analysis shows that, for a prominant buoyancy diffusivity, the magnetic stresses dominate in the angular momentum equation.

An axisymmetric, steady state disc is considered around a non-magnetic accreting star. Cylindrical polar coordinates (ϖ, ϕ, z) are used, centred on the star, with the disc's central plane corresponding to $z = 0$. The external medium is taken to be a vacuum. The steady state equations of momentum, induction and continuity are

$$(\mathbf{v} \cdot \nabla)\mathbf{v} = -\frac{1}{\rho}\nabla P - \nabla\psi + \frac{1}{\mu_0 \rho}(\nabla \wedge \mathbf{B}) \wedge \mathbf{B}, \tag{11.78}$$

$$\nabla \wedge (\mathbf{v} \wedge \mathbf{B}) + \nabla \wedge (\alpha B_\phi \hat{\boldsymbol{\phi}}) - \nabla \wedge (\eta \nabla \wedge \mathbf{B}) = \mathbf{0}, \tag{11.79}$$

$$\nabla \cdot (\rho \mathbf{v}) = 0, \tag{11.80}$$

where ψ is the stellar gravitational potential given by

$$\psi = -\frac{GM_{\mathrm{p}}}{(\varpi^2 + z^2)^{\frac{1}{2}}}, \tag{11.81}$$

with $M_{\rm p}$ the stellar mass. It is assumed that the turbulence, generated by the Balbus-Hawley instability, leads to the standard α-effect term in (11.79), with an azimuthal electric field αB_ϕ. It is also assumed that the viscous force is negligible in (11.78), as will be confirmed.

The ϖ, ϕ and z-components of (11.78) can be written as

$$\frac{v_\phi^2}{\varpi} = \frac{\partial \psi}{\partial \varpi} + \frac{\partial}{\partial \varpi}\left(\frac{v_\varpi^2}{2}\right) + v_z \frac{\partial v_\varpi}{\partial z} + \frac{1}{\rho}\frac{\partial P}{\partial \varpi} + \frac{1}{\varpi^2 \rho}\frac{\partial}{\partial \varpi}\left(\frac{\varpi^2 B_\phi^2}{2\mu_0}\right) - \frac{B_z J_\phi}{\rho}, \tag{11.82}$$

$$\frac{v_\varpi}{\varpi}\frac{\partial}{\partial \varpi}(\varpi^2 \Omega) + \frac{v_z}{\varpi}\frac{\partial}{\partial z}(\varpi^2 \Omega) = \frac{1}{\mu_0 \varpi^2 \rho}\frac{\partial}{\partial \varpi}(\varpi^2 B_\varpi B_\phi) + \frac{1}{\mu_0 \rho}\frac{\partial}{\partial z}(B_z B_\phi), \tag{11.83}$$

$$\frac{\partial \psi}{\partial z} + \frac{\partial}{\partial z}\left(\frac{v_z^2}{2}\right) + v_\varpi \frac{\partial v_z}{\partial \varpi} + \frac{1}{\rho}\frac{\partial P}{\partial z} + \frac{1}{\rho}\frac{\partial}{\partial z}\left(\frac{B_\phi^2}{2\mu_0}\right) + \frac{B_\varpi J_\phi}{\rho} = 0, \tag{11.84}$$

where $\Omega = v_\phi/\varpi$, and J_ϕ is the toroidal current density given by

$$J_\phi = \frac{1}{\mu_0}\left(\frac{\partial B_\varpi}{\partial z} - \frac{\partial B_z}{\partial \varpi}\right). \tag{11.85}$$

The poloidal and toroidal components of (11.79) give

$$v_\varpi B_z - v_z B_\varpi - \alpha B_\phi + \mu_0 \eta J_\phi = 0, \tag{11.86}$$

$$\frac{\eta}{\varpi}\left(\nabla^2 B_\phi - \frac{B_\phi}{\varpi^2}\right) + \frac{1}{\varpi^2}\frac{d\eta}{d\varpi}\frac{\partial}{\partial \varpi}(\varpi B_\phi) = -\mathbf{B}_p \cdot \nabla \Omega + \nabla \cdot \left(\frac{B_\phi}{\varpi}\mathbf{v}_p\right), \tag{11.87}$$

where subscripts p refer to poloidal components, and a radial dependence is allowed for η. Equation (11.80) is

$$\frac{1}{\varpi}\frac{\partial}{\partial \varpi}(\varpi \rho v_\varpi) + \frac{\partial}{\partial z}(\rho v_z) = 0, \tag{11.88}$$

where the vertical velocity component results because the disc surface h is ϖ-dependent. It follows that

$$\left|\frac{v_z}{v_\varpi}\right| \sim \frac{h}{\varpi} \ll 1. \tag{11.89}$$

The structure of the magnetic field must lead to an azimuthal force on rings of disc material, for net angular momentum advection. Since the exterior field is potential and axisymmetric, B_ϕ must vanish at the disc surface. After multiplying by ρ, the vertical integral of the last term in

(11.83) is consequently zero, contributing no net torque. A poloidal field with significant B_ϖ is therefore necessary, so a quadrupolar-type field must be generated. It was seen in the foregoing dynamo models that steady fields of this symmetry are indeed always favoured in discs, with essentially vacuum surroundings. If α is indeed negative for $0 < z < h$, as suggested by Brandenburg and Donner (1997), then oscillatory quadrupole fields may be favoured. Nevertheless, the steady solutions found here should have similar properties to the averaged oscillatory fields. The magnetic field, like P and ρ, varies on a vertical length-scale of $\sim h$ and on a horizontal scale of $\sim \varpi$. As shown in §2.3.2, the $\alpha\omega$-dynamo generates a dominant toroidal field with

$$\left|\frac{B_\phi}{B_\varpi}\right| \gg 1 \quad \text{and} \quad \left|\frac{B_\varpi}{B_z}\right| \sim \frac{\varpi}{h} \gg 1, \tag{11.90}$$

where the second relation follows from $\nabla \cdot \mathbf{B}_p = 0$ and a quadrupole field.

The magnetic diffusivity is taken to be due to buoyancy, with the form derived in §2.2.10 given by

$$\eta = \frac{\xi |B_{\phi c}|}{(\mu_0 \rho_c)^{\frac{1}{2}}} h, \tag{11.91}$$

where the subscript c denotes central values and

$$\left|\frac{B_\varpi}{B_\phi}\right|_c < \xi < 1. \tag{11.92}$$

It will be shown that $|B_\varpi/B_\phi|_c$ is $\sim v_\mathrm{T}/c_\mathrm{S}$, where v_T is the rms turbulent speed. This form for η will be used in both components of the induction equation. An alternative would be use the above form for toroidal diffusion, denoted η_ϕ, but take $\eta_\varpi = \xi |B_{\varpi c}| h/(\mu_0 \rho_c)^{1/2}$ for the diffusion of B_ϖ in (11.86). However, the resulting disc solutions for these two approaches have no significant differences, as will be shown.

The α function is taken to be the rms turbulent speed, v_T, which is expressed as a fraction of the sound speed in the central plane, so

$$\alpha = \tilde{\alpha} = \epsilon \left(\frac{P_c}{\rho_c}\right)^{\frac{1}{2}}, \quad 0 < z < h, \tag{11.93}$$

where $\epsilon \ll 1$ and α is antisymmetric in z. The numerical simulations of Brandenburg *et al* (1995), described in §11.2.4, indicate that this is a reasonable form to adopt for an α gererated by the Balbus-Hawley instability.

The thin nature of the disc, together with the smallness of the ratio $|B_\varpi/B_\phi|_c$, can be used to simplify the equations. Consider, first, the radial momentum, given by (11.82). Equation (11.81) yields

$$\frac{\partial \psi}{\partial \varpi} = \frac{GM_{\rm p}}{\varpi^2} = \frac{v_{\rm K}^2}{\varpi}, \tag{11.94}$$

where $v_{\rm K}$ is the Keplerian speed, and small terms of order $(z/\varpi)^2$ and above have been dropped. The perfect gas equation gives

$$\frac{1}{\rho}\left|\frac{\partial P}{\partial \varpi}\right| \sim \left(\frac{\Re T_c}{\mu}\right)\frac{1}{\varpi},$$

so, as with the viscous disc, a ratio of

$$\frac{|\partial P/\partial \varpi|}{\rho v_{\rm K}^2/\varpi} \lesssim 10^{-4} \tag{11.95}$$

follows for temperatures $\lesssim 10^5$ K and hence the radial pressure gradient is negligible.

The radial inflow speed can be estimated from (11.83) as

$$|v_\varpi| \sim \frac{B_\varpi B_\phi}{\mu_0 \rho v_\phi}.$$

For a steady state, the azimuthal induction equation (11.87) requires

$$\eta \frac{\partial^2 B_\phi}{\partial z^2} \sim \varpi B_\varpi \frac{\partial \Omega}{\partial \varpi},$$

yielding

$$B_\phi \sim \frac{B_\varpi v_\phi}{\eta/\varpi}\left(\frac{h}{\varpi}\right)^2. \tag{11.96}$$

It then follows that

$$|v_\varpi| \sim \frac{B_\varpi^2}{\mu_0 \rho(\eta/\varpi)}\left(\frac{h}{\varpi}\right)^2. \tag{11.97}$$

Equations (11.91), (11.92), (11.96) and (11.97) give

$$\frac{|v_\varpi|}{v_\phi} \sim \frac{1}{\xi^2}\left|\frac{B_\varpi}{B_\phi}\right|^3 \left(\frac{h}{\varpi}\right)^2 \ll 1. \tag{11.98}$$

It follows from (11.89) that the radial inertial terms are comparable, being

$$\frac{\partial}{\partial \varpi}\left(\frac{v_\varpi^2}{2}\right) \sim v_z \frac{\partial v_\varpi}{\partial z},$$

and (11.98) shows that these are small relative to v_ϕ^2/ϖ.

This leaves the two radial magnetic force terms to be considered. Since $|B_\varpi| \sim (\varpi/h)|B_z|$, the vertical derivative is dominant in (11.85) so

$$J_\phi = \frac{1}{\mu_0}\frac{\partial B_\varpi}{\partial z}. \tag{11.99}$$

Equations (11.90)–(11.92) and (11.96) then give

$$\frac{B_z J_\phi}{\rho(v_\phi^2/\varpi)} \sim \frac{1}{\xi^2}\left(\frac{B_\varpi}{B_\phi}\right)^4\left(\frac{h}{\varpi}\right)^2 \ll 1. \tag{11.100}$$

Finally (11.91) and (11.96) yield

$$\frac{B_\phi^2}{\mu_0\rho v_\phi^2} \sim \frac{1}{\xi^2}\left(\frac{B_\varpi}{B_\phi}\right)^2\left(\frac{h}{\varpi}\right)^2 \ll 1. \tag{11.101}$$

It follows that the gravitational term dominates in (11.82), giving

$$v_\phi = v_\mathrm{K} = \left(\frac{GM_\mathrm{p}}{\varpi}\right)^{\frac{1}{2}}. \tag{11.102}$$

The azimuthal velocity will only differ significantly from v_K close to the star, where a boundary layer forms with a radial length-scale $\ll \varpi$.

Consider, next, the components of the induction equation. In the poloidal component (11.86), equations (11.89)–(11.92), (11.97) and (11.99) give

$$\frac{v_\varpi B_z}{\mu_0\eta J_\phi} \sim \frac{v_z B_\varpi}{\mu_0\eta J_\phi} \sim \frac{1}{\xi^2}\left(\frac{B_\varpi}{B_\phi}\right)^2\left(\frac{h}{\varpi}\right)^2 \ll 1. \tag{11.103}$$

Had η been split as previously discussed, the above ratio would differ by a factor $\eta_\phi/\eta_\varpi = |B_\phi/B_\varpi|_c$ and so still be $\ll 1$. The poloidal advection terms are therefore small and (11.86) becomes

$$\mu_0\eta J_\phi - \alpha B_\phi = 0. \tag{11.104}$$

In the toroidal induction equation (11.87) the vertical derivative term dominates in the diffusion operator and (11.89), (11.91) and (11.97) yield

$$\frac{\partial(v_\varpi B_\phi)/\partial\varpi}{\eta(\partial^2 B_\phi/\partial z^2)} \sim \frac{\partial(v_z B_\phi)/\partial z}{\eta(\partial^2 B_\phi/\partial z^2)} \sim \frac{1}{\xi^2}\left(\frac{B_\varpi}{B_\phi}\right)^2\left(\frac{h}{\varpi}\right)^2 \ll 1. \tag{11.105}$$

The advection term is consequently negligible and hence vertical diffusion balances the radial shearing of B_ϖ giving

$$\eta\frac{\partial^2 B_\phi}{\partial z^2} = -\varpi B_\varpi\frac{d\Omega_\mathrm{K}}{d\varpi}. \tag{11.106}$$

This leaves the vertical momentum equation (11.84) to be considered. Equations (11.81), (11.99), (11.101) and (11.102) give the ratios of the magnetic terms to gravity as

$$\frac{B_\varpi J_\phi}{\rho(\partial\psi/\partial z)} \sim \frac{1}{\xi^2}\left(\frac{B_\varpi}{B_\phi}\right)^4 \ll 1, \tag{11.107}$$

and

$$\frac{\partial(B_\phi^2/2\mu_0)/\partial z}{\rho\partial\psi/\partial z} \sim \frac{1}{\xi^2}\left(\frac{B_\varpi}{B_\phi}\right)^2 \ll 1. \tag{11.108}$$

Equations (11.81) and (11.89) yield the ratios of the inertial terms to gravity, giving

$$\frac{v_\varpi \partial v_z/\partial\varpi}{\partial\psi/\partial z} \sim \frac{\partial(v_z^2/2)/\partial z}{\partial\psi/\partial z} \sim \left(\frac{v_\varpi}{v_\phi}\right)^2 \ll 1. \tag{11.109}$$

The vertical equilibrium therefore becomes

$$\frac{\partial\psi}{\partial z} + \frac{1}{\rho}\frac{\partial P}{\partial z} = 0. \tag{11.110}$$

The ratio $|B_\varpi/B_\phi|$ can be estimated in terms of the turbulent parameter ϵ. Equations (11.81), (11.91), (11.93), (11.99), (11.101), (11.104) and (11.110) give

$$\frac{\eta}{\alpha h} \sim \left|\frac{B_\phi}{B_\varpi}\right| \sim \frac{1}{\epsilon}\left|\frac{B_\varpi}{B_\phi}\right|,$$

and so

$$\left|\frac{B_\varpi}{B_\phi}\right| \sim \epsilon^{\frac{1}{2}}, \tag{11.111}$$

where (11.93) has $\epsilon \ll 1$. The effect of allowing for separate η_ϕ and η_ϖ is to change this ratio to $|B_\varpi/B_\phi| \sim \epsilon^{1/3}$.

Finally, the smallness of the viscous force can be shown. Denote the rms turbulent speed, the turbulent time-scale and the mixing length, in the absence of rotation, by $v_{\rm T}$, $\tau_{\rm T}$ and $\lambda_{\rm T}$. With $\lambda_{\rm T} \sim h$, the ratio of the Keplerian rotation time to $\tau_{\rm T}$ is

$$\frac{\tau_{\rm K}}{\tau_{\rm T}} \sim \frac{v_{\rm T}}{(h/\varpi)v_{\rm K}} \sim \frac{v_{\rm T}}{c_{\rm s}}, \tag{11.112}$$

by virtue of the vertical equilibrium. Hence $\tau_{\rm K} < \tau_{\rm T}$ for subsonic turbulence, and (2.170) gives a Rossby number $R_{\rm T} < 1$, so rotation will effect the convection. The simplest way of accounting for this is to multiply $v_{\rm T}$ and h by the factor $R_{\rm T} = \tau_{\rm K}/\tau_{\rm T}$. The α function is now the modified rms turbulent speed and hence

$$\tilde{\alpha} = \left(\frac{\tau_{\rm K}}{\tau_{\rm T}}\right) v_{\rm T} = \epsilon c_{\rm S}, \tag{11.113}$$

where the last equality is consistent with the definition of $\epsilon = \tilde{\alpha}/c_{\rm S}$. It follows from (11.112) and (11.113) that the rotationally modified ϵ is

$$\epsilon = \left(\frac{v_{\rm T}}{c_{\rm S}}\right)^2 = \left(\frac{\tau_{\rm K}}{\tau_{\rm T}}\right)^2, \tag{11.114}$$

being smaller than the unmodified value of $v_{\rm T}/c_{\rm S}$. The viscosity coefficient is now

$$\nu = \left(\frac{\tau_{\rm K}}{\tau_{\rm T}}\right) v_{\rm T} \left(\frac{\tau_{\rm K}}{\tau_{\rm T}}\right) h = \epsilon v_{\rm T} h. \tag{11.115}$$

Elimination of $v_{\rm T}$ using (11.114) therefore yields

$$\nu = \epsilon^{\frac{3}{2}} c_{\rm S} h. \tag{11.116}$$

The ratio of the viscous force to the azimuthal magnetic force is

$$\frac{F_{\rm v\phi}}{F_{\rm m\phi}} = \frac{\partial[\rho\nu\varpi^3(d\Omega_{\rm K}/d\varpi)]/\partial\varpi}{\partial(\varpi^2 B_\varpi B_\phi/\mu_0)/\partial\varpi},$$

and hence

$$\frac{F_{\rm v\phi}}{F_{\rm m\phi}} \sim \frac{\mu_0\rho\nu v_\phi}{\varpi B_\varpi B_\phi} \sim \xi^2 \frac{\nu}{\eta}\left(\frac{B_\phi}{B_\varpi}\right)^2, \tag{11.117}$$

where the last expression follows from (11.91) and (11.96). Equations (11.91), (11.108), (11.110) and (11.116) yield

$$\frac{\nu}{\eta} \sim \frac{\epsilon^{\frac{3}{2}}}{\xi}\left(\frac{\mu_0 P_c}{B_{\phi c}^2}\right)^{\frac{1}{2}} \sim \epsilon^{\frac{3}{2}}\left|\frac{B_\phi}{B_\varpi}\right|,$$

and so

$$\frac{F_{\rm v\phi}}{F_{\rm m\phi}} \sim \xi^2 \ll 1, \tag{11.118}$$

by virtue of (11.111). Using separate η_ϖ and η_ϕ, as previously defined, gives $F_{v\phi}/F_{m\phi} \sim \epsilon^{\frac{1}{2}}\xi^2$ so strengthening the inequality.

11.3.2. VERTICAL INTEGRALS

The equations for a thin magnetic disc, with buoyancy diffusivity, can now be gathered together as;

$$v_\phi = v_{\rm K} = \left(\frac{GM_{\rm p}}{\varpi}\right)^{\frac{1}{2}}, \tag{11.119}$$

$$v_\varpi \frac{\partial}{\partial \varpi}(\varpi^2 \Omega) + v_z \frac{\partial}{\partial z}(\varpi^2 \Omega) = \frac{1}{\mu_0 \varpi \rho}\frac{\partial}{\partial \varpi}(\varpi^2 B_\varpi B_\phi) + \frac{\varpi}{\mu_0 \rho}\frac{\partial}{\partial z}(B_z B_\phi), \tag{11.120}$$

$$\frac{z}{\varpi}\frac{v_{\rm K}^2}{\varpi} + \frac{1}{\rho}\frac{\partial P}{\partial z} = 0, \tag{11.121}$$

$$\eta \frac{\partial B_\varpi}{\partial z} = \alpha B_\phi, \tag{11.122}$$

$$\eta \frac{\partial^2 B_\phi}{\partial z^2} = -\varpi B_\varpi \frac{d\Omega_{\rm K}}{d\varpi}, \tag{11.123}$$

$$\frac{1}{\varpi}\frac{\partial}{\partial \varpi}(\varpi \rho v_\varpi) + \frac{\partial}{\partial z}(\rho v_z) = 0, \tag{11.124}$$

where

$$\eta = \frac{\xi |B_{\phi c}|}{(\mu_0 \rho_c)^{\frac{1}{2}}} h, \tag{11.125a}$$

and

$$\alpha = \begin{cases} \tilde{\alpha}(\varpi), & 0 < z < h, \\ 0, & z = 0, \\ -\tilde{\alpha}(\varpi), & -h < z < 0, \end{cases} \tag{11.125b}$$

with $\tilde{\alpha}$ given by (11.93).

Combining (11.120) and (11.124) gives

$$\frac{\partial}{\partial \varpi}(\varpi \rho v_\varpi \varpi^2 \Omega) + \frac{\partial}{\partial z}(\varpi \rho v_z \varpi^2 \Omega) = \frac{1}{\mu_0}\frac{\partial}{\partial \varpi}(\varpi^2 B_\varpi B_\phi) + \frac{1}{\mu_0}\frac{\partial}{\partial z}(\varpi^2 B_z B_\phi).$$

Integrating this vertically through the disc, using the vanishing of B_ϕ and ρ at $z = \pm h$ together with the even symmetry, yields

$$\frac{d}{d\varpi}\left[\frac{\dot{M}}{2\pi}\varpi^2\Omega_{\mathrm{K}} + \frac{2\varpi^2}{\mu_0}\int_0^h B_\varpi B_\phi dz\right] = 0, \tag{11.126}$$

where

$$\dot{M} = -4\pi\int_0^h \varpi\rho v_\varpi dz \tag{11.127}$$

is the mass transfer rate. As in the case of the viscous disc, a boundary layer will exist close to the stellar surface in which Ω changes rapidly from Ω_{K} to the angular velocity of the primary star, Ω_{p}. For accretion to occur, Ω_{p} must be less than Ω at the outer edge of the boundary layer, so $d\Omega/d\varpi = 0$ close to this radius. Integrating (11.126), taking $B_\varpi B_\phi$ to vanish at $R_{\mathrm{p}} + \delta$, where R_{p} is the stellar radius and δ the boundary layer width, gives

$$\int_0^h B_\varpi B_\phi dz = -\frac{\mu_0(GM_{\mathrm{p}})^{\frac{1}{2}}\dot{M}}{4\pi\varpi^{\frac{3}{2}}}\left[1 - \left(\frac{R_{\mathrm{p}}}{\varpi}\right)^{\frac{1}{2}}\right], \tag{11.128}$$

using $\delta/R_{\mathrm{p}} \ll 1$.

The radial magnetic field in the central plane can be related to the above integral. Equation (11.122) gives

$$B_\varpi B_\phi = \frac{\eta}{2\alpha}\frac{\partial}{\partial z}(B_\varpi^2).$$

The surface conditions on the quadrupolar magnetic field are

$$B_\phi(\varpi, h) = B_\varpi(\varpi, h) = 0. \tag{11.129}$$

The first condition is exact, resulting from axisymmetry and the potential nature of the outer field. The second condition follows because in the exterior vacuum B_ϖ and B_z are comparable, and in the quadrupole mode B_z is a higher order correction found by including the radial derivatives in the induction equation (see §11.2.1). It follows that

$$\frac{\eta}{2\tilde{\alpha}}B_\varpi^2(\varpi, 0) = -\int_0^h B_\varpi B_\phi dz,$$

so use of (11.128) gives

$$B_\varpi^2(\varpi, 0) = \frac{\mu_0(GM_{\mathrm{p}})^{\frac{1}{2}}\dot{M}\tilde{\alpha}}{2\pi\varpi^{\frac{3}{2}}\eta}\left[1 - \left(\frac{R_{\mathrm{p}}}{\varpi}\right)^{\frac{1}{2}}\right]. \tag{11.130}$$

Integrating (11.121), noting that the thermal pressure vanishes at the disc surface, yields

$$P_{\mathrm{c}} = \Omega_{\mathrm{K}}^2 \int_0^h z\rho dz \tag{11.131}$$

for the vertical equilibrium.

11.3.3. THERMAL ENERGY TRANSPORT

Magnetic flux is lost from the disc as a result of the buoyancy diffusivity. It is reasonable to assume that the flux is dissipated by turbulent reconnection processes. The dissipation rate, per unit volume, is

$$Q_{\mathrm{m}} = \mu_0 \eta |\mathbf{J}|^2. \tag{11.132}$$

The poloidal components of the current density are

$$J_{\varpi} = -\frac{1}{\mu_0}\frac{\partial B_\phi}{\partial z}, \qquad J_z = \frac{1}{\mu_0 \varpi}\frac{\partial}{\partial \varpi}(\varpi B_\phi).$$

It then follows from (11.99) for J_ϕ and (11.111) for $|B_\varpi/B_\phi|$ that J_ϖ is dominant, and so

$$Q_{\mathrm{m}} = \frac{\eta}{\mu_0}\left(\frac{\partial B_\phi}{\partial z}\right)^2. \tag{11.133}$$

The sum of the energy fluxes through the upper and lower disc faces is

$$2F_{+} = \int_{-h}^{h} Q_{\mathrm{m}} dz. \tag{11.134}$$

Integrating $(\partial B_\phi/\partial z)^2$ by parts, using (11.123) and (11.129) together with the central plane condition for a quadrupolar-type field of

$$\left(\frac{\partial B_\phi}{\partial z}\right)_{z=0} = 0, \tag{11.135}$$

gives

$$\int_0^h \left(\frac{\partial B_\phi}{\partial z}\right)^2 dz = \frac{\varpi}{\eta}\frac{d\Omega_{\mathrm{K}}}{d\varpi}\int_0^h B_\varpi B_\phi dz.$$

Equations (11.133) and (11.134) then yield

$$F_{+} = \frac{3GM_{\mathrm{p}}\dot{M}}{8\pi\varpi^3}\left[1 - \left(\frac{R_{\mathrm{p}}}{\varpi}\right)^{\frac{1}{2}}\right], \tag{11.136}$$

after use of (11.119) and (11.128).

For radiative diffusion in an optically thick disc, the vertical flux is

$$F = -\frac{4\sigma_{\rm B}}{3\kappa\rho}\frac{\partial}{\partial z}(T^4), \tag{11.137}$$

where $\sigma_{\rm B}$ is the Stefan-Boltzmann constant and κ is the Rosseland mean opacity. For free-free and bound-free transitions the opacity can be approximated by the Kramers form

$$\kappa = \bar{K}\rho T^{-\frac{7}{2}}, \tag{11.138}$$

where $\bar{K}$ is a constant (Cox and Giuli, 1968). Integrating (11.137) vertically, noting that the dominant variation comes from the T^4 factor and that F vanishes in the central plane, gives

$$T_c^4 = \frac{3\tau}{4\sigma_{\rm B}}F_+, \tag{11.139}$$

where the optical depth is

$$\tau = \bar{K}\rho_c^2 T_c^{-\frac{7}{2}}h. \tag{11.140}$$

The equation of state is

$$P = \frac{\Re}{\mu}\rho T, \tag{11.141}$$

so combining (11.139)–(11.141) relates P_c and ρ_c by

$$P_c^{\frac{15}{2}} = \frac{3\bar{K}}{4\sigma_{\rm B}}\left(\frac{\Re}{\mu}\right)^{\frac{15}{2}} hF_+\rho_c^{\frac{19}{2}}. \tag{11.142}$$

11.3.4. THE DYNAMO

The differential rotation in the disc creates toroidal magnetic field from the polidal component. The growth of B_ϕ is balanced by diffusion resulting from the dissipation of poloidal currents. The α-effect converts B_ϕ into B_ϖ, this process being balanced by the dissipation of toroidal currents.

Combining the components of the induction equation, given by (11.122) and (11.123), yields

$$\frac{\partial^3 B_\phi}{\partial z^3} + \frac{\varpi\Omega_{\rm K}'\tilde{\alpha}}{\eta^2}B_\phi = 0. \tag{11.143}$$

The toroidal magnetic field can be expressed in the separable form

$$B_\phi = \tilde{B}_\phi(\varpi) f_\phi\left(\frac{z}{h}\right). \tag{11.144}$$

Substitution into (11.143) gives

$$f_\phi'''(\zeta) + N f_\phi(\zeta) = 0, \tag{11.145}$$

where $\zeta = z/h$, primes denote differentiation, and

$$N = \frac{\varpi \Omega_{\rm K}' h^3 \tilde{\alpha}}{\eta^2}. \tag{11.146}$$

It was illustrated in §11.2.1 that a magnetic field of a global, rather than local, nature requires an essentially uniform dynamo number. Since a global field is needed here, the dynamo number N is taken to be constant and the fact that η depends on B_ϕ and ρ allows the disc structure to adjust accordingly. The quadrupolar boundary conditions, given by (11.129) and (11.135), together with (11.123) require

$$f_\phi''(1) = f_\phi(1) = f_\phi'(0) = 0. \tag{11.147}$$

The function $f_\phi(\zeta)$ is even, so (11.145) is solved for $0 < \zeta < 1$. Substituting $f_\phi = \exp(ikz)$, taking linear combinations of the three independent solutions and applying the boundary conditions, gives

$$f_\phi = A\left(\exp[-K(\zeta-1)] - 2\exp[K(\zeta-1)/2]\cos(\sqrt{3}K(\zeta-1)/2 - \pi/3)\right), \tag{11.148}$$

where A is a constant and the eigenvalues K are the solutions of

$$2\cos\left(\frac{\sqrt{3}K}{2}\right) + \exp\left(\frac{3K}{2}\right) = 0, \tag{11.149}$$

yielding negative values for K. The dynamo number $N = K^3$ so, using (11.119), and (11.146), gives

$$K^3 = -\frac{3\Omega_{\rm K} h^3 \tilde{\alpha}}{2\eta^2}. \tag{11.150}$$

Equations (11.119), (11.123) and (11.144) give the ratio of the field components as

$$\frac{B_\phi}{B_\varpi} = \frac{3\Omega_{\rm K} h^2}{2\eta} \frac{f_\phi}{f_\phi''}. \tag{11.151}$$

TABLE 11.1.

K	$f_\phi(0)/f_\phi''(0)$
-1.85	-2.82×10^{-1}
-5.44	-3.38×10^{-2}
-9.07	-1.22×10^{-2}
-12.70	-6.20×10^{-3}

It follows, using (11.150), that

$$B_\phi(\varpi, 0) = |K|^3 \frac{f_\phi(0)}{f_\phi''(0)} \frac{\eta}{\tilde{\alpha} h} B_\varpi(\varpi, 0), \tag{11.152}$$

where, from (11.148) and (11.149),

$$\frac{f_\phi(0)}{f_\phi''(0)} = \frac{[\exp K + 2\sqrt{3}\exp(-K/2)\sin(\sqrt{3}K/2)]}{K^2[\exp K - 2\sqrt{3}\exp(-K/2)\sin(\sqrt{3}K/2)]}. \tag{11.153}$$

Table 11.1 shows the first four values of K, obtained from (11.149), and the corresponding values of $f_\phi(0)/f_\phi''(0)$.

The central thermal pressure is given by (11.131). Approximating the integral as $\rho_c h^2/2$, relates the central pressure and density by

$$P_c = \frac{1}{2}\Omega_{\mathrm{K}}^2 h^2 \rho_c. \tag{11.154}$$

Then (11.93) becomes

$$\tilde{\alpha} = \frac{\epsilon}{\sqrt{2}}\Omega_{\mathrm{K}} h. \tag{11.155}$$

Equations (11.150) and (11.155) yield

$$\frac{\tilde{\alpha} h}{\eta} = \left(\frac{\sqrt{2}\epsilon}{3}\right)^{\frac{1}{2}} |K|^{\frac{3}{2}}, \tag{11.156}$$

so (11.130) and (11.152) give

$$B_{\varpi c} = \left(\frac{\mu_0}{2\pi}\right)^{\frac{1}{2}} \left(\frac{\sqrt{2}G}{3}\right)^{\frac{1}{4}} \epsilon^{\frac{1}{4}} |K|^{\frac{3}{4}} M_{\mathrm{p}}^{\frac{1}{4}} \dot{M}^{\frac{1}{2}} \frac{[1 - (R_{\mathrm{p}}/\varpi)^{\frac{1}{2}}]^{\frac{1}{2}}}{\varpi^{\frac{3}{4}} h^{\frac{1}{2}}}, \tag{11.157}$$

$$B_{\phi c} = \left(\frac{3\mu_0}{2\sqrt{2\pi}}\right)^{\frac{1}{2}} \left(\frac{\sqrt{2}G}{3}\right)^{\frac{1}{4}} \frac{|K|^{\frac{9}{4}}}{\epsilon^{\frac{1}{4}}} \frac{f_\phi(0)}{f''_\phi(0)} M_{\rm p}^{\frac{1}{4}} \dot{M}^{\frac{1}{2}} \frac{[1-(R_{\rm p}/\varpi)^{\frac{1}{2}}]^{\frac{1}{2}}}{\varpi^{\frac{3}{4}} h^{\frac{1}{2}}}. \quad (11.158)$$

It is noted that had separate η_ϕ and η_ϖ been used, these field component expressions would be multiplied by $(\eta_\phi/\eta_\varpi)^{1/4}$ and $(\eta_\varpi/\eta_\phi)^{1/4}$ respectively. This leaves their radial dependencies unaffected, just introducing constant factors of order unity. The product $B_\varpi B_\phi$ is unchanged.

11.3.5. MAGNETIC DISC STRUCTURE

The radial dependencies of disc quantities can now be found. Equations (11.125a), (11.155) and (11.156) give

$$B_{\phi c} = \left(\frac{3\mu_0}{2\sqrt{2}}\right)^{\frac{1}{2}} \frac{\epsilon^{\frac{1}{2}}}{\xi |K|^{\frac{3}{2}}} \Omega_{\rm K} \rho_c^{\frac{1}{2}} h. \quad (11.159)$$

The central density can be expressed in terms of the disc height, using (11.142) and (11.154), so

$$\rho_c = \frac{1}{2^{\frac{15}{4}}} \left(\frac{4\sigma_{\rm B}}{3\bar{K}}\right)^{\frac{1}{2}} \left(\frac{\mu}{\Re}\right)^{\frac{15}{4}} \frac{\Omega_{\rm K}^{\frac{15}{2}} h^7}{F_+^{\frac{1}{2}}}. \quad (11.160)$$

Substituting this in (11.159) and equating $B_{\phi c}$ to (11.158), using $\bar{K} = 1.4 \times 10^{19}\,\mathrm{m^2\,Kg^{-1}}$ and $\mu = 0.62$, yields the disc height as

$$h = 1.7 \times 10^6 \xi^{\frac{1}{5}} \frac{|K|^{\frac{3}{4}}}{\epsilon^{\frac{3}{20}}} \left|\frac{f_\phi(0)}{f''_\phi(0)}\right|^{\frac{1}{5}} \frac{\dot{M}_{-10}^{\frac{3}{20}}}{M_1^{\frac{3}{8}}} \varpi_8^{\frac{9}{8}} f^{\frac{3}{20}}\ \mathrm{m}, \quad (11.161)$$

where $\dot{M}_{-10} = \dot{M}/10^{-10} M_\odot \mathrm{yr}^{-1}$, $M_1 = M_{\rm p}/M_\odot$, $\varpi_8 = \varpi/10^8\,\mathrm{m}$ and

$$f = 1 - \left(\frac{R_{\rm p}}{\varpi}\right)^{\frac{1}{2}}. \quad (11.162)$$

The magnetic field components, given by (11.157) and (11.158), are then

$$B_{\varpi c} = 7.7 \times 10^{-2} \frac{\epsilon^{\frac{13}{40}} |K|^{\frac{3}{8}}}{\xi^{\frac{1}{10}}} \left|\frac{f_\phi(0)}{f''_\phi(0)}\right|^{-\frac{1}{10}} M_1^{\frac{7}{16}} \dot{M}_{-10}^{\frac{17}{40}} \frac{f^{\frac{17}{4}}}{\varpi_8^{\frac{21}{16}}}\ \mathrm{T}, \quad (11.163)$$

$$B_{\phi c} = -0.11 \frac{|K|^{\frac{15}{8}}}{\xi^{\frac{1}{10}} \epsilon^{\frac{7}{40}}} \left|\frac{f_\phi(0)}{f''_\phi(0)}\right|^{\frac{9}{10}} M_1^{\frac{7}{16}} \dot{M}_{-10}^{\frac{17}{40}} \frac{f^{\frac{17}{40}}}{\varpi_8^{\frac{21}{16}}}\ \mathrm{T}. \quad (11.164)$$

The central density, pressure and temperature are obtained from (11.136), (11.141), (11.142), (11.154), (11.160) and (11.161), giving

$$\rho_c = 2.4 \times 10^{-5} \frac{\xi^{\frac{7}{5}} |K|^{\frac{21}{4}}}{\epsilon^{\frac{27}{20}}} \left| \frac{f_\phi(0)}{f''_\phi(0)} \right|^{\frac{7}{5}} M_1^{\frac{5}{8}} \dot{M}_{-10}^{\frac{11}{20}} \frac{f^{\frac{11}{20}}}{\varpi_8^{\frac{15}{8}}} \,\mathrm{Kg\,m^{-3}}, \tag{11.165}$$

$$P_c = 4.9 \times 10^{3} \frac{\xi^{\frac{9}{5}} |K|^{\frac{27}{4}}}{\epsilon^{\frac{27}{20}}} \left| \frac{f_\phi(0)}{f''_\phi(0)} \right|^{\frac{9}{5}} M_1^{\frac{7}{8}} \dot{M}_{-10}^{\frac{17}{20}} \frac{f^{\frac{17}{20}}}{\varpi_8^{\frac{21}{8}}} \,\mathrm{Nm^{-2}}, \tag{11.166}$$

$$T_c = 1.5 \times 10^{4} \frac{\xi^{\frac{2}{5}} |K|^{\frac{3}{2}}}{\epsilon^{\frac{3}{10}}} \left| \frac{f_\phi(0)}{f''_\phi(0)} \right|^{\frac{2}{5}} M_1^{\frac{1}{4}} \dot{M}_{-10}^{\frac{3}{10}} \frac{f^{\frac{3}{10}}}{\varpi_8^{\frac{3}{4}}} \,\mathrm{K}, \tag{11.167}$$

where K and $|f_\phi(0)/f''_\phi(0)|$ are shown in Table 11.1 and $\epsilon^{\frac{1}{2}} < \xi < 1$. Equations (11.127), (11.161) and (11.165) give the inflow speed as

$$v_\varpi = -1.2 \times 10^{2} \frac{\epsilon^{\frac{6}{5}}}{\xi^{\frac{8}{5}} |K|^{6}} \left| \frac{f_\phi(0)}{f''_\phi(0)} \right|^{-\frac{8}{5}} \frac{\dot{M}_{-10}^{\frac{3}{10}}}{M_1^{\frac{1}{4}}} \frac{1}{\varpi^{\frac{1}{4}} f^{\frac{7}{10}}} \,\mathrm{m\,s^{-1}}, \tag{11.168}$$

where it should be remembered that the solutions are valid for $\varpi > R_\mathrm{p} + \delta$, with δ the boundary layer width.

11.3.6. CONCLUSIONS

The dimensionless parameter ϵ expresses the strength of the turbulence, and is related to the ratio $|B_\varpi/B_\phi|$ by (11.111). Equation (11.114) gives the rotationally modified rms turbulent speed as $v_\mathrm{T} = \epsilon^{1/2} c_\mathrm{S}$. The magnetic disc can operate with ϵ as low as 10^{-4}, corresponding to $v_\mathrm{T} = 10^{-2} c_\mathrm{S}$, and hence to weak turbulence. Such values of ϵ are in good agreement with the numerical simulation results of magnetic turbulence due to Brandenburg *et al* (1995), described in §11.2.4.

It is noted that the dynamo considered in §11.3.4 is not, of course, kinematic, since the induction equation is coupled to the dynamical and thermal equations via the dependence of η on B_ϕ and ρ.

Table 11.1 and equation (11.164) show that typical values of B_ϕ in the central plane are $\sim 10^3$ G. The radial structure is similar to that of the viscous disc, given by (2.297)–(2.303). However, unlike the viscous disc, the magnetic disc can operate in different eigenmodes. The lowest modes give the simplest vertical variation of the magnetic field and the lowest temperatures. It may be possible to relate transitions between such states to the outburst behaviour observed in binary star accretion discs.

11.4. Magneto-Viscous Discs

11.4.1. VERTICAL INTEGRALS

The foregoing disc solution is valid for $\eta_{\mathrm{B}} \gg \eta_{\mathrm{T}}$, leading to the magnetic stress dominating in the radial advection of angular momentum. The more general case with $\eta_{\mathrm{B}} \sim \eta_{\mathrm{T}}$ was considered by Campbell and Caunt (1997). Then η is given by

$$\eta = \eta_{\mathrm{B}} + \eta_{\mathrm{T}} = \frac{\xi |B_{\phi c}|}{(\mu_0 \rho_c)^{\frac{1}{2}}} h + \bar{\epsilon} c_{\mathrm{s}} h, \tag{11.169}$$

with $\bar{\epsilon} < 1$. The viscosity coefficient is

$$\nu = \epsilon^{\frac{3}{2}} c_{\mathrm{S}} h. \tag{11.170}$$

The radial magnetic force is still small, so the disc remains Keplerian up to an inner boundary layer at the stellar surface. The vertical and azimuthal momentum equations now become

$$\frac{z}{\varpi} \frac{v_{\mathrm{K}}^2}{\varpi} + \frac{1}{\rho} \frac{\partial}{\partial z} \left(P + \frac{B_\phi^2}{2\mu_0} \right) = 0, \tag{11.171}$$

$$\varpi \rho v_\varpi \frac{\partial}{\partial \varpi} (\varpi^2 \Omega) = \frac{1}{\mu_0} \frac{\partial}{\partial \varpi} (\varpi^2 B_\varpi B_\phi) + \frac{\varpi^2}{\mu_0} \frac{\partial}{\partial z} (B_\phi B_z) + \frac{\partial}{\partial \varpi} \left(\rho \nu \varpi^3 \frac{\partial \Omega}{\partial \varpi} \right), \tag{11.172}$$

noting $\partial \Omega / \partial z = 0$. The continuity and induction equations are unchanged.

Combining the continuity and azimuthal momentum equations and integrating vertically through the disc, using $B_\phi(\varpi, \pm h) = 0$, yields

$$\frac{d}{d\varpi} \left(\frac{\dot{M}}{2\pi} \varpi^2 \Omega + \nu \Sigma \varpi^3 \Omega' + \frac{2\varpi^2}{\mu_0} \int_0^h B_\varpi B_\phi dz \right) = 0.$$

Radial integration and application of the standard boundary layer condition then gives

$$\frac{\dot{M}}{2\pi} \left[\varpi^2 \Omega_{\mathrm{K}} - (G M_{\mathrm{p}} R_{\mathrm{p}})^{\frac{1}{2}} \right] + \nu \Sigma \varpi^3 \Omega_{\mathrm{K}}' + \frac{2\varpi^2}{\mu_0} \int_0^h B_\varpi B_\phi dz = 0, \tag{11.173}$$

valid for $\varpi > R_{\mathrm{p}} + \delta$, where δ is the boundary layer width. The vertical equilibrium (11.171) integrates to

$$P_c + \frac{B_{\phi c}^2}{2\mu_0} = \Omega_{\mathrm{K}}^2 \int_0^h z \rho dz. \tag{11.174}$$

Since viscous dissipation is now significant, the surface heat flux becomes

$$F_{+} = \frac{\eta}{\mu_0}\int_0^h \left(\frac{\partial B_\phi}{\partial z}\right)^2 dz + \frac{1}{2}\nu\Sigma(\varpi\Omega_{\mathrm{K}}')^2. \tag{11.175}$$

Combining (11.122), (11.173) and (11.175) gives (11.136) for F_{+}. Equation (11.142) relating P_c to ρ_c also still holds.

11.4.2. DISC SOLUTIONS

The vertical dynamo problem remains essentially unchanged, except that η is now given by (11.169). Equation (11.146), with $N = K^3$, yields

$$\eta^2 = (\eta_{\mathrm{T}} + \eta_{\mathrm{B}})^2 = \frac{3\Omega_{\mathrm{K}} h^3 \tilde{\alpha}}{2|K|^3}.$$

Then, noting from (11.93) and (11.169),

$$\tilde{\alpha} h = \epsilon c_{\mathrm{s}} h = \left(\frac{\epsilon}{\bar{\epsilon}}\right) \bar{\epsilon} c_{\mathrm{s}} h = \frac{\epsilon}{\bar{\epsilon}}\eta_{\mathrm{T}},$$

it follows that

$$(\eta_{\mathrm{T}} + \eta_{\mathrm{B}})^2 = \frac{3}{2|K|^3}\frac{\epsilon}{\bar{\epsilon}}\Omega_{\mathrm{K}} h^2 \eta_{\mathrm{T}}. \tag{11.176}$$

Using $\langle z\rho\rangle \simeq h\rho_c/2$ in (11.174), gives

$$\frac{P_c}{\rho_c} + \frac{B_{\phi c}^2}{2\mu_0\rho_c} = \frac{1}{2}h^2\Omega_{\mathrm{K}}^2.$$

Since $c_{\mathrm{s}}^2 = P_c/\rho_c$, this yields

$$\frac{1}{\bar{\epsilon}^2}\eta_{\mathrm{T}}^2 + \frac{1}{2\xi^2}\eta_{\mathrm{B}}^2 = \frac{1}{2}h^4\Omega_{\mathrm{K}}^2. \tag{11.177}$$

Dimensionless diffusivities can be defined by

$$\bar{x} = \frac{\eta_{\mathrm{T}}}{h^2\Omega_{\mathrm{K}}}, \qquad \bar{y} = \frac{\eta_{\mathrm{B}}}{h^2\Omega_{\mathrm{K}}}. \tag{11.178a,b}$$

Equations (11.176) and (11.177) then become

$$(\bar{x} + \bar{y})^2 = \frac{3}{2|K|^3}\frac{\epsilon}{\bar{\epsilon}}\bar{x}, \tag{11.179}$$

$$\frac{2}{\bar{\epsilon}^2}\bar{x}^2 + \frac{1}{\xi^2}\bar{y}^2 = 1. \tag{11.180}$$

It is noted that the coefficients in these equations are constants and hence $\bar{x}$ and $\bar{y}$ are spatially independent. Eliminating $(\bar{x}+\bar{y})$, and then $\bar{x}$, yields the quartic equation

$$(\bar{\epsilon}^2 + 2\xi^2)^2\bar{y}^4 - 12\frac{\bar{\epsilon}\epsilon\xi^2}{|K|^3}\bar{y}^3 + \left(\frac{9\epsilon^2\xi^2}{2K^6} - 2\bar{\epsilon}^2\xi^2(\bar{\epsilon}^2 + 2\xi^2)\right)\bar{y}^2$$
$$+ \frac{12\bar{\epsilon}\epsilon\xi^4}{|K|^3}\bar{y} + \left(\bar{\epsilon}^4\xi^4 - \frac{9\epsilon^2\xi^4}{2K^6}\right) = 0. \tag{11.181}$$

After solving this for $\bar{y}$, equation (11.180) gives $\bar{x}$. It follows from (11.178a,b) that the total diffusivity is

$$\eta = (\bar{x}+\bar{y})\Omega_K h^2. \tag{11.182}$$

The foregoing equations can be solved to obtain the radial disc structure. The solution is similar to (11.161)–(11.168), except with coefficients containing $\bar{x}(|K|,\epsilon,\bar{\epsilon},\xi)$ and $\bar{y}(|K|,\epsilon,\bar{\epsilon},\xi)$. Solutions are now possible with $F_{v\phi} \sim F_{m\phi}$, so the viscosity and magnetic stress can make similar contributions to the radial advection of angular momentum.

11.5. Discussion

It is most probable that binary star accretion discs have turbulent motions generated by the Balbus-Hawley instability. The turbulence and strong radial shear lead to the generation and maintenance of a large-scale magnetic field. Viscous and magnetic stresses cause radial advection of angular momentum via the azimuthal forces, $F_{v\phi}$ and $F_{m\phi}$. Provided these forces oppose the large-scale azimuthal motion, material will spiral in through the disc as angular momentum flows outwards. The force ratio $|F_{m\phi}|/|F_{v\phi}|$ depends on the nature of the turbulence and the diffusion mechanisms operating. It was seen in §11.3 that with η due to buoyancy, and a simple prescription for ν, $|F_{m\phi}| \gg |F_{v\phi}|$ so magnetic stresses account for the majority of angular mometum advection. The more general form of $\eta = \eta_T + \eta_B$ can give $F_{v\phi} \sim F_{m\phi}$.

So far, analytic and numerical work has indicated the importance of magnetic shear instabilities in angular momentum transport. Local simulations can generate small-scale turbulence, suggesting that mean-field theory is applicable. However, calculations of the effect of magnetic shearing instabilities beyond the present local box models are needed to analyse the details of the turbulent spectrum. Knowledge of this would then allow a calculation of α-effect terms and the transport effects due to viscosity and magnetic diffusion.

The dynamo models discussed in §11.2.1 illustrate that the nature of the disc's surroundings can have a significant effect on the type of magnetic field generated in the disc. It was seen that a low conductivity external region favours the generation of quadrupolar-type fields, whilst very highly conducting surroundings facilitates dipole-type fields. However, the latter case results in the disc field being excluded from the surroundings, and this situation is unlikely to be achieved in reality. A quadrupole field leads to the simplest mechanism of magnetic angular momentum transport in a radially sheared disc, via the $B_{\varpi}B_{\phi}$ stress. Since η depends on $\mathbf{B}$ and ρ, the disc can adjust to accommodate such a field.

Most disc models to date involve a vertically averaged structure. The future aim is to find solutions which self-consistently incorporate magnetic shear instabilities and vertical structure. This cannot be done until global simulations of magnetically generated turbulence are available.

It is likely that accretion discs have a corona which will interact with a magnetic field. Magnetic reconnection of buoyant fields in the lower density surface regions may supply the energy source for a hot corona. Heyvaerts and Priest (1989) considered the structure of a corona surrounding a Keplerian disc. The coronal magnetic field penetrates the disc and is stressed by its motions. Diffusive processes relax these stresses and the field is modelled as an essentially force-free structure. Coronal loops of a range of sizes were considered. The coronal interaction with the disc was found to be weak for finer scale magnetic fields and for a large-scale corona. Coronal heating becomes most efficient when the magnetic loops are of a size comparable to the disc's radius.

The existence of a disc corona raises the possibility of a wind flow. This would result in angular momentum transport away from the disc, which could have some influence on the inflow. Accretion disc winds are discussed in Chapter 12.

References

Balbus, S.A. and Hawley, J.F., 1991. *Astrophys. J.*, **376**, 214.

Blandford, R.D. and Payne, D.G., 1982. *Mon. Not. R. Astr. Soc.*, **199**, 883.

Brandenburg, A. and Donner, K.J., 1997. *Mon. Not. R. Astr. Soc.*, in press.

Brandenburg, A., Nordlund, A., Stein, R.F. and Torkelsson, U., 1995. *Astrophys. J.*, **446**, 741.

Brandenburg, A., Nordlund, A., Stein, R.F. and Torkelsson, U., 1996. *Astrophys. J. Lett.*, **458**, L45.

Campbell, C.G., 1992. *Geophys. Astrophys. Fluid. Dynam.*, **63**, 197.

Campbell, C.G. and Caunt, S., 1997. *Mon. Not. R. Astr. Soc.*, submitted.

Chandrasekhar, S., 1960. *Proc. Nat. Acad. Sci.*, **46**, 253.

Chandrasekhar, S., 1961. *Hydrodynamic and Hydromagnetic Stability*, Oxford University Press.

Coroniti, F.V., 1981. *Astrophys. J.*, **244**, 587.

Cox, J.P. and Giuli, T.G., 1968. *Principles of Stellar Structure*, Gordon and Breach.

Dubrulle, B. and Knoblock, E., 1993. *Astron. Astrophys.*, **274**, 667.

Eardley, D.M. and Lightman, A.P., 1975. *Astrophys. J.*, **200**, 181.

Galeev, A.A., Rosner, R. and Vaiana, G.S., 1979. *Astrophys. J.*, **229**, 318.

Gammie, C.F. and Balbus, S.A., 1994. *Mon. Not. R. Astr. Soc.*, **270**, 138.

Hawley, J.F. and Balbus, S.A., 1991. *Astrophys. J.*, **376**, 223.

Hawley, J.F. and Balbus, S.A., 1992. *Astrophys. J.*, **400**, 595.

Hawley, J.F., Gammie, C.F. and Balbus, S.A., 1995. *Astrophys. J.*, **440**, 743.

Heyvaerts, J.F. and Priest, E.R., 1989. *Astron. Astrophys.*, **216**, 230.

Ichimaru, S., 1977. *Astrophys. J.*, **214**, 840.

Lin, D.N.C. and Papaloizou, J.C.B., 1979. *Mon. Not. R. Astr. Soc.*, **188**, 191.

Lynden-Bell, D., 1969. *Nature*, **223**, 690.

Moffatt, H.K., 1978. *Magnetic Field Generation in Electrically Conducting Fluids*, Cambridge University Press.

Nordlund, A. and Stein, R.F., 1990. *Comp. Phys. Comm.*, **59**, 119.

Paczyński, B., 1977. *Astrophys. J.*, **216**, 822.

Papaloizou, J.C.B. and Lin, D.N.C., 1995. *Ann. Rev. Astron. Astrophys.*, **33**, 505.

Papaloizou, J.C.B. and Pringle, J.E., 1977. *Mon. Not. R. Astr. Soc.*, **181**, 441.

Papaloizou, J.C.B. and Szuszkiewicz, E., 1992. *Geophys. Astrophys. Fluid. Dynam.*, **66**, 223.

Parker, E.N., 1971. *Astrophy. J.*, **163**, 255.

Parker, E.N., 1972. *Astrophys. J.*, **174**, 499.

Pudritz, R.E., 1981. *Mon. Not. R. Astr. Soc.*, **195**, 897.

Rudiger, G., Elstner, D. and Stepinski, T.F., 1995. *Astron. Astrophys.*, **298**, 934.

Shakura, N.I. and Sunyaev, R.A., 1973. *Astron. Astrophys.*, **24**, 337.

Soward, A.M., 1992a. *Geophys. Astrophys. Fluid. Dynam.*, **64**, 163.

Soward, A.M., 1992b. *Geophys. Astrophys. Fluid. Dynam.*, **64**, 201.

Stella, L. and Rosner, R., 1984. *Astrophys. J.*, **277**, 312.

Stepinski, T.F. and Levy, E.H., 1990. *Astrophys. J.*, **362**, 318.

Stone, J.M., Hawley, J.F., Gammie, C.F. and Balbus, S.A., 1996. *Astrophys. J.*, **463**, 656.

Tout, C.A. and Pringle, J.E., 1992. *Mon. Not. R. Astr. Soc.*, **259**, 604.

Velikhov, E.P., 1959. *Soviet. J. Exp. Theor. Phys.*, **36**, 1398.

Vishniac, E.T. and Diamond, P., 1989. *Astrophys. J.*, **347**, 435.

Vishniac, E.T. and Diamond, P., 1992. *Astrophys. J.*, **398**, 561.

Vishniac, E.T., Jin, L. and Diamond, P., 1990. *Astrophys. J.*, **365**, 648.

Zhang, W., Diamond, P. and Vishniac, E.T., 1994. *Astrophys. J.*, **420**, 705.

CHAPTER 12

MAGNETIC WINDS

12.1. Introduction

The lobe-filling secondary stars in binaries with orbital periods $\lesssim 10\,\mathrm{hrs}$ have masses $\lesssim 1M_\odot$, and so will possess convective envelopes. Tidal synchronization means that these stars will be rotating rapidly. As seen in §2.3, turbulence and rapid rotation favour dynamo action, so such secondaries are likely to have a magnetic field. A hot expanding corona will lead to some mass loss, driven by thermal pressure gradients and centrifugal acceleration. A small coronal mass flux, and a moderate magnetic field, result in highly conducting material being channelled along field lines which are only slightly distorted by the flow. Field distortion becomes large when the kinetic energy density of the outflowing material becomes comparable to the poloidal magnetic energy density, equality occurring at the Alfvén speed

$$v_{\mathrm{A}} = \left(\frac{B_p^2}{\mu_0 \rho} \right)^{\frac{1}{2}} . \tag{12.1}$$

If this speed is reached before a magnetic coronal loop starts to close, then the field is dragged out with the flow. The Alfvénic points, defined by (12.1), constitute the Alfvén surface S_{A}.

Within S_{A} the stellar magnetic field tries to bring the gas into corotation with the star. The magnetic torque imparts angular momentum to the flowing material, at the expense of the stellar angular momentum, causing a braking torque on the star. The open field lines constitute the wind zone, which accounts for the main loss of stellar angular momentum. A magnetic wind from a lobe-filling secondary star in a binary system leads to a loss of orbital angular momentum, via tidal torques. This process is invoked to explain mass transfer in systems with orbital periods $\gtrsim 3\,\mathrm{hrs}$, since gravitational radiation losses are not sufficient to account for the larger accretion rates observed at these periods.

It has been suggested that magnetic winds are also generated by accretions discs, and that these may be a significant means of angular momentum advection. A poloidal field must be maintained in the disc so, in a diffusive axisymmetric case, dynamo action is required or there must be an external source of poloidal field. Most of the work done in this area has been aimed at explaining bipolar outflows and jets in *T* Tauri stars and active galactic nuclei, which are believed to have large accretion discs (e.g. Basri and Bertout, 1989; Pringle, 1993). The possibility of magnetically influenced winds from discs in binary stars is discussed here.

In §12.2 stellar magnetic wind theory is formulated. A simple field model is presented and results are derived for the fast rotator regime. Section 12.3 applies this wind theory to secondary stars as a means of driving mass transfer, and a range of dynamo laws are considered. The origin of the period gap, and winds from AM Herculis stars are addressed. The generation and stability of winds from accretion discs are considered in §12.4, and the results are discussed in §12.5.

12.2. Stellar Magnetic Wind Theory

The basic wind theory for a rotating, magnetic star is formulated here. Parker (1963) showed that a hot stellar corona cannot be contained by the pressure of the surrounding interstellar medium, and so expands to generate a wind. Schatzman (1962) pointed out that if a strong stellar magnetic field keeps the wind corotating with the star by magnetic torques out to large distances, then far more angular momentum per unit mass will be carried off than in a non-magnetic wind, in which the gas conserves its angular momentum. The theory of a steady, axisymmetric magnetic wind was formulated by Mestel (1967, 1968) and Weber and Davis (1967). Further work was done by many authors, including Pneuman and Kopp (1971), Okamoto (1974) and Sakurai (1985). The braking of late-type stars was considered by Mestel and Spruit (1987), including the fast rotator regime which is of particular relevance to corotating secondary stars in binaries. While fundamental results are common to these papers, specific models vary. The formulation presented here is that of Mestel (1968) and Mestel and Spruit (1987).

12.2.1. ANGULAR MOMENTUM TRANSPORT

Consider a star rotating with angular velocity Ω_s about the z-axis in a cylindrical coordinate system (ϖ, ϕ, z), with its origin at the stellar centre. The unperturbed magnetic field is taken to be dipolar with its moment along the z-axis, so the poloidal field $\mathbf{B}_p$ is axisymmetric. The dead zone will consist of field lines emanating from the stellar surface nearer to the equator than the poles. The associated flux tubes form closed loops trapping within them hot gas. These loops close sufficiently close to the star so the magnetic pressure exceeds the gas pressure. The wind zone consists of field lines emerging nearer the poles which extend further and are unable to trap the hot gas. These flux tubes are pulled open by the flowing gas before it reaches the Alfvén surface S_A, so the outer poloidal field adopts a more radial structure. Angular momentum transport occurs in the wind zone. The star is taken to be a spherically symmetric gravity source, so in the absence of magnetic torques the wind material would conserve its specific angular momentum $\varpi^2\Omega$.

The steady state momentum, induction and continuity equations are

$$(\mathbf{v}\cdot\nabla)\mathbf{v} = -\frac{1}{\rho}\nabla P - \nabla\psi + \frac{1}{\mu_0\rho}(\nabla\wedge\mathbf{B})\wedge\mathbf{B}, \tag{12.2}$$

$$\nabla\wedge(\mathbf{v}\wedge\mathbf{B}) = \mathbf{0}, \tag{12.3}$$

$$\nabla\cdot(\rho\mathbf{v}) = 0, \tag{12.4}$$

where perfect conductivity is assumed, and ψ is the stellar gravitational potential given by

$$\psi = -\frac{GM_s}{r}. \tag{12.5}$$

The poloidal and toroidal components of the induction equation (12.3) yield

$$\mathbf{v}_p = \kappa\mathbf{B}_p, \tag{12.6}$$

$$\Omega - \frac{\kappa B_\phi}{\varpi} = \alpha, \tag{12.7}$$

where κ is a scalar function of position and α is a constant on each field-streamline. These equations combine to give

$$\mathbf{v} = \kappa\mathbf{B} + \varpi\alpha\hat{\boldsymbol{\phi}}. \tag{12.8}$$

Since $\nabla\cdot\mathbf{B}_p = 0$, equations (12.4) and (12.6) lead to

$$\mathbf{B}_p\cdot\nabla(\rho\kappa) = 0, \tag{12.9}$$

so

$$\rho\kappa = \frac{\rho v_p}{B_p} = \epsilon, \tag{12.10}$$

where ϵ is constant on a field line, being the rate of mass flow per unit poloidal flux. The azimuthal component of (12.2), together with (12.4), yields

$$\nabla \cdot (\varpi^2 \Omega \rho \mathbf{v}_p) = \frac{\varpi}{\mu_0} [(\nabla \wedge \mathbf{B}) \wedge \mathbf{B}]_\phi = \frac{1}{\mu_0} \mathbf{B}_p \cdot \nabla(\varpi B_\phi), \qquad (12.11)$$

relating the advection of angular momentum to the magnetic torque. Use of (12.6) and (12.10) in (12.11) gives

$$\epsilon \varpi^2 \Omega - \frac{\varpi B_\phi}{\mu_0} = -\beta, \qquad (12.12)$$

where β is constant on a field line. This represents the total rate of transport of angular momentum per unit poloidal flux tube, carried jointly by the gas and magnetic stresses.

Equations (12.7), (12.10) and (12.12) combine to yield

$$B_\phi = \frac{\mu_0 \beta / \varpi + \mu_0 \epsilon \alpha \varpi}{1 - \mu_0 \epsilon^2 / \rho}, \qquad (12.13)$$

and

$$\Omega = \frac{\alpha + \mu_0 \epsilon \beta / \varpi^2 \rho}{1 - \mu_0 \epsilon^2 / \rho}. \qquad (12.14)$$

It follows from (12.10) that

$$\frac{\mu_0 \epsilon^2}{\rho} = \frac{\mu_0 \rho v_p^2}{B_p^2} = \frac{v_p^2}{v_{\mathrm{A}}^2}, \qquad (12.15)$$

where v_{A} is the Alfvén speed. For expected mass loss rates and surface magnetic fields, near the star $v_p \ll v_{\mathrm{A}}$. The density decreases outwards so $\mu_0 \epsilon^2 / \rho$ increases until it reaches unity at the Alfvénic point P_{A} where

$$v_p = v_{\mathrm{A}} = \frac{B_p}{(\mu_0 \rho)^{\frac{1}{2}}}, \qquad (12.16)$$

and

$$\rho = \rho_{\mathrm{A}} = \mu_0 \epsilon^2. \qquad (12.17)$$

Equations (12.13) and (12.14) then shows that B_ϕ and Ω will only be non-singular at P_{A} if

$$-\beta = \frac{\alpha \varpi_{\mathrm{A}}^2 \rho_{\mathrm{A}}}{\mu_0 \epsilon} = \epsilon \alpha \varpi_{\mathrm{A}}^2 = \left(\frac{\rho v_p}{B_p} \right) \alpha \varpi_{\mathrm{A}}^2. \qquad (12.18)$$

This shows that the angular momentum transport per unit flux tube, carried jointly by the gas and field, is equivalent to that which would be carried by the gas if it were kept corotating with angular velocity α out to the Alfvén point $P_{\rm A}$.

Use of (12.17) and (12.18) in (12.13) and (12.14) gives

$$B_\phi = -\mu_0 \epsilon \varpi_{\rm A}^2 \alpha \frac{(1 - \varpi^2/\varpi_{\rm A}^2)}{\varpi(1 - \rho_{\rm A}/\rho)}, \tag{12.19}$$

and

$$\Omega = \frac{(1 - \varpi_{\rm A}^2 \rho_{\rm A}/\varpi^2 \rho)}{(1 - \rho_{\rm A}/\rho)} \alpha = \frac{[1 - (\varpi_{\rm A}^2 B_{\rm A}/\varpi^2 B_p)(v_p/v_{\rm A})]}{(1 - \rho_{\rm A}/\rho)} \alpha, \tag{12.20}$$

where $B_{\rm A}$ is B_p at $\varpi_{\rm A}$. For any realistic poloidal field, the quantity $\varpi^2 B_p$ is slowly varying along a field-streamline, while $v_p/v_{\rm A}$ and $\rho_{\rm A}/\rho$ decrease rapidly towards the coronal base. It therefore follows from (12.20) that well inside the Alfvénic point the rotational shear is small, since $\Omega \simeq \alpha$. At the cornal base, $r = r_0$, and $\Omega = \Omega_{\rm s}$ so

$$\alpha = \frac{(1 - \rho_{\rm A}/\rho_0)}{[1 - (\varpi_{\rm A}^2 B_{\rm A}/\varpi_0^2 B_0)(v_0/v_{\rm A})]} \Omega_{\rm s}, \tag{12.21}$$

showing that the constant α is very close to $\Omega_{\rm s}$. Hence, near to the star, $\Omega \simeq \Omega_{\rm s}$ and (12.18) and (12.19) show that the magnetic term in (12.12) dominates the material term by a factor $(\varpi_{\rm A}/\varpi)^2$. Well beyond $P_{\rm A}$, where $\rho_{\rm A}/\rho \gg 1$ and $v_p/v_{\rm A} \gg 1$, (12.20) gives $\varpi^2 \Omega \simeq \varpi_{\rm A}^2 \Omega_{\rm s}$, and material angular momentum transport dominates.

The outward transport of angular momentum requires $-\beta B_p > 0$ and so, from (12.18),

$$\rho v_p \alpha \varpi_{\rm A}^2 > 0.$$

For outflows this implies $\alpha > 0$ and hence $\Omega_{\rm s} > 0$. Equations (12.20) and (12.21) show that beyond the coronal base Ω lags $\Omega_{\rm s}$. The consequent shear generates B_ϕ and the resulting $B_p B_\phi$ stress gives a torque which acts to increase Ω. The reaction to this torque causes a decrease in the angular velocity of the star. The star spins down on the braking time-scale which far exceeds the wind flow time-scale of $\sim \varpi/v_p$. The flow can therefore be treated as evolving on the braking time-scale through quasi-steady states, in which the azimuthal magnetic torque acting on the matter in a volume element at a fixed point is balanced by a divergence of angular momentum flux, as stated in (12.11).

12.2.2. THE WIND FLOW

The wind speed is analysed by considering the component of the equation of motion along the flow. The wind zone is taken to be isothermal with sound speed a_{w}, so

$$P = a_{\mathrm{w}}^2 \rho. \tag{12.22}$$

Using this and (A2), (12.2) can be written

$$(\nabla \wedge \mathbf{v}) \wedge \mathbf{v} = -\nabla \left(\frac{1}{2} v^2 + a_{\mathrm{w}}^2 \ln \rho - \frac{GM_{\mathrm{s}}}{r} \right) + \frac{1}{\mu_0 \rho} (\nabla \wedge \mathbf{B}) \wedge \mathbf{B}.$$

Taking the scalar product of this with $\mathbf{v}$ gives

$$\mathbf{v} \cdot \nabla \left(\frac{1}{2} v^2 + a_{\mathrm{w}}^2 \ln \rho - \frac{GM_{\mathrm{s}}}{r} \right) = \frac{1}{\mu_0 \rho} \mathbf{v} \cdot [(\nabla \wedge \mathbf{B}) \wedge \mathbf{B}]. \tag{12.23}$$

Use of (12.8) for $\mathbf{v}$ yields

$$\frac{1}{\mu_0 \rho} \mathbf{v} \cdot [(\nabla \wedge \mathbf{B}) \wedge \mathbf{B})] = \frac{\alpha \varpi}{\mu_0 \rho} [(\nabla \wedge \mathbf{B}) \wedge \mathbf{B})]_{\phi} = \alpha \mathbf{v} \cdot \nabla (\varpi^2 \Omega), \tag{12.24}$$

which is the rate of work done by the magnetic torque, with the last equality following from the angular momentum equation (12.11). Equations (12.23) and (12.24) show that

$$\frac{1}{2} v_p^2 + \frac{1}{2} \varpi^2 \Omega^2 + a_{\mathrm{w}}^2 \ln \rho - \frac{GM_{\mathrm{s}}}{r} - \alpha \varpi^2 \Omega = E \tag{12.25}$$

is constant on a streamline, where E is the total energy per unit mass. This is a generalized Bernoulli integral for an isothermal flow. The continuity equation enables (12.25) to be written as

$$\nabla \cdot \left[\rho \mathbf{v} \left(\frac{1}{2} v_p^2 + \frac{1}{2} \varpi^2 \Omega^2 + a_{\mathrm{w}}^2 \ln \rho - \frac{GM_{\mathrm{s}}}{r} - \alpha \varpi^2 \Omega \right) \right] = 0,$$

where the quantity in square brackets is the total energy flux. Equation (12.11), together with $\mathbf{v}_p \cdot \nabla \alpha = \mathbf{B}_p \cdot \nabla \alpha = 0$, gives the last term as

$$\begin{aligned} -\nabla \cdot (\rho \mathbf{v}_p \alpha \varpi^2 \Omega) &= -\nabla \cdot \left(\frac{\alpha}{\mu_0} \varpi B_{\phi} \mathbf{B}_p \right) \\ &= -\nabla \cdot \left(\frac{1}{\mu_0} (\mathbf{v} \wedge \mathbf{B}) \wedge \mathbf{B} \right) = \nabla \cdot (\mathbf{E} \wedge \mathbf{H}), \end{aligned}$$

where $\mathbf{H} = \mathbf{B}/\mu_0$ and the second and last equalities follow from (12.8) and Ohm's law with perfect conductivity, respectively. Hence

$$\nabla \cdot \left[\rho \mathbf{v} \left(\frac{1}{2} v_p^2 + \frac{1}{2} \varpi^2 \Omega^2 + a_{\mathrm{w}}^2 \ln \rho - \frac{GM_{\mathrm{s}}}{r} \right) + \mathbf{E} \wedge \mathbf{H} \right] = 0, \tag{12.26}$$

where $\mathbf{E} \wedge \mathbf{H}$ is the Poynting flux of electromagnetic energy.

If the structure of $\mathbf{B}_p$ is supposed known, substitution of (12.10) and (12.20) in (12.25) gives a relation

$$H(\varpi, \rho) = E \tag{12.27}$$

between ρ and ϖ on a field-streamline. The choice (12.18) ensures that B_ϕ and Ω are non-singular, and it then follows that all non-singular solutions for ρ pass through the Alfvénic point P_{A}. The function $H(\varpi, \rho_{\mathrm{A}})$ is only finite at $\varpi = \varpi_{\mathrm{A}}$. Equation (12.27) yields

$$\frac{d\rho}{d\varpi} = -\frac{\partial H/\partial \varpi}{\partial H/\partial \rho}. \tag{12.28}$$

Use of (12.7) and (12.10) to eliminate Ω and v_p in the Bernoulli integral (12.25) gives

$$H = \frac{\epsilon^2 B_p^2}{2\rho^2} + a_{\mathrm{w}}^2 \ln \rho - \frac{GM_{\mathrm{s}}}{r} + \frac{1}{2}\left(\alpha\varpi + \frac{\epsilon B_\phi}{\rho}\right)^2 - \alpha\varpi\left(\alpha\varpi + \frac{\epsilon B_\phi}{\rho}\right).$$

Taking $\partial H/\partial \rho$, employing (12.19) for B_ϕ/ρ, yields

$$\rho\frac{\partial H}{\partial \rho} = -v_p^2 + a_{\mathrm{w}}^2 - \frac{\epsilon^2 B_\phi^2}{\rho^2(1 - \rho/\rho_{\mathrm{A}})}.$$

Then noting, from (12.17), that

$$\frac{\rho_{\mathrm{A}}}{\rho} = \frac{\mu_0 \epsilon^2}{\rho} = \frac{\mu_0 \rho v_p^2}{B_p^2} = \left(\frac{v_p}{v_{\mathrm{A}}}\right)^2,$$

gives

$$\rho\frac{\partial H}{\partial \rho} = \frac{v_p^4 - (a_{\mathrm{w}}^2 + v_{\mathrm{A}p}^2 + v_{\mathrm{A}\phi}^2)v_p^2 + a_{\mathrm{w}}^2 v_{\mathrm{A}p}^2}{(v_{\mathrm{A}p}^2 - v_p^2)}, \tag{12.29}$$

where $v_{\mathrm{A}\phi} = B_\phi/(\mu_0\rho)^{1/2}$. The condition $\partial H/\partial \rho = 0$ gives v_p equal to v_s or v_f, the slow and fast magnetosonic wave speeds (see §2.2.6). At these values of v_p it follows from (12.28) that $\partial H/\partial \varpi = 0$ is necessary for smoothly varying ρ, defining the slow and fast critical points ϖ_s and ϖ_f. Then (12.27) gives

$$H(\varpi_s, \rho_s) = H(\varpi_f, \rho_f) = E. \tag{12.30}$$

These conditions suffice to fix the solution along each field-streamline in terms of a non-dimensionalized coronal temperature, stellar rotation rate and magnetic flux. These are, respectively,

$$\ell_{\mathrm{w}} = \frac{GM_{\mathrm{s}}}{r_0 a_{\mathrm{w}}^2}, \quad \nu = \frac{\alpha^2 r_0^3}{GM_{\mathrm{s}}}, \quad \zeta_{\mathrm{w}} = \frac{B_0^2}{2\mu_0(\rho_0)_{\mathrm{w}} a_{\mathrm{w}}^2}, \tag{12.31}$$

evaluated at the coronal base $r = r_0$. Usually, the wind speed at the slow point is close to the sound speed. The fast speed is approximately the local Alfvén speed determined by the total magnetic field.

12.2.3. THE BRAKING TORQUE

The total rate of angular momentum transport is found by integrating $-\beta B_p$ over the coronal base surface $r = r_0$, from the poles to the limiting field lines. Near the stellar surface $\mathbf{B}_p$ will be close to its unperturbed dipole structure, having components

$$B_r = B_0 \left(\frac{r_0}{r}\right)^3 \cos\theta, \quad B_\theta = \frac{B_0}{2}\left(\frac{r_0}{r}\right)^3 \sin\theta, \tag{12.32}$$

with the poles at $\theta = 0$ and $\theta = \pi$. The poloidal field strength is therefore

$$B_p = B_0 \left(\frac{r_0}{r}\right)^3 \left(1 - \frac{3}{4}\sin^2\theta\right)^{\frac{1}{2}}. \tag{12.33}$$

The rate of angular momentum transport is

$$-\dot{J} = -\int_{S_0} \beta \mathbf{B}_p \cdot \hat{\mathbf{r}} dS_0,$$

where $\dot{J}$ is the stellar torque, and S_0 is the coronal base surface. Hence

$$-\dot{J} = -4\pi \int_0^{\bar{\theta}_0} \left(\frac{B_r}{B_p}\right)_0 \beta B_p(r_0, \theta_0) r_0^2 \sin\theta_0 d\theta_0, \tag{12.34}$$

where the limiting field line cuts the stellar surface at $(r_0, \bar{\theta}_0)$, and the symmetry of $-\beta B_p$ about the equatorial plane has been used. Equations (12.18) and (12.10) give

$$-\beta = \epsilon\alpha\varpi_{\mathrm{A}}^2 = \frac{\rho_0 v_0 \alpha \varpi_{\mathrm{A}}^2}{B_p(r_0, \theta_0)},$$

where $v_0 = v_p(r_0, \theta_0)$, so (12.32)–(12.34) yield

$$-\dot{J} = 4\pi\Omega_{\mathrm{s}} r_0^4 \int_0^{\bar{\theta}_0} \rho_0 v_0 \left(\frac{\alpha}{\Omega_{\mathrm{s}}}\right)\left(\frac{\varpi_{\mathrm{A}}}{r_0}\right)^2 \frac{\sin\theta_0 \cos\theta_0 d\theta_0}{(1 - \frac{3}{4}\sin^2\theta_0)^{\frac{1}{2}}}. \tag{12.35}$$

The factor $(\varpi_A/r_0)^2$, which increases with field strength B_0, represents the enhanced braking due to the magnetic torque. The magnetic field also plays a part in determining the upper limit $\bar{\theta}_0$ which reflects the extent of the dead zone.

The associated mass loss rate from the stellar surface is given by

$$\dot{M}_w = -4\pi r_0^2 \int_0^{\bar{\theta}_0} \rho_0 v_0 \left(\frac{B_r}{B_p}\right)_0 \sin\theta_0 d\theta_0,$$

and hence

$$\dot{M}_w = -4\pi r_0^2 \int_0^{\bar{\theta}_0} \rho_0 v_0 \frac{\sin\theta_0 \cos\theta_0 d\theta_0}{\left(1-\frac{3}{4}\sin^2\theta_0\right)^{\frac{1}{2}}}. \tag{12.36}$$

It follows from (12.35) and (12.36) that

$$\frac{\dot{J}}{r_0^2 \Omega_s \dot{M}_w} = \frac{I}{K} = \Gamma, \tag{12.37}$$

where

$$I = \int_0^{\bar{s}} \rho_0 v_0 \left(\frac{\varpi_A}{r_0}\right)^2 \left(1 - \frac{3}{4}s\right)^{-\frac{1}{2}} ds,$$

$$K = \int_0^{\bar{s}} \rho_0 v_0 \left(1 - \frac{3}{4}s\right)^{-\frac{1}{2}} ds,$$

$s = \sin^2\theta_0$, and $\alpha = \Omega_s$ has been used. The quantity Γ measures the effect of the magnetic field in increasing the angular momentum loss for a given mass loss. For an unconstrained spherically symmetric wind, $\Gamma = 2/3$. Magnetic channelling can increase Γ by factors of up to $\sim 10^3$.

12.2.4. A SIMPLE FIELD MODEL

Well within the Alfvénic surface S_A the magnetic energy density dominates the kinetic energy density of the wind. The poloidal field $\mathbf{B}_p$ will therefore remain close to its unperturbed structure, and channel the wind flow. Well beyond S_A the kinetic energy density of the flow dominates the energy density of the magnetic field. In this region the field will be passive, being pulled out to follow the nearly radial wind.

The detailed construction of $\mathbf{B}_p$ requires solution of the poloidal trans-$\mathbf{B}_p$ component of the equation of motion. This poses a formidable problem

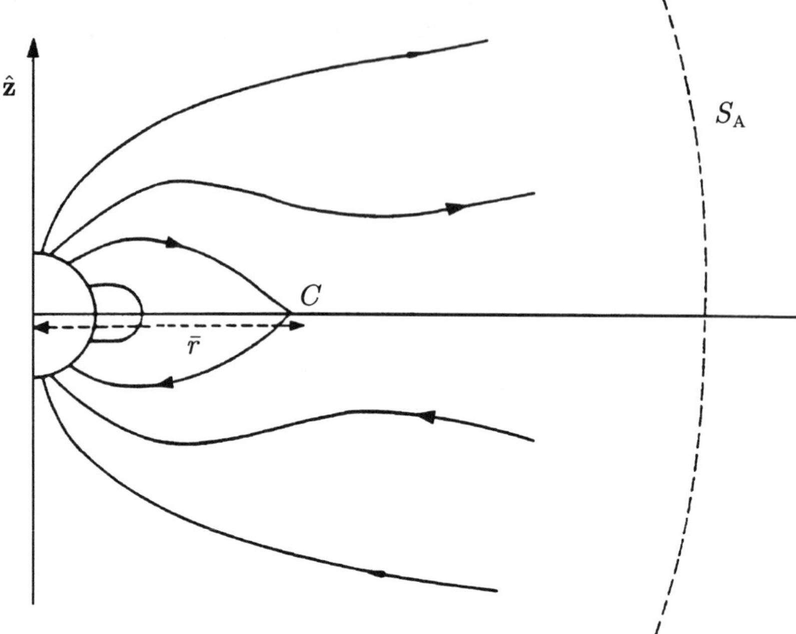

Figure 12.1. Schematic magnetic field model. From Mestel and Spruit (1987).

(e.g. Sakurai, 1990). A simple field model was used by Mestel and Spruit (1987) in considering the braking of late-type stars. The field structure, shown in Figure 12.1, represents the basic properties of the distortion in the flow regions. The poloidal field is taken to be dipolar, with components given by (12.32), out to a radius $\bar{r}$ which corresponds to the equatorial boundary of the dead zone. Beyond $\bar{r}$, $\mathbf{B}_p$ is taken as radial, so

$$\mathbf{B}_p = \bar{B}\left(\frac{\bar{r}}{r}\right)^2 \hat{\mathbf{r}}. \tag{12.38}$$

The field has a cusp on the equator at $\bar{r}$ and $B_p(r, \pi/2) = 0$ for $r > \bar{r}$. Close to $\theta = \pi/2$, B_r varies rapidly with θ and changes sign across the equator. Away from the equator B_r is expected to vary slowly with θ.

Material in the dead zone is in hydrostatic balance and is taken to be isothermal, with sound speed a_{d}. Hence

$$P = a_{\mathrm{d}}^2 \rho, \tag{12.39}$$

and the force balance is

$$\nabla\left(\frac{GM_{\mathrm{s}}}{r} + \frac{1}{2}\alpha^2 r^2 \sin^2\theta - a_{\mathrm{d}}^2 \ln\rho\right) + \frac{1}{\mu_0 \rho}(\nabla \wedge \mathbf{B}) \wedge \mathbf{B} = \mathbf{0}. \tag{12.40}$$

Taking the scalar product with $\mathbf{B}$ gives

$$\frac{GM_{\rm s}}{r} + \frac{1}{2}\alpha^2 r^2 \sin^2\theta - a_{\rm d}^2 \ln\rho = U, \tag{12.41}$$

where U is a constant along the field. It follows that the density along a field line is

$$\rho = (\rho_0)_{\rm d} \exp\left[-\frac{GM_{\rm s}}{R_{\rm s}a_{\rm d}^2}\left(1 - \frac{R_{\rm s}}{r}\right) + \frac{\alpha^2 R_{\rm s}^2}{2a_{\rm d}^2}\left(\frac{r^2\sin^2\theta}{R_{\rm s}^2} - \sin^2\theta_0\right)\right], \tag{12.42}$$

with the coronal base $r_0 = R_{\rm s}$, the stellar radius. Since the wind is an energy sink, a latitude-independent heat input at the coronal base should give a significantly lower temperature for the wind zone than for the dead zone. When the cusp C is well within $S_{\rm A}$, then $|B_\phi| \ll |B_p|$, so the vertical continuity of total pressure at C is

$$\left(\frac{B_p^2}{2\mu_0}\right)_{\rm w} = P_{\rm d}, \tag{12.43}$$

using $P_{\rm d} \gg P_{\rm w}$. Taking $\bar{B} = (B_0/2)(R_{\rm s}/\bar{r})^3$, except at $\theta = \pi/2$ where $\bar{B} = 0$, equation (12.38) becomes

$$\mathbf{B}_p = \begin{cases} \pm\frac{B_0}{2}\left(\frac{R_{\rm s}}{\bar{r}}\right)^3\left(\frac{\bar{r}}{r}\right)^2\hat{\mathbf{r}}, & \theta \neq \pi/2, \\ 0, & \theta = \pi/2, \end{cases} \tag{12.44}$$

with the positive sign applying above the equatorial plane. Equations (12.39) and (12.42)–(12.44) then give

$$\left(\frac{\bar{r}}{R_{\rm s}}\right)^6 = \frac{B_0^2}{8\mu_0(\rho_0)_{\rm d}a_{\rm d}^2}\exp\left[\frac{GM_{\rm s}}{R_{\rm s}a_{\rm d}^2}\left(1 - \frac{R_{\rm s}}{\bar{r}}\right)\right]\exp\left[-\frac{\alpha^2 R_{\rm s}^2}{2a_{\rm d}^2}\left(\frac{\bar{r}^2}{R_{\rm s}^2} - \frac{R_{\rm s}}{\bar{r}}\right)\right], \tag{12.45}$$

which determines $\bar{r}$, the extent of the dead zone.

As previously seen, the wind solution is fixed by considering the slow and fast magnetosonic points on a given field-streamline. The condition (12.30) then determines the solution. The fast magnetosonic point is beyond $S_{\rm A}$ and so lies in the radial field domain, with $r^2 B_r$ =constant on a given field-streamline. The Bernoulli integral (12.25) can then be written

$$H(r,\rho) = \frac{\epsilon^2 B_{\rm A}^2 r_{\rm A}^4}{2\rho^2 r^4} + \frac{1}{2}\alpha^2 r^2 \sin^2\theta\left[\frac{\Omega}{\alpha}\left(\frac{\Omega}{\alpha} - 2\right)\right] - \frac{GM_{\rm s}}{r} + a_{\rm w}^2 \ln\frac{\rho}{\rho_{\rm A}}, \tag{12.46}$$

and (12.20) gives

$$\frac{\Omega}{\alpha} = \frac{1 - \rho_{\rm A} r_{\rm A}^2/\rho r^2}{1 - \rho_{\rm A}/\rho}. \tag{12.47}$$

The critical point conditions $\partial H/\partial r = 0$, $\partial H/\partial \rho = 0$ at (r_f, ρ_f) yield

$$2V^2 = -\gamma X^2 \frac{(V^2 - 2VX^2 + 1)}{(VX^2 - 1)^2}, \tag{12.48}$$

and

$$V^2 - \delta = \gamma V^2 X^2 \frac{(X^2 - 1)^2}{(VX^2 - 1)^3}, \tag{12.49}$$

with dimensionless quantities,

$$X = \frac{r_f}{r_\mathrm{A}}, \quad V = \frac{\rho_\mathrm{A}}{\rho_f}\frac{1}{X^2} = \frac{v_f}{v_\mathrm{A}}, \quad \gamma = \frac{\alpha^2 r_\mathrm{A}^2}{v_\mathrm{A}^2}\sin^2\theta, \quad \delta = \frac{a_\mathrm{W}^2}{v_\mathrm{A}^2}. \tag{12.50}$$

The gravitational term has been dropped in (12.48) since it is small compared with v_f^2 term.

In the case of rapid rotation, which is appropriate to the secondary stars in close binaries, centrifugal driving forces v_A to well above a_W, so that $\delta \ll 1$. Dividing (12.48) by (12.49) gives a quadratic equation for X^2 containing the small quantity δ in its coefficients. To leading order, this yields the finite root

$$X \simeq \left[\frac{1}{2\delta} V(V-1)(3-V)\right]^{\frac{1}{2}}, \tag{12.51}$$

which substituted in (12.48) gives $V \simeq \gamma^{\frac{1}{3}}$ and hence, from (12.50),

$$v_f^3 \simeq \alpha^2 \varpi_\mathrm{A}^2 v_\mathrm{A}. \tag{12.52}$$

Equation (12.51) shows that $V = v_f/v_\mathrm{A}$ lies between 1 and 3 and, since $\delta \ll 1$, then $X \gg 1$. It therefore follows from (12.50) that

$$\frac{\rho_\mathrm{A}}{\rho_f} = VX^2 \gg 1. \tag{12.53}$$

Along a radial field-streamline the continuity equation gives

$$\frac{v_\mathrm{A}}{v_f} = \frac{\rho_f}{\rho_\mathrm{A}}\left(\frac{\varpi_f}{\varpi_\mathrm{A}}\right)^2. \tag{12.54}$$

Use of these results in (12.19) and (12.20) yields

$$\varpi_f (B_\phi)_f \simeq -4\pi\epsilon\alpha\varpi_\mathrm{A}^2\left(\frac{v_\mathrm{A}}{v_f}\right), \tag{12.55}$$

$$\varpi_f^2 \Omega_f \simeq \alpha \varpi_{\rm A}^2 \left(1 - \frac{v_{\rm A}}{v_f}\right). \tag{12.56}$$

In the rapid rotation regime, the gravitational and thermal terms can be dropped in the Bernoulli integral (12.25). Noting that $\varpi_{\rm A}/\varpi_f \ll 1$, (12.25), (12.52) and (12.56) give

$$H(\varpi_f, \rho_f) = v_{\rm A}^2 \left[\frac{3}{2}\left(\frac{v_f}{v_{\rm A}}\right)^2 - \left(\frac{v_f}{v_{\rm A}}\right)^3\right]. \tag{12.57}$$

To determine the solution for rapid rotators, the flow must be made to pass through the slow magnetosonic point. Except for very rapid rotators, the Ω-terms can be dropped in (12.25) in the region closer to the axis in which the slow point ϖ_s will occur. Hence

$$H = \frac{\epsilon^2 B_p^2}{2\rho^2} + a_{\rm w}^2 \ln \rho - \frac{GM_{\rm s}}{r}. \tag{12.58}$$

In this region $\mathbf{B}_p$ is the dipole field (12.32), and on a field-streamline

$$\sin^2\theta = \frac{r}{R_{\rm s}} \sin^2\theta_0. \tag{12.59}$$

The conditions $\partial H/\partial \rho = 0$, $\partial H/\partial r = 0$ then yield

$$v_p = a_{\rm w}, \qquad r_s = \frac{GM_{\rm s}}{3a_{\rm w}^2}, \tag{12.60}$$

and so from (12.58)

$$H(\varpi_s, \rho_s) = a_{\rm w}^2 \left[-\frac{5}{2} + \ln \frac{\rho_s}{\rho_{\rm A}}\right]. \tag{12.61}$$

Finally, the condition

$$H(\varpi_f, \rho_f) = H(\varpi_s, \rho_s)$$

must be applied. Since $a_{\rm w}^2 = \delta v_{\rm A}^2$, (12.57) and (12.61) yield

$$\frac{3}{2}\left(\frac{v_f}{v_{\rm A}}\right)^2 - \left(\frac{v_f}{v_{\rm A}}\right)^3 = \delta\left[-\frac{5}{2} + \ln\left(\frac{\rho_s}{\rho_{\rm A}}\right)\right].$$

Noting that $V = v_f/v_{\rm A}$, for $\delta \ll 1$ it follows that $V = 3/2$ which, together with (12.51) for $X = r_f/r_{\rm A}$, gives

$$\frac{v_f}{v_{\rm A}} = \frac{3}{2}, \qquad \frac{r_f}{r_{\rm A}} = \frac{3}{4}\frac{v_{\rm A}}{a_{\rm w}}. \tag{12.62}$$

Then (12.52) and (12.56) give

$$v_{\mathrm{A}} = 0.54\alpha\varpi_{\mathrm{A}}, \qquad \Omega_f = \frac{\alpha}{3}\left(\frac{\varpi_{\mathrm{A}}}{\varpi_f}\right)^2. \tag{12.63a, b}$$

Hence, for a fast rotator, centrifugal force drives the wind speed at the Alfvénic point up to about one half of the corotation speed.

This leaves the rate of braking to be found. The Alfvén speed is given to a reasonable approximation, applicable to slow and fast rotators, by

$$\frac{v_{\mathrm{A}}}{a_{\mathrm{w}}} = \left[\left(\frac{v_{\mathrm{th}}}{a_{\mathrm{w}}}\frac{r_{\mathrm{A}}}{R_{\mathrm{s}}}\right)^2 + \frac{\alpha^2 r_{\mathrm{A}}^2}{3a_{\mathrm{w}}^2}\sin^2\theta\right]^{\frac{1}{2}}, \tag{12.64}$$

where v_{th} is Parker's thermally driven wind. Then noting from (12.10), (12.17) and (12.44) that

$$\mu_0\epsilon = \mu_0\frac{\rho_0 v_0}{(B_p)_0} = \frac{B_{\mathrm{A}}}{v_{\mathrm{A}}} = \frac{B_0}{2v_{\mathrm{A}}}\frac{R_{\mathrm{s}}}{\bar{r}}\left(\frac{R_{\mathrm{s}}}{r_{\mathrm{A}}}\right)^2,$$

(12.64) gives

$$\left(\frac{r_{\mathrm{A}}}{R_{\mathrm{s}}}\right)^3 = \frac{\zeta_{\mathrm{w}}}{(v_0/a_{\mathrm{w}})}\left(\frac{R_{\mathrm{s}}}{\bar{r}}\right)\left[\left(\frac{v_{\mathrm{th}}}{a_{\mathrm{w}}}\right)^2 + \frac{\alpha^2 R_{\mathrm{s}}^2}{3a_{\mathrm{w}}^2}\sin^2\theta\right]^{-\frac{1}{2}}, \tag{12.65}$$

where typically $v_0/a_{\mathrm{w}} \sim 0.15$, and ζ_{w} is the parameter defined in (12.31).

From (12.18) the flux of angular momentum along a field-streamline can be expressed as

$$-\beta B_p = \rho v_p \alpha \varpi_{\mathrm{A}}^2. \tag{12.66}$$

The distance $\varpi_{\mathrm{A}} = r_{\mathrm{A}}\sin\theta$ will vary between field lines. However, a lower limit to the braking torque can be found by taking the flow beyond $\bar{r}$ to be spherically symmetric, with $v_p(r)$ having its equatorial variation. The Alfvénic surface S_{A} is therefore approximated as a sphere of radius $r_{\mathrm{A}} = \varpi_{\mathrm{A}}$ at $\theta = \pi/2$. Equation (12.65) shows that r_{A} varies slowly with θ, its most rapid variation being near $\theta = 0$ where the contribution to angular momentum transport is least. The integral of (12.66) over S_{A} is then

$$-\dot{J} = 4\pi\alpha\int_0^{\frac{\pi}{2}} \rho_{\mathrm{A}} v_{\mathrm{A}} r_{\mathrm{A}}^2 \sin^2\theta\, r_{\mathrm{A}}^2 \sin\theta\, d\theta,$$

giving

$$-\dot{J} = \frac{8\pi}{3}(\rho_{\mathrm{A}} v_{\mathrm{A}} r_{\mathrm{A}}^2)\alpha r_{\mathrm{A}}^2. \tag{12.67}$$

Since $\rho_A v_A = B_A^2/\mu_0 v_A$, $-\dot{J}$ can also be written as

$$-\dot{J} = \frac{\Phi^2 \alpha}{6\pi\mu_0 v_A}, \tag{12.68}$$

where $\Phi = 4\pi r_A^2 B_A$ is the magnitude of the flux of open magnetic field lines that form the wind zone.

It follows from (12.21) that α is very close to the stellar rotation rate Ω_s and so, for most purposes, $\alpha = \Omega_s$ can be used. The authors considered rotation rates $\Omega_s = k\Omega_\odot$ with $1 \leq k \leq 80$, where $\Omega_\odot = 2.5 \times 10^{-6}\,\mathrm{s}^{-1}$ corresponds to a period of $P_\odot = 29$ days. The magnetic field is assumed to be dynamo generated, with pole strength B_0 related to Ω_s by

$$B_0 = (B_0)_\odot \left(\frac{\Omega_s}{\Omega_\odot}\right)^n. \tag{12.69}$$

The simplest model has $(\rho_0)_d$ and a_d both independent of Ω_s. However, it is difficult to see how this could be consistent with X-ray observations of late-type stars, which show a luminosity $L_X \propto \Omega_s^2$ (e.g. Pallavicini *et al*, 1981). The variation of coronal emission is determined mostly by the variation in density, with $L_X \propto \rho_0^2$ so $\rho_0 \propto \Omega_s$. The X-ray luminosity is expected to depend on B_0, since this is a measure of the degree of activity of the star. Consistency with (12.69) and $L_X \propto \Omega_s^2$ requires $L_X \propto B_0^{2/n}$. The choice $n = 1$ is made, leading to

$$\zeta_d = (\zeta_d)_\odot \frac{[B_0/(B_0)_\odot]^2}{(\rho_0)_d/(\rho_0)_{d\odot}} = (\zeta_d)_\odot \left(\frac{\Omega_s}{\Omega_\odot}\right), \tag{12.70}$$

where

$$\zeta_d = \frac{B_0^2}{2\mu_0(\rho_0)_d a_d^2} \tag{12.71}$$

occurs in (12.45), which determines the dead zone radius $\bar{r}$.

The cases $\zeta_d \propto \Omega_s$ and $\zeta_d \propto \Omega_s^2$ are considered, corresponding to $(\rho_0)_d \propto B_0$ and $(\rho_0)_d$ independent of B_0, respectively. The rotation rate Ω_s is varied from the solar value ($P = 29$ days) to a value of $80\Omega_\odot$ ($P = 9$ hrs), for the cases $(\zeta_d)_\odot = 4$, 60, taking stars of solar mass. In all cases the value $\Omega_s \simeq 15\Omega_\odot$ marks a turn-over in the behaviour of $\bar{r}/R_s$. For $\Omega_s < 15\Omega_\odot$ the centrifugal effect on the pressure in the dead zone, described by (12.39) and (12.42), is weak, so $\bar{r}$ increases with Ω_s because the stronger field traps more gas. This effect occurs in (12.45) through the $B_0^2/(\rho_0)_d$ term, which increases with Ω_s. For $\Omega_s > 15\Omega_\odot$ the centrifugal effect, described by the second exponential in (12.45), dominates and $\bar{r}$ begins to shrink. The results are summarized in Table 12.1.

TABLE 12.1. The dead zone extent

	$(\xi_d)_\odot = 4$		$(\xi_d)_\odot = 60$	
$\Omega_S/\Omega_\odot$	$\bar{r}/R_S\ (\xi_d \propto \Omega_S)$	$\bar{r}/R_S\ (\xi_d \propto \Omega_S^2)$	$\bar{r}/R_S\ (\xi_d \propto \Omega_S)$	$\bar{r}/R_S\ (\xi_d \propto \Omega_S^2)$
1	2.5	2.5	5.0	5.0
2	3.1	3.5	5.8	6.6
3	3.4	4.1	6.2	7.8
5	3.9	5.5	6.8	9.2
7	4.1	6.2	7.2	10.0
10	4.4	6.9	7.4	10.7
20	4.6	7.2	6.9	9.6
30	4.4	6.5	6.1	8.2
60	3.4	4.7	4.3	5.4
80	3.0	3.9	3.6	4.5

The Alfvén radius r_A is found from (12.65), using $v_0 \simeq 0.15a_w$ and the previously determined value of $\bar{r}/R_s$, for each Ω_s. The parameter ζ_w is defined in (12.31), where $B_0^2/(\rho_0)_w$ is taken to have the same Ω_s dependencies as $B_0^2/(\rho_0)_d$ occurring in ζ_d. Since the wind zone has a lower density and temperature that the dead zone, $\zeta_w > \zeta_d$ is expected. The authors take $\zeta_w = 120$ for the sun so that (12.65) yields $r_A \simeq 12R_s$, in agreement with observation (Pizzo *et al*, 1983). The wind zone temperature is taken to be one half that of the dead zone. The results are shown in Table 12.2.

TABLE 12.2. Equatorial values of the ratio r_A/R_S, with $(\xi_d)_\odot = 60,\ (\xi_w)_\odot = 120$.

$\Omega_S/\Omega_\odot$	$(r_A/R_S)_{eq}\ (\xi_w \propto \Omega_S)$	$(r_A/R_S)_{eq}\ (\xi_w \propto \Omega_S^2)$
1	12	12
2	15	19
5	20	36
10	25	52
20	29	72
30	31	88
60	35	127
80	37	149

The angular momentum loss rate is given by (12.67). The continuity equation for radial flow gives $\rho_A v_A r_A^2 = \bar{\rho}\bar{v}\bar{r}^2$, while constancy of $\rho v_p/B_p$

TABLE 12.3. The function $K(\Omega_S/\Omega_\odot)$

	$q=2$, $(\rho_0)_W \propto B_0$	$q=1$, $(\rho_0)_W$ independent of B_0
$\Omega_S/\Omega_\odot$	$K(\Omega_S/\Omega_\odot)$	$K(\Omega_S/\Omega_\odot)$
1	30.5	30.5
2	1.6×10^2	1.1×10^2
5	1.5×10^3	7.0×10^2
10	8.4×10^3	2.5×10^3
20	4.8×10^4	1.1×10^4
30	1.4×10^5	2.8×10^4
60	1.0×10^6	1.8×10^5
80	2.4×10^6	4.0×10^5

along a field-streamline yields

$$\frac{\bar{\rho}\bar{v}}{(\rho_0)_W v_0} = \frac{\bar{B}}{B_0} = \frac{1}{2}\left(\frac{R_S}{\bar{r}}\right)^3,$$

where the last equality follows from (12.44). Use of these results allows $-\dot{J}$ to be expressed as

$$-\dot{J} = \frac{4\pi}{3}\left[(\rho_0)_\odot v_0 R_S^2\right]\Omega_\odot R_S^2 K(\Omega_S/\Omega_\odot), \tag{12.72a}$$

where

$$K(\Omega_S/\Omega_\odot) = \left(\frac{R_S}{\bar{r}}\right)\left(\frac{r_A}{R_S}\right)^2\left(\frac{\Omega_S}{\Omega_\odot}\right)^q, \tag{12.72b}$$

with $q=1$ or 2. The case $q=2$ corresponds to $(\rho_0)_W \propto \Omega_S \propto B_0$, while $q=1$ occurs for $(\rho_0)_W$ independent of Ω_S. Table 12.3 shows values of $K(\Omega_S/\Omega_\odot)$. Braking is more efficient when $(\rho_0)_W \propto B_0$, primarily due to the increase in mass loss which more than compensates for the reduction in the Alfvén radius. The smaller extent of the dead zone in this case, illustrated in Table 12.1, also enhances the braking.

The braking laws predicted by (12.72) will be discussed in the next section, in the context of the secondary stars in binaries. The fast rotator theory will be relevant here, since for orbital periods of $P \lesssim 9$ hrs synchronized secondaries have $\Omega_S \gtrsim 80\Omega_\odot$.

12.3. Stellar Winds in Close Binaries

12.3.1. ORBITAL ANGULAR MOMENTUM LOSS AND MASS TRANSFER

A lobe-filling secondary star in a close binary system is expected to be kept near to orbital corotation, by the tidal field of the primary star. Asynchronous motions result in internal dissipations which cause the tidal bulge to lag the motion of the line of stellar centres. The resulting gravitational torque spins the star towards synchronism. It follows from (2.226) that for orbital periods of $P \lesssim 7\,\text{hrs}$ a main sequence secondary will have a mass $M_s \lesssim 0.7 M_\odot$, and hence possess a deep convective envelope. The star is fully convective for $M_s \lesssim 0.3 M_\odot$, corresponding to $P \lesssim 3\,\text{hrs}$.

A rapidly rotating star with a convective envelope is likely to have a dynamo-generated magnetic field. As seen in §2.3, an $\alpha\omega$-dynamo requires differential rotation, as well as turbulence, to operate. The secondary may retain some rotational shearing motions if a small radiative core is not completely synchronized, while the convective envelope is. Some differential rotation may remain in the convective region if the turbulent stress tensor has a non-standard form, allowing turbulent motions to generate non-uniform rotation (Campbell and Papaloizou, 1983). Alternatively, an α^2-dynamo may operate.

A magnetic secondary with a wind, generated by an expanding corona, would be subject to the braking effect formulated in the previous section. There will, of course, be some differences from the case of a single star. The presence of the primary means that, even if the secondary had a magnetic field symmetric about its rotation axis, the wind would not be axisymmetric due to the ϕ-dependence of the total gravitational field. The presence of a strongly magnetic primary would introduce a further complication, this being discussed in the §12.3.7.

Despite the non-axisymmetric gravity field, the fundamental effect of a magnetically influenced wind from the secondary in a binary system should be the same as it is for a single star, namely the removal of stellar angular momentum. However, the spin evolution of the secondary star is affected by its tidal interaction with the primary. The synchronization process involves the exchange of angular momentum between the secondary and the orbital motion, via the tidal torque. The wind causes a spin down of the secondary, so driving Ω_s towards an under-synchronous state. However, tidal torques then become operable and act to increase Ω_s back to the orbital rate. The net effect is the continuous removal of orbital angular momentum.

The orbital angular momentum can be expressed as

$$L_{\text{orb}} = M_s M_p \left(\frac{GD}{M}\right)^{\frac{1}{2}}, \qquad (12.73)$$

where $M = M_s + M_p$ and D is the stellar separation. For a mass ratio range of $0.1 \leq M_s/M_p \leq 0.8$, the mean radius of the secondary's Roche lobe obeys

$$\frac{M}{M_s}\left(\frac{R_L}{D}\right)^3 \simeq 0.1. \tag{12.74}$$

Taking M as constant, (12.73) and (12.74) give an evolution equation for R_L as

$$\frac{\dot{R}_L}{R_L} = 2\frac{\dot{L}_{orb}}{L_{orb}} - 2\frac{\dot{M}_s}{M_s}\left(\frac{5}{6} - \frac{M_s}{M_p}\right), \tag{12.75}$$

where $\dot{M}_s$ is the mass loss rate from the secondary due to Roche lobe overflow. The orbital evolution time-scale is $L_{orb}/\dot{L}_{orb}$ which is comparable to the mass transfer time-scale

$$\tau_M = \frac{M_s}{|\dot{M}_s|}. \tag{12.76}$$

Validity of the condition of constant M over the binary lifetime therefore requires

$$|\dot{M}_w| \ll |\dot{M}_s|, \tag{12.77}$$

where $\dot{M}_w$ is the wind mass loss rate.

A lower main sequence secondary star, in thermal equilibrium, has a mean radius given approximately by

$$\frac{R_s}{R_\odot} = q\left(\frac{M_s}{M_\odot}\right), \tag{12.78}$$

with $q \simeq 1.1$. For the secondary to continuously transfer mass to the primary, R_L must shrink at least as fast as R_s to keep matter in contact with the L_1 region. Equation (12.75) shows that for $M_p > (6/5)M_s$ a continuous decrease in R_L requires $\dot{L}_{orb} < 0$, since the last term is positive. Also, $|\dot{L}_{orb}|/L_{orb}$ must be sufficiently large to yield $\dot{R}_L < 0$ and $|\dot{R}_L| \geq |\dot{R}_s|$. The secondary will remain in thermal equilibrium provided $\tau_M \gg \tau_{th}$, where τ_{th} is the time-scale for thermal adjustment.

A loss of orbital angular momentum is caused by magnetic braking in conjunction with strong tidal coupling, as outlined above, and by gravitational radiation losses. The latter process leads to the mass transfer rate

$$\dot{M}_s = -1.9 \times 10^{-10}\frac{(1-\mu)^2}{(4-7\mu)\mu^{\frac{2}{3}}} M_\odot\,\mathrm{yr}^{-1}, \tag{12.79}$$

where $\mu = M_\mathrm{s}/M$ and $R_\mathrm{s} = R_\mathrm{L}$ is taken (see §2.4.2). Over the range of mass ratios of interest (12.79) gives $|\dot{M}_\mathrm{s}| \sim 10^{-10} M_\odot \mathrm{yr}^{-1}$. Observations of optical and X-ray luminosities imply mass transfer rates of this magnitude for systems with periods $P \lesssim 3\,\mathrm{hrs}$. However, longer period close binaries are believed to have $|\dot{M}_\mathrm{s}| \sim 10^{-9} M_\odot \mathrm{yr}^{-1}$ which cannot be explained by gravitational radiation losses. The foregoing magnetic braking mechanism can provide a way of removing orbital angular momentum fast enough to explain the higher values of $|\dot{M}_\mathrm{s}|$. Verbunt and Zwaan (1981) estimated the magnetic braking torque by using the Skumanich (1972) spin-down law for single stars, assumed to be due to magnetic braking. By taking this law to apply to a binary secondary star, they derived (2.244) for $\dot{M}_\mathrm{s}$. This yields sufficiently high values of $|\dot{M}_\mathrm{s}|$ for $P > 3\,\mathrm{hrs}$ but, since its derivation makes no direct use of braking theory, the expression cannot be related to the field structure of the secondary or to the details of the wind flow.

Campbell (1997) applied magnetic braking theory to find expressions for $\dot{M}_\mathrm{s}$. The foregoing fast rotator theory is relevant here, since for $P \lesssim 7\,\mathrm{hrs}$ it follows that $\Omega_\mathrm{s} \gtrsim 100\Omega_\odot$. Firstly, the effect of orbital evolution on the magnetic braking rate must be considered. If the secondary is kept close to orbital corotation by tidal forces then $\dot{L}_\mathrm{orb} = \dot{J}$ can be used, where $\dot{J}$ is the loss rate of stellar angular momentum via magnetic braking. The tidal torque is dissipative and so will only be finite when there is some asynchronism. For a turbulent viscosity the tidal torque can be comparable to the braking torque at small degrees of asynchronism, so Ω_s can be taken to be essentially the same as the orbital angular velocity Ω. Hence

$$\Omega_\mathrm{s}^2 = \Omega^2 = \frac{GM}{D^3}. \tag{12.80}$$

The stellar angular momentum loss rate is given by (12.68) as

$$\dot{J} = -\frac{\Phi^2 \Omega_\mathrm{s}}{6\pi\mu_0 v_\mathrm{A}}, \tag{12.81a}$$

where

$$\Phi = 4\pi\bar{r}^2\bar{B} = 2\pi R_\mathrm{s}^2 \left(\frac{R_\mathrm{s}}{\bar{r}}\right) B_0, \tag{12.81b,c}$$

with the last equality following from (12.44). Since the Alfvénic points occur in the radial field region, the magnitude of the wind mass loss rate can be written $|\dot{M}_\mathrm{w}| = 4\pi r_\mathrm{A}^2 \rho_\mathrm{A} v_\mathrm{A}$. Use of this in (12.67) yields

$$\dot{J} = -\frac{2}{3}|\dot{M}_\mathrm{w}| r_\mathrm{A}^2 \Omega_\mathrm{s}. \tag{12.82}$$

Equating this to (12.81a) gives

$$r_{\mathrm{A}}^2 = \frac{\Phi^2}{4\pi\mu_0|\dot{M}_{\mathrm{w}}|v_{\mathrm{A}}}. \qquad (12.83)$$

The fast rotator theory result (12.63a) gives $v_{\mathrm{A}} \simeq 0.5 r_{\mathrm{A}}\Omega_{\mathrm{s}}$, so (12.82) and (12.83) lead to

$$\dot{J} = -\frac{2}{3(2\pi\mu_0)^{\frac{2}{3}}}|\dot{M}_{\mathrm{w}}|^{\frac{1}{3}}\Phi^{\frac{4}{3}}\Omega_{\mathrm{s}}^{\frac{1}{3}}. \qquad (12.84)$$

Equations (12.81c) and (12.84) then yield

$$\dot{L}_{\mathrm{orb}} = \dot{J} = -\frac{2}{3}\left(\frac{2\pi}{\mu_0}\right)^{\frac{2}{3}} B_{\odot}^{\frac{4}{3}}\dot{M}_{\odot}^{\frac{1}{3}}\left(\frac{R_{\mathrm{s}}}{\bar{r}}\right)^{\frac{4}{3}}\left(\frac{B_0}{B_{\odot}}\right)^{\frac{4}{3}}\left(\frac{\dot{M}_{\mathrm{w}}}{\dot{M}_{\odot}}\right)^{\frac{1}{3}} R_{\mathrm{s}}^{\frac{8}{3}}\Omega_{\mathrm{s}}^{\frac{1}{3}}, \qquad (12.85)$$

where $B_{\odot}$ is the coronal base polar value of the sun's magnetic field, and $\dot{M}_{\odot}$ is the solar wind mass loss rate. For a more complicated field than dipolar, $B_{\odot}$ and B_0 would be essentially the mean surface values. As the orbit evolves, with a nearly synchronous secondary having $\Omega_{\mathrm{s}} = \Omega$, the stellar quantities R_{s}, B_0, $\dot{M}_{\mathrm{w}}$ and $\bar{r}$ will change as M_{s} and Ω_{s} evolve. The total mass M can be taken as constant provided that $|\dot{M}_{\mathrm{w}}| \ll |\dot{M}_{\mathrm{s}}|$.

The ratio $\bar{r}/R_{\mathrm{s}}$ is determined by the solution of (12.45). This equation can be written as

$$\frac{\bar{r}}{R_{\mathrm{s}}} = \left(\frac{\zeta_{\mathrm{d}}}{4}\right)^{\frac{1}{6}} \exp\left[\frac{GM_{\mathrm{s}}}{6R_{\mathrm{s}}a_{\mathrm{d}}^2}\left(1 - \frac{R_{\mathrm{s}}}{\bar{r}}\right)\right] \exp\left[-\frac{R_{\mathrm{s}}^2\Omega_{\mathrm{s}}^2}{12a_{\mathrm{d}}^2}\left(\frac{\bar{r}^2}{R_{\mathrm{s}}^2} - \frac{R_{\mathrm{s}}}{\bar{r}}\right)\right], \qquad (12.86a)$$

where

$$\zeta_{\mathrm{d}} = \frac{B_0^2}{2\mu_0(\rho_0)_{\mathrm{d}}a_{\mathrm{d}}^2}. \qquad (12.86b)$$

Eliminating D between (12.74) and (12.80) gives

$$R_{\mathrm{s}}^2\Omega_{\mathrm{s}}^2 = 0.1G\left(\frac{M_{\mathrm{s}}}{R_{\mathrm{s}}}\right). \qquad (12.87)$$

Equation (12.78) shows that $M_{\mathrm{s}}/R_{\mathrm{s}}$ is constant and hence $R_{\mathrm{s}}\Omega_{\mathrm{s}}$ is also constant as the orbit evolves. The virial theorem gives approximate scalings for characteristic temperatures and densities in a star as $T \propto M_{\mathrm{s}}/R_{\mathrm{s}}$ and $\rho \propto M_{\mathrm{s}}/R_{\mathrm{s}}^3$. Taking the corona to follow these scalings yields

$$a_{\mathrm{d}}^2 \propto \frac{M_{\mathrm{s}}}{R_{\mathrm{s}}}, \quad (\rho_0)_{\mathrm{d}} \propto \frac{M_{\mathrm{s}}}{R_{\mathrm{s}}^3}. \qquad (12.88\mathrm{a,b})$$

For T_0 independent of Ω_s, (12.78) and (12.88a) lead to a_d being invariant during evolution. The exponentiated coefficients $GM_s/R_s a_d^2$ and $(R_s\Omega_s)^2/a_d^2$ in (12.86a) are therefore also invariant. This leaves the factor $\zeta_d^{1/6}$ to be considered in the determination of $\bar{r}/R_s$. Using the dynamo law (12.69), with $1 \le n < 2$, together with $(\rho_0)_d \propto \Omega_s$ and $\propto M_s/R_s^3$, (12.86b) and (12.88a) give

$$\zeta_d^{\frac{1}{6}} \propto \left[\frac{(R_s\Omega_s)^2\Omega_s^{2n-3}}{(M_s/R_s)^2}\right]^{\frac{1}{6}} \propto \Omega_s^{(2n-3)/6},$$

where the last proportionality follows from the constancy of $R_s\Omega_s$ and M_s/R_s. For the above range of n, $\zeta_d^{1/6}$ is therefore weakly dependent on Ω_s. Since this factor occurs outside the exponentials in (12.86a), and the exponentiated coefficients are invariant, the solution $\bar{r}/R_s$ will be very slowly varying as mass transfer occurs and can therefore be taken as constant in (12.85).

Consider, now, the variation of the wind mass loss rate. Since the radial field region begins at $r = \bar{r}$, this can be written

$$\dot{M}_w = -4\pi\bar{r}^2\bar{\rho}\bar{v}.$$

Conservation of mass and flux along a poloidal flux tube yield

$$\bar{\rho}\bar{v} = \rho_0 v_0 \left(\frac{\bar{B}}{B_0}\right) = \frac{1}{2}\rho_0 v_0 \left(\frac{R_s}{\bar{r}}\right)^3,$$

and hence

$$\dot{M}_w = -2\pi\left(\frac{R_s}{\bar{r}}\right) R_s^2 \rho_0 v_0. \tag{12.89}$$

The coronal base flow speed in the wind zone is taken as

$$v_0 = 0.15 a_w. \tag{12.90}$$

Then, since $R_s/\bar{r}$ is constant,

$$|\dot{M}_w| \propto R_s^2\left(\frac{M_s}{R_s^3}\right)\Omega_s\left(\frac{M_s}{R_s}\right)^{\frac{1}{2}} \propto \left(\frac{M_s}{R_s}\right)^{\frac{3}{2}}\Omega_s,$$

using the foregoing scalings. Because M_s/R_s is constant, $\dot{M}_w$ evolves with the orbit as

$$\frac{\dot{M}_w}{\dot{M}_\odot} = \frac{\Omega_s}{\Omega_\odot}. \tag{12.91}$$

The wind zone equivalent of ζ_{d} is

$$\zeta_{\mathrm{w}} = \frac{B_0^2}{2\mu_0(\rho_0)_{\mathrm{w}} a_{\mathrm{w}}^2}, \tag{12.92}$$

giving the ratio of the magnetic energy density to the thermal pressure at the coronal base. Use of (12.90) and (12.92) in (12.89) gives the solar mass loss rate

$$\dot{M}_{\odot} = -\frac{0.15\pi}{\mu_0} \frac{R_{\odot}^2 B_{\odot}^2}{a_{\mathrm{w}}(\zeta_{\mathrm{w}})_{\odot}} \left(\frac{R_{\mathrm{s}}}{\bar{r}}\right). \tag{12.93}$$

The orbital angular momentum follows from (12.73), (12.74) and (12.78) as

$$L_{\mathrm{orb}} = 10^{\frac{1}{6}} q^{\frac{1}{2}} \left(\frac{G R_{\odot}}{M_{\odot}}\right)^{\frac{1}{2}} \frac{M_{\mathrm{p}} M_{\mathrm{s}}^{\frac{4}{3}}}{M^{\frac{1}{3}}}. \tag{12.94}$$

A lobe-filling secondary has $R_{\mathrm{s}} = R_{\mathrm{L}}$ which, together with (12.75) and (12.78), yields

$$\frac{\dot{L}_{\mathrm{orb}}}{L_{\mathrm{orb}}} = \left(\frac{4}{3} - \frac{M_{\mathrm{s}}}{M_{\mathrm{p}}}\right) \frac{\dot{M}_{\mathrm{s}}}{M_{\mathrm{s}}}. \tag{12.95}$$

Equations (12.85), (12.94) and (12.95) enable expressions to be derived for $\dot{M}_{\mathrm{s}}$, for various dynamo laws. Consistency requires $|\dot{M}_{\mathrm{w}}| \ll |\dot{M}_{\mathrm{s}}|$ and thermal equilibrium holds for $\tau_{\mathrm{th}} \ll M_{\mathrm{s}}/|\dot{M}_{\mathrm{s}}|$. Hence it is necessary to derive expressions for $\dot{M}_{\mathrm{w}}$ and τ_{th}.

12.3.2. THE WIND MASS LOSS RATE AND THE THERMAL TIME-SCALE

The wind mass loss rate follows from (12.78), (12.87), (12.91) and (12.93) as

$$\dot{M}_{\mathrm{w}} = -\frac{1.9 \times 10^{-11}}{q^{\frac{3}{2}}(\zeta_{\mathrm{w}})_{\odot}} \left(\frac{B_{\odot}}{1\,\mathrm{G}}\right)^2 \left(\frac{R_{\mathrm{s}}}{\bar{r}}\right) \left(\frac{M_{\odot}}{M}\right) \frac{1}{\mu} M_{\odot}\,\mathrm{yr}^{-1}, \tag{12.96}$$

where $\mu = M_{\mathrm{s}}/M$. There is some suggestion that stellar dynamos may saturate above a critical rotation rate Ω_c. Assuming ρ_0, and hence $\dot{M}_{\mathrm{w}}$, also saturates at Ω_c, gives

$$\dot{M}_{\mathrm{w}} = -\frac{2.4 \times 10^{-13}}{(\zeta_{\mathrm{w}})_{\odot}} \left(\frac{B_{\odot}}{1\,\mathrm{G}}\right)^2 \left(\frac{R_{\mathrm{s}}}{\bar{r}}\right) \left(\frac{\Omega_c}{\Omega_{\odot}}\right) M_{\odot}\,\mathrm{yr}^{-1}, \tag{12.97}$$

applicable for $\Omega_{\mathrm{s}} > \Omega_c$.

The Kelvin time-scale for thermal adjustment of the secondary is given by $\tau_{\text{th}} \sim E_{\text{th}}/L_{\text{S}}$, where E_{th} and L_{S} are the star's thermal energy and surface luminosity. The virial theorem can be used to find E_{th}, where, since the contributions of rotational and magnetic energies are negligible,

$$E_{\text{th}} = -\frac{1}{2}E_{\text{G}} \simeq \frac{3}{4}\frac{GM_{\text{S}}^2}{R_{\text{S}}}.$$

Taking the approximate main sequence mass-luminosity relation

$$\frac{L_{\text{S}}}{L_\odot} = \left(\frac{M_{\text{S}}}{M_\odot}\right)^5, \tag{12.98}$$

together with (12.78), yields

$$\tau_{\text{th}} = \frac{2.3 \times 10^7}{q}\left(\frac{M_\odot}{M}\right)^4 \frac{1}{\mu^4}\,\text{yrs}. \tag{12.99}$$

The self-consistency condition $|\dot{M}_{\text{w}}| \ll |\dot{M}_{\text{S}}|$, and the thermal equilibrium condition $\tau_{\text{th}} \ll \tau_{\text{M}}$, are investigated for each dynamo law.

12.3.3. LINEAR DYNAMO LAW

In the case of a linear dynamo law $B_0(\Omega_{\text{S}})$ is given by

$$\frac{B_0}{B_\odot} = \frac{\Omega_{\text{S}}}{\Omega_\odot}. \tag{12.100}$$

Equations (12.85) and (12.91) then lead to

$$\dot{L}_{\text{orb}} = -\frac{2}{3}\left(\frac{2\pi}{\mu_0}\right)^{\frac{2}{3}} \frac{B_\odot^{\frac{4}{3}}\dot{M}_\odot^{\frac{1}{3}}}{\Omega_\odot^{\frac{5}{3}}}\left(\frac{R_{\text{S}}}{\bar{r}}\right)^{\frac{4}{3}} R_{\text{S}}^{\frac{8}{3}}\Omega_{\text{S}}^2. \tag{12.101}$$

The mass transfer rate follows from (12.78), (12.87), (12.93)–(12.95) and (12.101) as

$$\dot{M}_{\text{S}} = -\frac{7.7 \times 10^{-10}}{q^{\frac{5}{6}}(\zeta_{\text{w}})_\odot^{\frac{1}{3}}}\left(\frac{B_\odot}{1\,\text{G}}\right)^2 \left(\frac{R_s}{\bar{r}}\right)^{\frac{5}{3}} \left(\frac{M_\odot}{M}\right)^{\frac{1}{3}} \frac{\mu^{\frac{1}{3}}}{(4-7\mu)} M_\odot\text{yr}^{-1}, \tag{12.102}$$

where $\mu = M_{\text{S}}/M$, and values of $\Omega_\odot = 2.5 \times 10^{-6}\,\text{s}^{-1}$ and $a_{\text{w}} = 1.2 \times 10^5\,\text{m}\,\text{s}^{-1}$ have been used. Since the gas can expand freely in the wind zone, its temperature should be lower than that of gas in the dead zone. Taking $a_{\text{d}}^2 = 2a_{\text{w}}^2$, (12.86b) and (12.92) give

$$\zeta_{\text{w}} = 2\zeta_{\text{d}}, \tag{12.103}$$

for uniform ρ_0. For $(\zeta_d)_\odot = 8$, the solution of (12.86a) is $\bar{r}/R_s = 1.5$. Then, using $q = 1.1, B_\odot = 10^{-4}$ tesla, $M_p = 1 M_\odot$ and $M_s = 0.4 M_\odot$, (12.102) and (12.103) yield $|\dot{M}_s| = 4.2 \times 10^{-11} M_\odot \mathrm{yr}^{-1}$. Using the same masses in (12.79) gives $|\dot{M}_s|_{gr} = 1.1 \times 10^{-10} M_\odot \mathrm{yr}^{-1}$, so the effect of gravitational radiation is stronger than that of magnetic braking in this case. Field saturation reduces $|\dot{M}_s|$.

Equations (12.96) and (12.102) give

$$\frac{|\dot{M}_w|}{|\dot{M}_s|} = \frac{2.5 \times 10^{-2}}{q^{\frac{2}{3}} (\zeta_w)_\odot^{\frac{2}{3}}} \left(\frac{\bar{r}}{R_s}\right)^{\frac{2}{3}} \left(\frac{M_\odot}{M}\right)^{\frac{2}{3}} \frac{(4 - 7\mu)}{\mu^{\frac{4}{3}}}. \tag{12.104}$$

Above the period gap, the range $3\,\mathrm{hrs} \lesssim P \lesssim 6\,\mathrm{hrs}$ is of interest for close binaries and, by (2.226), this corresponds to $0.3\,M_\odot \lesssim M_s \lesssim 0.6\,M_\odot$. Over this range of M_s, with $M = 1.4 M_\odot$, (12.104) yields $9 \times 10^{-3} < |\dot{M}_w|/|\dot{M}_s| < 8 \times 10^{-2}$, so justifying treating M as constant.

The ratio of the mass transfer time-scale to the thermal time-scale follows from (12.76), (12.99) and (12.102) as

$$\frac{\tau_M}{\tau_{th}} = 55 q^{\frac{11}{6}} (\zeta_w)_\odot^{\frac{1}{3}} \left(\frac{B_\odot}{1\,\mathrm{G}}\right)^{-2} \left(\frac{\bar{r}}{R_s}\right)^{\frac{5}{3}} \left(\frac{M}{M_\odot}\right)^{\frac{16}{3}} (4 - 7\mu)\mu^{\frac{14}{3}}. \tag{12.105}$$

For $(\zeta_w)_\odot = 16$ and the above mass range, this gives $4 \lesssim \tau_M/\tau_{th} \lesssim 38$, with $\tau_M \sim 3\tau_{th}$ corresponding to $M_s \sim 0.3 M_\odot$. Hence, for a linear dynamo law, the secondary is in marginal thermal equilibrium close to the top of the period gap. However, explanations of the gap require the secondary to be out of thermal equilibrium at $M_s \sim 0.4 M_\odot$, while (12.105) yields $\tau_M \sim \tau_{th}$ at this mass. Also, (12.102) gives values of $|\dot{M}_s|$ that are at most comparable to $|\dot{M}_s|_{gr}$, and hence cannot explain the higher mass transfer rates believed to occur above the period gap. It is noted that $|\dot{M}_s|$ scales with $B_\odot^2$ and $B_\odot = 1\,\mathrm{G}$ ($= 10^{-4}$tesla) has been used. There is some uncertainty in $B_\odot$, but values significantly higher than 2 G are unlikely. A value of $\bar{r}/R_s = 1.5$ has been taken here. An essentially synchronized secondary in a close binary has $\Omega_s \gtrsim 100\Omega_\odot$, and observations of rapidly rotating stars indicate that matter can escape from the dead zone (Collier Cameron and Robinson, 1989). Presumably the high centrifugal force allows matter near $\bar{r}$ to escape from the magnetic constraint. This suggests that for the secondary star values of $\bar{r}/R_s$ nearer to 1 than to 10 are likely, supporting the value of 1.5 used here. With the presently available braking theory, it is therefore necessary to consider stronger dependencies of B_0 on Ω_s than linear if magnetic winds are to explain mass transfer rates above the period gap.

12.3.4. INVERSE ROSSBY NUMBER LAW

As the secondary star transfers mass to the primary its spectral type will change as $M_{\rm S}$ decreases. The depth of the convective envelope will increase as $M_{\rm S}$ decreases towards $\sim 0.3\,M_\odot$, at which mass the star is fully convective. This change may affect the value of B_0, in addition to the effect of a change in $\Omega_{\rm S}$. There is some observational evidence to suggest that B_0 scales as the inverse turbulent Rossby number $R_{\rm T}^{-1} \propto \tau_{\rm T}\Omega_{\rm S}$, where $\tau_{\rm T}$ is the convective turnover time-scale (Saar, 1991).

The rms convective speed in a region with no energy sources is given by simple mixing length theory as

$$v_{\rm T} = \left(\frac{|g|\lambda_{\rm T} L_{\rm S}}{2\pi r^2 \rho c_P T}\right)^{\frac{1}{3}}, \qquad (12.106)$$

where $\lambda_{\rm T}$ is the mixing length, and $L_{\rm S}$ is the surface luminosity. Taking $\lambda_{\rm T}$ as a pressure scale height gives

$$\lambda_{\rm T} = \lambda_P = \frac{c_{\rm S}^2}{\gamma|g|}, \qquad (12.107)$$

where $c_{\rm S}$ is the adiabatic sound speed. Using $P = (\Re/\mu)\rho T$, with μ the mean molecular weight and $\tau_{\rm T} \sim \lambda_{\rm T}/v_{\rm T}$, (12.106) and (12.107) yield

$$\tau_{\rm T} \propto \frac{r^2 T}{M(r)}\left(\frac{r^2\rho}{L_{\rm S}}\right)^{\frac{1}{3}}, \qquad (12.108)$$

where $M(r)$ is the mass within a radius r (noting tidal distortion will be small deep in the convective envelope). Expression (12.108) gives the radial variation of $\tau_{\rm T}$, so to compare characteristic values between stars of different masses take the simple homologous scalings $r^2/M(r) \propto R_{\rm S}^2/M_{\rm S}, T \propto M_{\rm S}/R_{\rm S}$ and $\rho \propto M_{\rm S}/R_{\rm S}^3$ at a given point. This gives

$$\tau_{\rm T} \propto \left(\frac{R_{\rm S}^2 M_{\rm S}}{L_{\rm S}}\right)^{\frac{1}{3}}. \qquad (12.109)$$

Using the mass-luminosity relation (12.98) then leads to

$$\frac{B_0}{B_\odot} = \frac{\tau_{\rm T}\Omega_{\rm S}}{(\tau_{\rm T})_\odot\Omega_\odot} = \left(\frac{M_\odot^4}{R_\odot^2}\right)^{\frac{1}{3}}\left(\frac{R_{\rm S}^2}{M_{\rm S}^4}\right)^{\frac{1}{3}}\frac{\Omega_{\rm S}}{\Omega_\odot}. \qquad (12.110)$$

The rate of loss of orbital angular momentum follows from (12.85), (12.91) and (12.110) as

$$\dot{L}_{\rm orb} = -\frac{2}{3}\left(\frac{2\pi}{\mu_0}\right)^{\frac{2}{3}} q^{\frac{16}{9}}\,\frac{\dot{M}_\odot^{\frac{1}{3}} B_\odot^{\frac{4}{3}} R_\odot^{\frac{8}{9}}}{\Omega_\odot^{\frac{5}{3}}}\left(\frac{R_{\rm S}}{\bar{r}}\right)^{\frac{4}{3}} R_{\rm S}^{\frac{16}{9}}\,\Omega_{\rm S}^2. \qquad (12.111)$$

The resulting mass transfar rate is

$$\dot{M}_{\rm s} = -7.7 \times 10^{-10} \frac{q^{\frac{1}{18}}}{(\zeta_{\rm w})_{\odot}^{\frac{1}{3}}} \left(\frac{B_{\odot}}{1\,{\rm G}}\right)^2 \left(\frac{R_{\rm s}}{\bar{r}}\right)^{\frac{5}{3}} \left(\frac{M_{\odot}}{M}\right)^{\frac{11}{9}} \frac{1}{(4-7\mu)\mu^{\frac{5}{9}}} M_{\odot}{\rm yr}^{-1}. \tag{12.112}$$

The foregoing parameters give $|\dot{M}_{\rm s}| = 10^{-10} M_{\odot}{\rm yr}^{-1}$ for this case, which is comparable to $|\dot{M}_{\rm s}|_{\rm gr}$, at an orbital period of 4 hrs. Field saturation reduces $|\dot{M}_{\rm s}|$. Hence the inverse Rossby number scaling produces higher mass transfer rates then a simple linear scaling of B_0 with $\Omega_{\rm s}$, but still not high enough to explain values of $|\dot{M}_{\rm s}| \gtrsim 10^{-9} M_{\odot}{\rm yr}^{-1}$.

Equations (12.96) and (12.112) give

$$\frac{|\dot{M}_{\rm w}|}{|\dot{M}_{\rm s}|} = \frac{2.5 \times 10^{-2}}{q^{\frac{14}{9}} (\zeta_{\rm w})_{\odot}^{\frac{2}{3}}} \left(\frac{\bar{r}}{R_{\rm s}}\right)^{\frac{2}{3}} \left(\frac{M}{M_{\odot}}\right)^{\frac{2}{9}} \frac{(4-7\mu)}{\mu^{\frac{4}{9}}}. \tag{12.113}$$

For $M = 1.4 M_{\odot}$ and $0.3 M_{\odot} \leq M_{\rm s} \leq 0.6 M_{\odot}$, together with $(\zeta_{\rm w})_{\odot} = 16$, (12.113) yields $6.6 \times 10^{-3} < |\dot{M}_{\rm w}|/|\dot{M}_{\rm s}| < 2.2 \times 10^{-2}$, so justifying the constancy of M.

The time-scale ratio follows from (12.99) and (12.112) as

$$\frac{\tau_M}{\tau_{\rm th}} = 55 q^{\frac{17}{18}} (\zeta_{\rm w})_{\odot}^{\frac{1}{3}} \left(\frac{B_{\odot}}{1\,{\rm G}}\right)^{-2} \left(\frac{\bar{r}}{R_{\rm s}}\right)^{\frac{5}{3}} \left(\frac{M}{M_{\odot}}\right)^{\frac{56}{9}} (4-7\mu)\mu^{\frac{50}{9}}. \tag{12.114}$$

The above mass range gives $1 < \tau_M/\tau_{\rm th} < 22$, so at $M_{\rm s} = 0.3 M_{\odot}$ the secondary is marginally in thermal equilibrium. For $B_{\odot} = 2\,{\rm G}$ this case is close to having the required properties, but only without field saturation.

12.3.5. A NON-LINEAR LAW

Consider the scaling

$$\frac{B_0}{B_{\odot}} = \left(\frac{\Omega_{\rm s}}{\Omega_{\odot}}\right)^{\frac{7}{4}}. \tag{12.115}$$

Equation(12.85) then gives

$$\dot{L}_{\rm orb} = \dot{J} = -\frac{2}{3} \left(\frac{2\pi}{\mu_0}\right)^{\frac{2}{3}} \frac{\dot{M}_{\odot}^{\frac{1}{3}} B_{\odot}^{\frac{4}{3}}}{\Omega_{\odot}^{\frac{8}{3}}} \left(\frac{R_{\rm s}}{\bar{r}}\right)^{\frac{4}{3}} R_{\rm s}^{\frac{8}{3}} \Omega_{\rm s}^3. \tag{12.116}$$

For a single star, of essentially constant mass, this corresponds to the braking equation

$$I\dot{\Omega}_{\rm s} = -K\Omega_{\rm s}^3,$$

where I is the moment of inertia and K is a positive constant. The solution is

$$\Omega_{\rm s} = \left(\frac{1}{\Omega_0^2} + \frac{2K}{I}t\right)^{-\frac{1}{2}},$$

and for $\Omega_{\rm s} \ll \Omega_0$,

$$\Omega_{\rm s} = \left(\frac{I}{2K}\right)^{\frac{1}{2}} \frac{1}{t^{\frac{1}{2}}}, \tag{12.117}$$

which is the Skumanich (1972) law. This case can therefore be compared with the work of Verbunt and Zwaan (1981), which employed such a braking law and gave the mass transfer rate (2.244).

Equations (12.95) and (12.116) lead to the mass transfer rate

$$\dot{M}_{\rm s} = -\frac{6.2\times 10^{-8}}{q^{\frac{7}{3}}(\zeta_{\rm w})_\odot^{\frac{1}{3}}}\left(\frac{B_\odot}{1\,{\rm G}}\right)^2\left(\frac{R_{\rm s}}{\bar{r}}\right)^{\frac{5}{3}}\left(\frac{M_\odot}{M}\right)^{\frac{4}{3}}\frac{1}{(4-7\mu)\mu^{\frac{2}{3}}}\,M_\odot\,{\rm yr}^{-1}. \tag{12.118}$$

At a period of 4 hrs, corresponding to $M_{\rm s} = 0.4M_\odot$, this gives $|\dot{M}_{\rm s}| = 7.5\times 10^{-9}M_\odot{\rm yr}^{-1}$. This is an order of magnitude higher than that derived from (2.244). The dependencies on μ differ in these expressions, since (2.244) was derived using the observational fit (2.243) for $R_{\rm s}\Omega_{\rm s}$, which is independent of stellar parameters.

The forms for $|\dot{M}_{\rm w}|/|\dot{M}_{\rm s}|$ and $\tau_M/\tau_{\rm th}$ corresponding to (12.118) are

$$\frac{|\dot{M}_{\rm w}|}{|\dot{M}_{\rm s}|} = 3\times 10^{-4}\frac{q^{\frac{5}{6}}}{(\zeta_{\rm w})_\odot^{\frac{2}{3}}}\left(\frac{\bar{r}}{R_s}\right)^{\frac{2}{3}}\left(\frac{M}{M_\odot}\right)^{\frac{1}{3}}\frac{(4-7\mu)}{\mu^{\frac{1}{3}}}, \tag{12.119}$$

and

$$\frac{\tau_M}{\tau_{\rm th}} = 0.7q^{\frac{10}{3}}(\zeta_{\rm w})_\odot^{\frac{1}{3}}\left(\frac{B_\odot}{1\,{\rm G}}\right)^{-2}\left(\frac{\bar{r}}{R_{\rm s}}\right)^{\frac{5}{3}}\left(\frac{M}{M_\odot}\right)^{\frac{19}{3}}(4-7\mu)\mu^{\frac{17}{3}}. \tag{12.120}$$

For the mass range $0.3\,M_\odot \le M_{\rm s} \le 0.6\,M_\odot$ these expressions yield the ratios $1.1\times 10^{-4} < |\dot{M}_{\rm w}|/|\dot{M}_{\rm s}| < 3.5\times 10^{-4}$ and $1.6\times 10^{-2} < \tau_M/\tau_{\rm th} < 0.4$, so M is conserved to high accuracy, and the secondary is well out of thermal equilibrium over the lower part of this mass range.

The values of $|\dot{M}_{\rm s}|$ given by (12.118) are rather large, so the case of field saturation at Ω_c is considered. The resulting mass transfer rate is

$$\dot{M}_{\rm s} = -\frac{5.3}{10^{13}}\frac{q^{\frac{5}{3}}}{(\zeta_{\rm w})_\odot^{\frac{1}{3}}}\left(\frac{B_\odot}{1\,{\rm G}}\right)^2\left(\frac{R_{\rm s}}{\bar{r}}\right)^{\frac{5}{3}}\left(\frac{\Omega_c}{\Omega_\odot}\right)^{\frac{8}{3}}\left(\frac{M}{M_\odot}\right)^{\frac{4}{3}}\frac{\mu^2}{(4-7\mu)}\,M_\odot{\rm yr}^{-1}. \tag{12.121}$$

TABLE 12.4. The ratio $\tau_M/\tau_{\rm th}$

$M_{\rm s}/M_\odot$	$\tau_M/\tau_{\rm th}$	$\|\dot{M}_{\rm s}\|/M_\odot\,{\rm yr}^{-1}$
0.3	0.27	4.2×10^{-10}
0.4	0.51	9.5×10^{-10}
0.5	0.74	2.0×10^{-9}
0.6	0.86	4.2×10^{-9}

This, together with (12.97), gives

$$\frac{|\dot{M}_{\rm w}|}{|\dot{M}_{\rm s}|} = \frac{0.46}{q^{\frac{5}{3}}(\zeta_{\rm w})_\odot^{\frac{2}{3}}}\left(\frac{\bar{r}}{R_{\rm s}}\right)^{\frac{2}{3}}\left(\frac{M_\odot}{M}\right)^{\frac{4}{3}}\left(\frac{\Omega_\odot}{\Omega_c}\right)^{\frac{5}{3}}\frac{(4-7\mu)}{\mu^2}, \tag{12.122}$$

which yields $5 \times 10^{-4} < |\dot{M}_{\rm w}|/|\dot{M}_{\rm s}| < 3.6 \times 10^{-3}$. The time-scale ratio is

$$\frac{\tau_M}{\tau_{\rm th}} = 8 \times 10^4 \frac{(\zeta_{\rm w})_\odot^{\frac{1}{3}}}{q^{\frac{2}{3}}}\left(\frac{B_\odot}{1\,{\rm G}}\right)^{-2}\left(\frac{\bar{r}}{R_{\rm s}}\right)^{\frac{5}{3}}\left(\frac{M}{M_\odot}\right)^{\frac{11}{3}}\left(\frac{\Omega_\odot}{\Omega_c}\right)^{\frac{8}{3}}(4-7\mu)\mu^3. \tag{12.123}$$

Table 12.4 shows values of $|\dot{M}_{\rm s}|/M_\odot{\rm yr}^{-1}$ and $\tau_M/\tau_{\rm th}$ above the period gap, using the foregoing parameters and $\Omega_c = 80\Omega_\odot$ in (12.121) and (12.123). The mass transfer rates are in good agreement with those implied by observations of systems with $P \gtrsim 3$ hrs. The values of $\tau_M/\tau_{\rm th}$ show that at $M_{\rm s} = 0.6M_\odot$ the secondary is marginally in thermal equilibrium. As $M_{\rm s} = 0.3M_\odot$ is approached the star starts to be driven out of thermal equilibrium, since it cannot fully adjust thermally as it losses mass. This case therefore has the required properties for explanation of the period gap.

12.3.6. THE PERIOD GAP

The cataclysmic variables are binaries containing a white dwarf primary and a lobe-filling main sequence secondary. Their orbital periods lie in the range $80\,{\rm min} \lesssim P \lesssim 7\,{\rm hrs}$ and their mass transfer rates span $5 \times 10^{-11} \lesssim |\dot{M}_{\rm s}|/M_\odot{\rm yr}^{-1} \lesssim 3 \times 10^{-9}$. There is an almost complete absence of systems with periods in the range 2 to 3 hrs. As seen in the foregoing discussion, gravitational radiation losses can account for the lower mass transfer rates occurring at periods $\lesssim 3$ hrs, and hence for systems below the period gap. The foregoing analysis showed that magnetic braking can explain the larger values of $|\dot{M}_{\rm s}|$ observed at $P > 3$ hrs, provided the dynamo law relating B_0 to $\Omega_{\rm s}$ is nearer quadratic than linear. The gap is believed to contain systems in which the secondary star is detached from its Roche lobe, so mass transfer, which is the main source of luminosity via accretion onto the primary, is absent. This poses the problem of why the secondary becomes detached.

The current explanation for the origin of the period gap involves magnetic braking. It was seen in §12.3.5 that for $0.3 \lesssim M_s/M_\odot \lesssim 0.7$ the mass transfer time-scale $M_s/|\dot{M}_s|$ can become $\lesssim 10^9$yrs and hence be comparable to the thermal time-scale τ_{th}. The secondary will then not have time to fully adjust thermally as it loses mass. This results in an over-luminous and oversized star for its mass. If, at around a period of 3 hrs, something happens to sharply reduce magnetic braking, then the orbital angular momentum loss will become controlled by gravitational radiation. The time-scale for Roche lobe shrinkage due to this weaker process is $\gg \tau_{th}$, so the star contracts back to its thermal equilibrium radius faster than its lobe shrinks and therefore becomes detached. The resulting low luminosity system subsequently evolves through the period gap, driven by gravitational radiation, until the secondary fills it lobe and resumes mass transfer at a period of ~ 2 hrs.

Various mechanisms have been proposed to account for a sudden decrease in the strength of magnetic braking at a period of $\sim$ 3 hrs. This period corresponds to a secondary mass of $\sim 0.3M_\odot$, which is the mass at which a main sequence star is expected to become fully convective. It has been suggested that when this state is reached the stellar dynamo begins to operate in a different way, causing a reduction in field strength and/or a change in field configuration, which results in a lower magnetic torque (e.g. Spruit and Ritter, 1983; Rappaport, Verbunt and Joss, 1983). If the dynamo mainly operates at the interface between the radiative core and convective envelope, existing when $M_s > 0.3M_\odot$, then the disappearance of this interface at $M_s \sim 0.3M_\odot$ would cause a reduction in field strength. However, there is no compelling theoretical reason that this should be the case. Observational evidence in support of this hypothesis would require magnetic activity to decline abruptly in stars later than spectral type M5. The most favourable evidence for this is that X-ray emission from stars later than M5 has maximum luminosities that are at least an order of magnitude less than those of M stars of earlier spectral type (Golub, 1983). However, present data are not extensive enough to prove that a sharp decrease in magnetic activity occurs close to M5.

It should be noted that a reduction by a factor of 10 in the angular momentum loss rate is sufficient to produce a period gap of the observed width (e.g. Spruit and Ritter, 1983). Equation (12.85) has $|\dot{J}| \propto B_0^{4/3}|\dot{M}_w|^{1/3}$ so, with a coronal base density $(\rho_0)_w \propto B_0$ and ignoring the small change in $\bar{r}/R_s$ generated in (12.86), it follows that $|\dot{J}| \propto B_0^{5/3}$. This theory therefore only requires a decrease in B_0 by a factor of 4 to cause $|\dot{J}|$ to drop by the

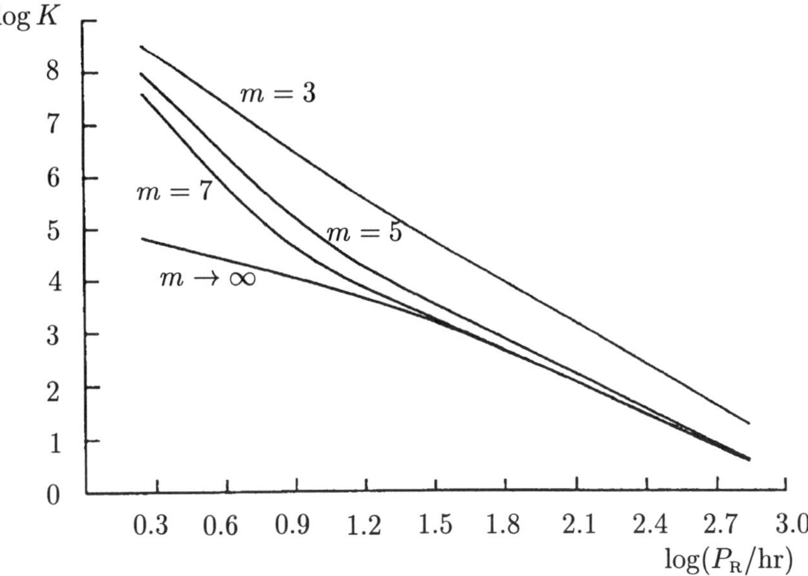

Figure 12.2. The dimensionless angular momentum loss rate, $K(\Omega_S/\Omega_\odot)$, as a function of stellar rotation period for different multipole orders. Based on Taam and Spruit (1989).

required factor of 10.

Taam and Spruit (1989) argue that a sudden decline in magnetic braking when the secondary becomes fully convective is more likely to be due to a change in field configuration, rather than a sudden decrease in field strength. They point out that the spectroscopic data, taken as a whole, indicate a steady decline in magnetic activity through the M5 spectral type, rather than an abrupt decrease. They note that at high rotation rates the irregular variability that is observed may be evidence for the dynamo operating in higher modes, generating fields with a smaller scale structure than those occurring in more slowly rotating stars.

The authors hypothesise that when the secondary becomes fully convective its dynamo process changes to produce fields of higher multipole structure. The foregoing braking theory is used, except with the field inside $\bar{r}$ having a radial dependence of r^{-m}. Equation (12.72) is used to calculate the angular momentum loss rate, for several values of m corresponding to different multipoles. The function $K(\Omega_s/\Omega_\odot)$ is shown in Figure 12.2. The angular momentum loss rate declines by a factor of 2000 between a dipole field and an asymptotically high order field at an orbital period of 3hrs. This effect is mainly due to a decrease in the fraction of the magnetic surface flux that extends out to the Alfvén radius. This fraction of open field lines is

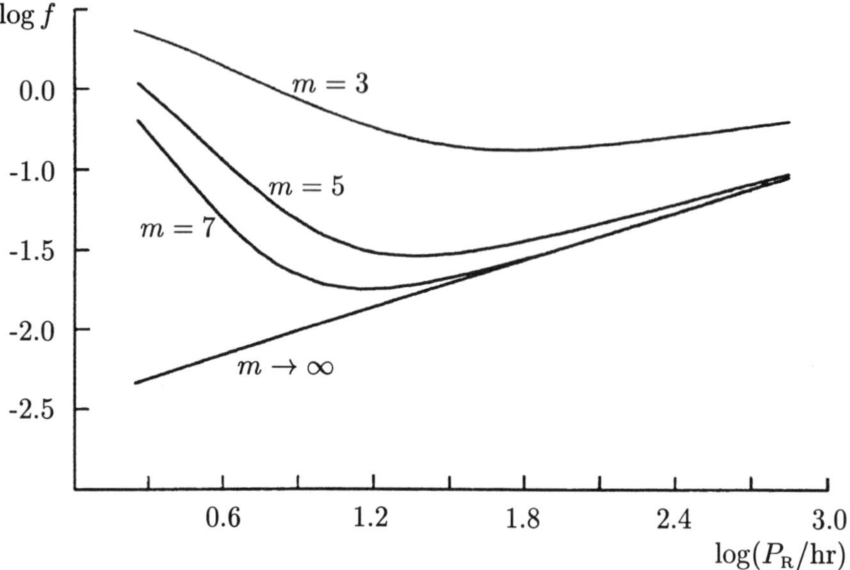

Figure 12.3. The fraction of open field lines, f, as a function of stellar rotation period for different multipole orders. Based on Taam and Spruit (1989).

$$f = \left(\frac{\bar{r}^2}{R_s^2}\right)\frac{B(\bar{r})}{B(R_s)}. \tag{12.124}$$

The variation of f with period is shown in Figure 12.3. With increasing multipole order, $\bar{r}$ moves closer to the stellar surface because the field strength decreases more rapidly with distance. At high rotation rates, the pressure (and hence the density) in the dead zone is strongly influenced by the centrifugal acceleration, which increases rapidly with distance from the star. For higher order multipoles, the effect of the centrifugal acceleration on f is therefore smaller than at low order, and fewer field lines are open. It should be noted, however that at high rotation rates the point where gravity and centrifugal force cancel is inside the dead zone. Hence the density of gas in hydrostatic equilibrium increases outwards in the outer part of this zone. Instabilities could then lead to the gas escaping from the field in the outer regions, so contributing to the wind and increasing $|\dot{J}|$ to some extent.

It should be noted that the mass transfer rates $\dot{M}_s$ discussed in this section are average rates. Short-time scale variations in $\dot{M}_s$ can occur; for example AM Her binaries have low states, lasting typically 2 to 3 months, during which $|\dot{M}_s|$ is well below its average value.

12.3.7. AM HERCULIS SYSTEMS

Recent observations suggest that the period gap is not as significant for AM Her systems as it is for binaries with essentially non-magnetic accretors. About one sixth of the presently known AM Her systems have periods in the 2–3 hr gap. Li, Wu and Wickramasinghe (1994) proposed an explanation for this. They noted that at a distance of 2–3 stellar radii from the secondary the primary's magnetic field significantly affects the wind dynamics, assuming typical binary parameters and a secondary polar surface field of 140 G. Both stars were taken to have dipolar fields with moments perpendicular to the orbital plane. For stellar masses of $M_{\rm p} = 0.7M_\odot$ and $M_{\rm s} = 0.4M_\odot$, the wind and dead zones were found using the criteria of §12.2.4, so a dead zone occurs when $B_p^2/2\mu_0 > P$ and a wind zone of open fields lines when $P > B_p^2/2\mu_0$. Figure 12.4 shows the wind and dead zones in the plane containing the dipole moments. This illustrates that the wind zones of the secondary are greatly reduced by the presence of the primary's magnetic field.

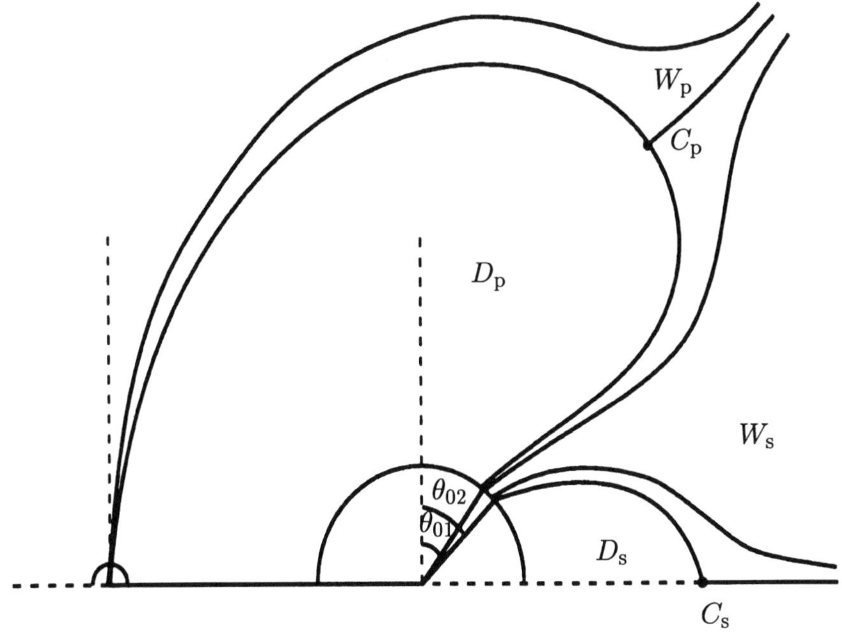

Figure 12.4. The wind and dead zones in the plane of the dipole moments. Large dead zones, $D_{\rm P}$ and $D_{\rm S}$, occur. Based on Li, Wu and Wickramasinghe (1994).

By considering a simple ring model, in which the surface segment between θ_{01} and θ_{02} is extended around the stellar surface, an estimate can be

TABLE 12.5. Reduction in magnetic braking

$(B_p)_0$ (MG)	κ	$\lvert\dot{J}_{mb}\rvert/\lvert\dot{J}_{gr}\rvert$
0	1.0	8.2×10^{-2}
10	0.1	6.1×10^{-3}
30	3.2×10^{-2}	6.7×10^{-4}
50	1.0×10^{-2}	4.5×10^{-6}
70	2.4×10^{-4}	2.1×10^{-10}

made for the magnetic braking. The foregoing braking theory was applied, using a linear dynamo law and a range of primary fields, to find $-\dot{J}$. The results are shown in Table 12.5, where $(B_p)_0$ is the primary star's surface polar field and κ is an estimate of the ratio of the total open flux of the secondary to the flux of a ring of angular width $\Delta\theta = \theta_{02} - \theta_{01}$. It is noted that even for $(B_p)_0 = 0$ the effect of magnetic braking is weaker than that of gravitational radiation, as shown in §12.3.3 for such a dynamo law. However, the reduction in $|\dot{J}_{mb}|$ for $(B_p)_0 > 30\,\mathrm{MG}$, illustrated in Table 12.5, is sufficiently large that magnetic braking would still become negligible in this model even for the stronger dynamo laws considered in §12.3. It follows that magnetic braking would not be sufficient to drive the secondary out of thermal equilibrium in the AM Her systems, which would therefore evolve via gravitational radiation through the period gap with mass transfer continuing. This analysis assumes, of course, that the dead zones remain effective at high rotation rates.

12.4. Accretion Disc Winds

12.4.1. THE WIND SOURCE

If an accretion disc has a poloidal magnetic field, and a hot corona, then the possibility of a magnetically-influenced wind arises. Consider a cylindrical coordinate system (ϖ, ϕ, z) centred on the accretor. The footpoint of an infinitesimal poloidal flux tube, at $\varpi = \varpi_0$, $z = 0$, rotates with the local angular velocity $\Omega_0 = \Omega_K(\varpi_0)$. The effective potential at a point on a field line, assumed corotating, is then

$$\psi = -\frac{GM_p}{r} - \frac{1}{2}\Omega_0^2\varpi^2, \qquad (12.125)$$

where M_p is the mass of the accreting star. Near-corotation will occur at points well inside the Alfvén surface. Blandford and Payne (1982) showed that there is a critical launching angle for the wind flow, above which the mass flux rapidly decreases due to the presence of a potential barrier above the disc. They illustrated this numerically, but the critical angle can be derived analytically.

For a Keplerian disc, the effective potential (12.125) can be written

$$\psi = -\frac{GM_{\rm p}}{\varpi_0}\left[\frac{\varpi_0}{(\varpi^2+z^2)^{\frac{1}{2}}}+\frac{1}{2}\left(\frac{\varpi}{\varpi_0}\right)^2\right]. \tag{12.126}$$

Consider a straight poloidal field line making an angle i with the disc midplane $z=0$. The equation of this line is

$$z = \tan i(\varpi - \varpi_0). \tag{12.127}$$

A unit vector along the field is then

$$\hat{\mathbf{s}} = \cos i\,\hat{\boldsymbol{\varpi}} + \sin i\,\hat{\mathbf{z}}. \tag{12.128}$$

The force per unit mass along $\mathbf{B}_p$ is

$$F_s = -\hat{\mathbf{s}}\cdot\nabla\psi = -\frac{d\psi}{ds}. \tag{12.129}$$

Use of (12.126) and (12.128) yields

$$-\frac{d\psi}{ds} = -\frac{GM_{\rm p}}{(\varpi^2+z^2)^{\frac{1}{2}}}(\varpi\cos i + z\sin i) + \frac{GM_{\rm p}}{\varpi_0^3}\varpi\cos i, \tag{12.130}$$

giving force balance along the field when

$$\frac{\varpi\cos i + z\sin i}{(\varpi^2+z^2)^{\frac{3}{2}}} = \frac{\varpi\cos i}{\varpi_0^3}. \tag{12.131}$$

This is satisfied, for any value of i, in the disc midplane where $\varpi=\varpi_0$, $z=0$. However, the stability of the balance does depend on i. The existence of equilibrium points above the disc midplane, and their nature, must also be investigated.

Solutions of (12.131), together with the field line equation (12.127), can be sought assuming $z/\varpi_0 \ll 1$ and their consistency checked. Expansion of these equations to second order in z/ϖ_0 yields

$$\left(\frac{z}{\varpi_0}\right)\left[(\tan^2 i+3)\left(\frac{z}{\varpi_0}\right) - \tan i(\tan^2 i - 3)\right] = 0.$$

This gives the midplane solution $z=0$, together with the solution

$$\frac{z}{\varpi_0} = \frac{\tan i(\tan^2 i - 3)}{\tan^2 i + 3}. \tag{12.132}$$

For $i > 0$ the field line points must have $z \geq 0$. It follows from (12.132) that a critical angle exists, given by $i_c = \tan^{-1}\sqrt{3} = 60°$, below which the solution is invalid since it then yields $z < 0$. For $i \leq i_c$ the only equilibrium solution is at $z = 0$. For $i > i_c$ force balance along $\mathbf{B}_p$ is also possible above the disc's midplane.

The stability of the equilibrium points can be determined by considering the sign of $d^2\psi/ds^2$. It follows from (12.128) and (12.129) that

$$\frac{d^2\psi}{ds^2} = \cos i \frac{\partial}{\partial \varpi}\left(\frac{d\psi}{ds}\right) + \sin i \frac{\partial}{\partial z}\left(\frac{d\psi}{ds}\right).$$

This leads to

$$\left(\frac{d^2\psi}{ds^2}\right)_{\rm eq} = \frac{GM_{\rm p}}{\varpi_0^3}\left[\left(\frac{\varpi_0^2}{\varpi^2+z^2}\right)^{\frac{3}{2}} - \left\{1+3\left(\frac{\varpi}{\varpi_0}\right)^2\left(\frac{\varpi^2+z^2}{\varpi_0^2}\right)^{\frac{1}{2}}\right\}\cos^2 i\right], \tag{12.133}$$

where an equilibrium point is a solution of (12.127) and (12.131), satisfying

$$\frac{\varpi^2+z^2}{\varpi_0^2} = \left[1+\frac{\tan^2 i(z/\varpi_0)}{\tan i + (z/\varpi_0)}\right]^{\frac{2}{3}}, \tag{12.134a}$$

$$\frac{\varpi}{\varpi_0} = 1 + \frac{1}{\tan i}\left(\frac{z}{\varpi_0}\right). \tag{12.134b}$$

At the midplane point,

$$\left(\frac{d^2\psi}{ds^2}\right)_{z=0} = \frac{GM_{\rm p}}{\varpi_0^3}\cos^2 i(\tan^2 i - 3). \tag{12.135}$$

It follows that for $i < i_c$ this point is a maximum for ψ along $\mathbf{B}_p$, and is therefore unstable. For $i > i_c$ the midplane point is stable.

Consider, now, equilibrium solutions above the midplane, which exist when $i > i_c$. When $i_c < i \lesssim 62°$ expansion in z/ϖ_0 is valid and (12.132)–(12.134) give

$$\left(\frac{d^2\psi}{ds^2}\right)_{\rm eq} = -\frac{GM_{\rm p}}{\varpi_0^3}\frac{\cos^2 i(8\tan^2 i - 9)(\tan^2 i - 3)}{(\tan^2 i + 3)}, \tag{12.136}$$

showing that such points are unstable. For $i > 62°$, z/ϖ_0 is no longer small but it can be shown numerically that, for a given value of i, there is only one unstable equilibrium point above the disc.

The foregoing results have significance for wind flow from the disc. For $i > 60^\circ$ there is a potential barrier to the flow, and material in the disc plane must have sufficient thermal energy to climb out of the potential well and surmount this barrier. For $i \gtrsim 65^\circ$ the unstable point lies significantly beyond the disc surface so, for typical coronal temperatures, the mass loss flux decreases rapidly with increasing values of i. For $i < 60^\circ$ there is no barrier to the flow, since the only equilibrium point lies in the disc midplane and is unstable. Hot material can therefore freely leave the disc, and be accelerated outwards by the strong centrifugal force. This differs from the stellar case, in that in a disc wind flow the centrifugal force at the coronal base already balances gravity, with the radial pressure gradient being negligible. Since most stars rotate well below break-up speed, the possibility of free mass loss does not arise. For $i \lesssim 50^\circ$ the mass loss rate becomes so high that the disc evaporates on a dynamical time-scale (i.e. $\sim 2\pi/\Omega_0$). The rapid changes in mass flux either side of the critical angle are relevant to disc wind stability, as discussed below.

12.4.2. WIND FLOW STABILITY

The stability of magnetic wind-driven accretion discs was considered by Lubow, Papaloizou and Pringle (1994). The wind removes angular momentum from the disc, which may be sufficient to drive mass inflow. For an axisymmetric disc the fundamental results of stellar wind theory, derived in §12.2, are applicable. The mass inflow rate through the disc is

$$\dot{M}(\varpi_0) = -4\pi\varpi_0\rho_c v_\varpi h, \tag{12.137}$$

where ρ_c is the central density, v_ϖ the inflow velocity and h the disc scale height. Unlike the viscous disc, considered in §2.4.3, $\dot{M}$ now depends on ϖ_0 since the wind removes some mass from each ring of width $d\varpi_0$. Mass conservation gives

$$\frac{d\dot{M}}{d\varpi_0} = 4\pi\varpi_0\dot{m}, \tag{12.138}$$

where $\dot{m}(\varpi_0) \simeq \rho v_p$ is the mass loss rate per unit area from one side of the disc due to the wind. If most of the matter is to be accreted, rather than lost in the wind, then

$$\dot{m} \ll \frac{\dot{M}}{4\pi\varpi_0^2} \tag{12.139}$$

is required, which is equivalent to $d\dot{M}/d\varpi_0 \simeq 0$.

For a dipole-type disc magnetic field,

$$B_\phi(\varpi_0, h) = -B_\phi(\varpi_0, -h) = B_\phi^s. \tag{12.140}$$

For a thin disc, the vertical magnetic field will be independent of z, to first order in z/ϖ_0, and so can be denoted $B_z^d(\varpi_0)$. Equation (12.11) gives the advection of angular momentum due to a magnetic torque. For inflow through the disc this yields

$$\frac{d}{d\varpi_0}(\dot{M}\varpi_0^2\Omega_{\mathrm{K}}) = 4\pi\varpi_0 T_{\mathrm{m}}, \tag{12.141}$$

where

$$T_{\mathrm{m}} = \frac{\varpi_0}{\mu_0} B_\phi^s B_z^d \tag{12.142}$$

is the magnetic torque per unit area on one side of the disc, and the angular velocity is taken as Keplerian with

$$\Omega_{\mathrm{K}} = \left(\frac{GM_{\mathrm{p}}}{\varpi_0^3}\right)^{\frac{1}{2}}, \tag{12.143}$$

where M_{p} is the mass of the accreting primary star. It follows from (12.137), (12.141) and (12.142) that

$$v_\varpi \simeq \frac{\varpi_0}{h}\frac{v_{\mathrm{A}z}^2}{v_{\mathrm{K}}}\frac{B_\phi^s}{B_z^d}, \tag{12.144}$$

where

$$v_{\mathrm{A}z} = \frac{B_z^d}{(\mu_0\rho_c)^{\frac{1}{2}}}. \tag{12.145}$$

As in the case of an axisymmetric stellar wind, the total rate of transport of angular momentum per unit poloidal flux is conserved along field-streamlines, as expressed by (12.12). This, together with (12.10) and (12.18), gives

$$\rho v_p \varpi^2 \Omega - \frac{\varpi}{\mu_0} B_\phi B_p = \rho v_p \varpi_{\mathrm{A}}^2 \Omega_{\mathrm{K}}, \tag{12.146}$$

since $\alpha \simeq \Omega_{\mathrm{K}}$, the rotation rate near the coronal base. Since ϖ_0 refers to a point in the disc, while ϖ_{A} denotes the Alfvén point above the disc, (12.146) yields

$$T_{\mathrm{m}} = \frac{\varpi_0}{\mu_0} B_\phi^s B_z^d \simeq -\dot{m}\Omega_{\mathrm{K}}(\varpi_0)(\varpi_{\mathrm{A}}^2 - \varpi_0^2), \tag{12.147}$$

if the approximation of corotation is applied all the way to ϖ_A. It is noted that, since $\dot{m} > 0$ and $\varpi_A > \varpi_0$, (12.147) gives $T_m < 0$ as required to drive inflow through the disc. Using this torque expression in (12.141) gives

$$\frac{\dot{M}}{4\pi\varpi_0^2} \simeq \dot{m}\left[\left(\frac{\varpi_A}{\varpi_0}\right)^2 - 1\right], \tag{12.148}$$

so the small mass loss condition (12.139) requires

$$\varpi_A \gg \varpi_0. \tag{12.149}$$

The magnetic wind is taken to be self-similar with

$$B_p^A = B_z^d(\varpi_0)\left(\frac{\varpi_0}{\varpi_A}\right)^n, \tag{12.150}$$

where $n > 1$. The poloidal speed at the Alfvén point is defined by

$$v_A = \frac{B_p^A}{(\mu_0 \rho_A)^{\frac{1}{2}}}. \tag{12.151}$$

Equation (12.63a) gave the result from fast rotator theory that the Alfvén speed $v_A \simeq 0.5\varpi_A\Omega_s$, where Ω_s was the stellar rotation rate. The present authors use

$$v_A \simeq \varpi_A \Omega_K(\varpi_0), \tag{12.152}$$

consistent with their approximation of corotation out to ϖ_A. Since $\rho v_p/B_p$ is constant along a field-streamline, it follows that

$$\frac{\dot{m}}{B_z^d} \simeq \frac{\rho_A v_A}{B_p^A}. \tag{12.153}$$

Combining (12.150)–(12.153) gives

$$\dot{m} \simeq \frac{(B_z^d)^2}{\mu_0 v_K}\left(\frac{\varpi_0}{\varpi_A}\right)^{n+1}. \tag{12.154}$$

Eliminating $\dot{m}$ between this and (12.147) yields

$$\frac{B_\phi^s}{B_z^d} \simeq \left(\frac{\varpi_0}{\varpi_A}\right)^{n-1}. \tag{12.155}$$

If magnetic stresses do not affect the vertical equilibrium significantly, then

$$\frac{v_K}{c_s} \simeq \frac{\varpi_0}{h}. \tag{12.156}$$

Using (12.154)–(12.156), together with (12.144) for v_ϖ, gives

$$\dot{m} \simeq \rho_0 c_\mathrm{s} \left(\frac{h}{\varpi_0}\right) \left(\frac{v_\varpi}{c_\mathrm{s}}\right)^{\frac{n+1}{n-1}} \left(\frac{v_{\mathrm{A}z}}{c_\mathrm{s}}\right)^{-\frac{4}{n-1}} . \tag{12.157}$$

This equation gives the mass loss rate causing the inflow; the magnetic stresses associated with the wind remove disc angular momentum and energy allowing matter to spiral inwards.

For a dipole-type field, the ϖ-component is antisymmetric about the midplane, so

$$B_\varpi(\varpi_0, h) = -B_\varpi(\varpi_0, -h) = B^s_\varpi. \tag{12.158}$$

For a thin disc, the poloidal component of the induction equation shows that

$$v_\varpi B_z + \eta \frac{\partial B_\varpi}{\partial z} \simeq 0,$$

implying

$$\frac{B^s_\varpi}{B^d_z} \simeq \frac{h v_\varpi}{\eta}, \tag{12.159}$$

where η is the magnetic diffusivity in the disc.

As seen from the foregoing magnetic wind theory of §12.2.2, the flow is fixed by ensuring that the solution passes through the critical points. The position of the sonic point is approximately determined by the point on the field line at which the net tangential force vanishes, with an unstable equilibrium. As seen in §12.4.1, the height z_s of this point governs the wind mass flux $\dot{m}$. Hence

$$\dot{m} \simeq (\rho c_\mathrm{s})_{z_s}. \tag{12.160}$$

The disc is assumed to be locally isothermal, so

$$\rho(\varpi_0, z) = \rho(\varpi_0) \exp\left(-\frac{z^2}{2h^2}\right). \tag{12.161}$$

As shown in §12.4.1, if the poloidal field inclination to the disc's symmetry plane is $i \ll i_c$, where $i_c = 60°$, then $z_s = 0$. There is then no barrier to the wind flow which consequently evaporates the disc on a dynamical time-scale of $\sim 1/\Omega_\mathrm{K}$, with $\dot{m} \sim \dot{M}/4\pi\varpi_0^2$. Conversely, for $i \gg i_c$, the sonic point z_s lies well above the disc surface and there is a stable equilibrium point at $z = 0$. The resulting high potential barrier leads to small values of

$\dot{m}$. Values of $\dot{m}$ sufficiently large to generate the rate of angular momentum advection necessary to drive the inflow $\dot{M}$, but still satisfy $\dot{m} \ll \dot{M}/4\pi\varpi_0^2$, require $i_c < i \lesssim 62°$.

A dipole-type field has $i = 90°$ at $z = 0$, but its inclination decreases with height. Provided the field is sufficiently curved in the vicinity of the disc, the sonic point will occur approximately where the field inclination reaches i_c, with field components given by

$$B_\varpi = B_\varpi^s(\varpi_0)\left(\frac{z}{h}\right), \qquad B_z = B_z^d(\varpi_0), \tag{12.162}$$

to first order in z/ϖ_0. The sonic point then occurs where $B_z/B_\varpi \simeq \tan i_c = \sqrt{3}$ and hence

$$z_s \simeq \frac{h}{\sqrt{3}}\frac{B_z^d}{B_\varpi^s}. \tag{12.163}$$

A second relationship between $\dot{m}$ and v_ϖ now follows from (12.159)–(12.161) and (12.163) as

$$\dot{m} \simeq \rho_0 c_s \exp\left(-\frac{\eta^2}{6h^2v_\varpi^2}\right). \tag{12.164}$$

This represents the mass loss caused by the radial inflow, since v_ϖ is related to B_ϖ^s/B_z^d, and hence to the sonic point height z_s, via (12.159) and (12.163) respectively.

Figure 12.5 shows the self-consistent solutions resulting from simultaneously satisfying (12.157) and (12.164) for $\dot{m}$ and v_ϖ. Curve A represents $\dot{m}/\rho_0 c_s$, given by (12.164), as a function of $|v_\varpi|/c_s$. Curves B and C show two possible variations for (12.157), corresponding to different values of η. In general, three types of solution are possible. Solution 1 arises when the values of $\dot{m}$ given by (12.157) exceed those given by (12.164) for all v_ϖ, the curves only meeting to yield the trivial solution S1 at $\dot{m} = v_\varpi = 0$. The approximate condition for this case is that when $|v_\varpi| \sim \eta/h$ then (12.157) yields $\dot{m} > \rho_0 c_s$ which exceeds the value of $\dot{m} \sim 0.8\rho_0 c_s$ given by (12.164). This requires

$$\eta > c_s h \left(\frac{\varpi_0}{h}\right)^{\frac{n-1}{n+1}} \left(\frac{v_{Az}}{c_s}\right)^{\frac{4}{n+1}}. \tag{12.165}$$

In this case the magnetic diffusivity in the disc is so large that the poloidal field cannot be bent sufficiently by the inflow to give a finite self-consistent wind solution. As shown in §12.4.1, at such high inclinations of the poloidal field the unstable point lies well above the disc midplane, leading to low $\dot{m}$.

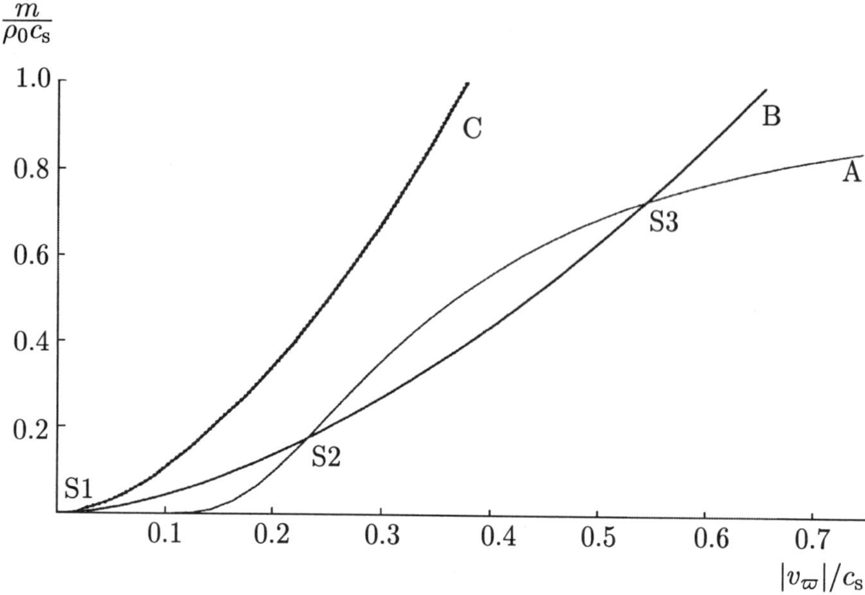

Figure 12.5. The dimensionless mass loss rate $\dot{m}/\rho_0 c_s$ versus dimensionless disc radial velocity $|v_\varpi|/c_s$. The three solutions are denoted S1, S2 and S3. Based on Lubow, Papaloizou and Pringle (1994).

When η is small enough to satisfy the reverse of (12.165), two non-trivial solutions are possible. Solution S2 corresponds to a value of $\dot{m}$ in the mid part of the range $0 < \dot{m} < \rho_0 c_s$, giving field lines with tangents near the disc surface making angles of $\sim 60°$ with the midplane. This solution corresponds to the type discussed by Blandford and Payne (1982). However, it is clear from Figure 12.5 that the solution S2 is unstable; a small increase in $|v_\varpi|$ from its equilibrium value bends $\mathbf{B}_p$ leading to a much higher mass loss rate from the disc (curve A) than that which is required to maintain the slightly increased $|v_\varpi|$ (curve B). A negative perturbation also grows.

Solution 3 corresponds to $\dot{m} \sim \rho_0 c_s$, with a large $|v_\varpi|$ strongly distorting $\mathbf{B}_p$, which consequently makes an angle with the horizontal of significantly less than 60° near the midplane. Large values of $\dot{m}$ arise because, as shown in §12.4.1, there is no potential barrier to the wind flow for such angles and hence the disc evaporates locally on a time-scale of $\sim \Omega_K^{-1}$, so being rapidly destroyed. Such a solution clearly violates the condition $\dot{m} \ll \dot{M}/4\pi\varpi_0^2$. Equations (12.137) and (12.157) give

$$\frac{|v_\varpi|}{c_s} \simeq \epsilon^{\frac{n-1}{2}} \left(\frac{v_{Az}}{c_s}\right)^2, \qquad (12.166)$$

where $\epsilon = 4\pi\varpi_0^2\dot{m}/\dot{M}$. Hence, for $n > 1$, the condition $\epsilon \ll 1$ requires

$$\frac{|v_\varpi|}{c_\mathrm{S}} \ll \left(\frac{v_{\mathrm{A}z}}{c_\mathrm{S}}\right)^2 . \tag{12.167}$$

The foregoing solution is local, and the authors emphasise the need for a global solution. They note that, under circumstances in which the sonic and Alfvénic points are closer together, stable wind solutions may be possible. Taking account of the magnetic field in the vertical equilibrium, and using a diffusivity dependent on $\mathbf{B}$ and ρ, may allow the disc to regulate wind launching in a stable way.

Detailed models have been constructed for magnetically driven jets and outflows from accretion discs (e.g. Lovelace, Berk and Contopoulos, 1991; Lovelace, Romanova and Contopoulos, 1993). These have been aimed at explaining the high velocity bipolar outflows and optical jets observed in star forming systems, which are believed to contain Keplerian accretion discs (e.g. Basri and Bertout, 1989). Such models may also be relevant to the jets observed in radio galaxies (e.g. Pringle, 1993). These models are axisymmetric, and hence subject to the Cowling theorem discussed in §2.3.1. The poloidal magnetic field must therefore either be generated by a dynamo α-effect in the disc, or supplied by an external source. The models to date rely on poloidal field being captured from the environment of the disc, and advected inward by the accretion flow. Such a mechanism may be operable in the large discs surrounding protostars and active galactic nuclei, for which the models are constructed. They seem unlikely to be applicable to the much smaller accretion discs occurring in binary stars which are supplied by a narrow accretion stream. The required magnetic fields in binary star accretion discs need to be dynamo generated.

12.5. Discussion

The foregoing sections illustrate the importance of magnetic wind theory in close binary stars, and hence the need for future work in this area. Although the simple field models discussed are qualitatively reasonable, the formidable problem of the self-consistent determination of the poloidal magnetic field remains to be solved.

So far, axisymmetric wind theory has been applied to secondary stars. However, since the total gravitational field surrounding such stars is non-axisymmetric, an appropriate ϕ-dependent theory needs to be developed.

A better understanding of magnetic field generation in stars is needed to determine the dependence of the surface field on the rotation rate, and in relation to the coronal structure. In discs, self-consistent global wind models incorporating field generation and disc structure must be considered, to assess whether stable flows are possible.

References

Basri, G. and Bertout, C., 1989. *Astrophys. J.*, **341**, 340.
Blandford, R.D. and Payne, D.G., 1982. *Mon. Not. R. Astr. Soc.*, **199**, 883.
Campbell, C.G., 1997. *Mon. Not. R. Astr. Soc.*, in press.
Campbell, C.G. and Papaloizou, J.C.B., 1983. *Mon. Not. R. Astr. Soc.*, **204**, 433.
Collier Cameron, A. and Robinson, R.D., 1989. *Mon. Not. R. Astr. Soc.*, **238**, 657.
Golub, L., 1983. In *IAU Colloquium 71., Activity in Red Stars*, eds., Byrne, P.B. and Rodono, M., Reidel.
Li, J., Wu, K. and Wickramasinghe, D.T., 1994. *Mon. Not. R. Astr. Soc.*, **268**, 61.
Lovelace, R.V.E., Berk, H.L. and Contopoulos, J., 1991. *Astrophys. J.*, **379**, 696.
Lovelace, R.V.E., Romanova, M.M. and Bisnovatyi-Kogan, G.S., 1995. *Mon. Not. R. Astr. Soc.*, **275**, 244.
Lovelace, R.V.E., Romanova, M.M. and Contopoulos, J., 1993. *Astrophys. J.*, **403**, 158.
Lubow, S.H., Papaloizou, J.C.B. and Pringle, J.E., 1994. *Mon. Not. R. Astr. Soc.*, **268**, 1010.
Mestel, L., 1967. In *Plasma Astrophysics*, ed. Sturrock, P.A., Academic Press, London.
Mestel, L., 1968. *Mon. Not. R. Astr. Soc.*, **138**, 359.
Mestel, L. and Spruit, H.C., 1987. *Mon. Not. R. Astr. Soc.*, **226**, 57.
Okamoto, I., 1974. *Mon. Not. R. Astr. Soc.*, **166**, 683.
Pallavicini, R., Golub, L., Rosner, R., Vaiana, G.S., Ayres, T. and Linsky, J.L., 1981. *Astrophys. J.*, **248**, 279.
Parker, E.N., 1963. *Interplanetary Dynamical Processes*, Interscience, New York.
Pizzo, V., Schwenn, R., Marsch, E., Rosenbrauer, H., Muhlhauser, K.H. and Neubrauer, F.M., 1983. *Astrophys. J.*, **271**, 335.
Pneuman, G.W. and Kopp, R.A., 1971. *Solar. Phys.*, **18**, 258.
Pringle, J.E., 1993. In *Astrophysical Jets*, eds Burgarella, D., Livio, M. and O'Dea, C., Cambridge University Press.
Rappaport, S., Verbunt, F. and Joss. P.C., 1983. *Astrophys. J.*, **275**, 713.
Saar, S.H., 1991. In *The sun and cool stars: activity, magnetism, dynamos.*, eds Tuominen, I., Moss, D. and Rudiger, G., Springer, Berlin.
Sakurai, T., 1985. *Astron. Astrophys.*, **152**, 121.
Sakurai, T., 1990. *Computer Physics Reports*, **12**, 247.
Schatzman, E., 1962. *Ann. Astrophys.*, **25**, 18.
Skumanich, A., 1972. *Astrophys. J.*, **171**, 565.
Spruit, H.C., 1994. In *Cosmical Magnetism*, ed. Lynden-Bell, D., Kluwer Academic Publishers.
Spruit, H.C. and Ritter, H., 1983. *Astron. Astrophys.*, **124**, 267.
Taam, R.E. and Spruit. H.C., 1989. *Astrophys. J.*, **345**, 972.
Verbunt, F. and Zwaan, C., 1981. *Astron. Astrophys.*, **100**, L7.
Weber, E.J. and Davis, L., 1967. *Astrophys. J.*, **148**, 217.

APPENDIX

A. Physical Constants and Solar Parameters

Speed of light $c = 2.998 \times 10^8 \, \mathrm{m\,s^{-1}}$
Electron charge $e = 1.602 \times 10^{-19} \, \mathrm{C}$
Electron mass $m_{\mathrm{e}} = 9.109 \times 10^{-31} \, \mathrm{Kg}$
Proton mass $m_{\mathrm{p}} = 1.673 \times 10^{-27} \, \mathrm{Kg}$
Boltzmann constant $k = 1.381 \times 10^{-23} \, \mathrm{J\,K^{-1}}$
Gas constant $\Re = 8.314 \times 10^3 \, \mathrm{m^2\,s^{-2}\,K^{-1}}$
Stefan-Boltzmann constant $\sigma_{\mathrm{B}} = 5.669 \times 10^{-8} \, \mathrm{J\,K^{-4}\,m^{-2}\,s^{-1}}$
Gravitational constant $G = 6.67 \times 10^{-11} \, \mathrm{N\,m^2\,Kg^{-2}}$
Permittivity of free space $\epsilon_0 = 8.854 \times 10^{-12}$ farad $\mathrm{m^{-1}}$
Permeability of free space $\mu_0 = 4\pi \times 10^{-7}$ henry $\mathrm{m^{-1}}$
Solar mass $M_\odot = 1.989 \times 10^{30} \, \mathrm{Kg}$
Solar radius $R_\odot = 6.960 \times 10^8 \, \mathrm{m}$
Solar luminosity $L_\odot = 3.90 \times 10^{26} \, \mathrm{J\,s^{-1}}$

B. Vector Identities

$$\mathbf{E} \wedge (\mathbf{F} \wedge \mathbf{G}) = (\mathbf{E} \cdot \mathbf{G})\mathbf{F} - (\mathbf{E} \cdot \mathbf{F})\mathbf{G} \tag{A1}$$

$$\nabla(\mathbf{F} \cdot \mathbf{G}) = (\mathbf{F} \cdot \nabla)\mathbf{G} + (\mathbf{G} \cdot \nabla)\mathbf{F} + \mathbf{F} \wedge (\nabla \wedge \mathbf{G}) + \mathbf{G} \wedge (\nabla \wedge \mathbf{F}) \tag{A2}$$

$$\nabla \cdot (\Psi \mathbf{F}) = \Psi \nabla \cdot \mathbf{F} + \mathbf{F} \cdot \nabla \Psi \tag{A3}$$

$$\nabla \cdot (\mathbf{F} \wedge \mathbf{G}) = \mathbf{G} \cdot (\nabla \wedge \mathbf{F}) - \mathbf{F} \cdot (\nabla \wedge \mathbf{G}) \tag{A4}$$

$$\nabla \wedge (\Psi \mathbf{F}) = \Psi \nabla \wedge \mathbf{F} + (\nabla \Psi) \wedge \mathbf{F} \tag{A5}$$

$$\nabla \wedge (\mathbf{F} \wedge \mathbf{G}) = (\mathbf{G} \cdot \nabla)\mathbf{F} - (\mathbf{F} \cdot \nabla)\mathbf{G} + \mathbf{F}\nabla \cdot \mathbf{G} - \mathbf{G}\nabla \cdot \mathbf{F} \tag{A6}$$

$$\nabla \wedge (\nabla \wedge \mathbf{F}) = \nabla(\nabla \cdot \mathbf{F}) - \nabla^2 \mathbf{F} \tag{A7}$$

C. Operators in Orthogonal Coordinates

C.1. SPHERICAL POLAR COORDINATES

$$\mathbf{B} = B_r\hat{\mathbf{r}} + B_\theta\hat{\boldsymbol{\theta}} + B_\phi\hat{\boldsymbol{\phi}}. \tag{A8}$$

$$\nabla\Psi = \frac{\partial\Psi}{\partial r}\hat{\mathbf{r}} + \frac{1}{r}\frac{\partial\Psi}{\partial\theta}\hat{\boldsymbol{\theta}} + \frac{1}{r\sin\theta}\frac{\partial\Psi}{\partial\phi}\hat{\boldsymbol{\phi}}. \tag{A9}$$

$$\nabla\cdot\mathbf{B} = \frac{1}{r^2}\frac{\partial}{\partial r}\left(r^2 B_r\right) + \frac{1}{r\sin\theta}\frac{\partial}{\partial\theta}(\sin\theta B_\theta) + \frac{1}{r\sin\theta}\frac{\partial B_\phi}{\partial\phi}. \tag{A10}$$

$$\begin{aligned}
(\nabla\wedge\mathbf{B})_r &= \frac{1}{r\sin\theta}\left[\frac{\partial}{\partial\theta}(\sin\theta B_\phi) - \frac{\partial B_\theta}{\partial\phi}\right],\\
(\nabla\wedge\mathbf{B})_\theta &= \frac{1}{r\sin\theta}\frac{\partial B_r}{\partial\phi} - \frac{1}{r}\frac{\partial}{\partial r}(rB_\phi),\\
(\nabla\wedge\mathbf{B})_\phi &= \frac{1}{r}\frac{\partial}{\partial r}(rB_\theta) - \frac{1}{r}\frac{\partial B_r}{\partial\theta}.
\end{aligned} \tag{A11}$$

C.2. CYLINDRICAL COORDINATES

$$\mathbf{B} = B_\varpi\hat{\boldsymbol{\varpi}} + B_\phi\hat{\boldsymbol{\phi}} + B_z\hat{\mathbf{z}}. \tag{A12}$$

$$\nabla\Psi = \frac{\partial\Psi}{\partial\varpi}\hat{\boldsymbol{\varpi}} + \frac{1}{\varpi}\frac{\partial\Psi}{\partial\phi}\hat{\boldsymbol{\phi}} + \frac{\partial\Psi}{\partial z}\hat{\mathbf{z}}. \tag{A13}$$

$$\nabla\cdot\mathbf{B} = \frac{1}{\varpi}\frac{\partial}{\partial\varpi}(\varpi B_\varpi) + \frac{1}{\varpi}\frac{\partial B_\phi}{\partial\phi} + \frac{\partial B_z}{\partial z}. \tag{A14}$$

$$\begin{aligned}
(\nabla\wedge\mathbf{B})_\varpi &= \frac{1}{\varpi}\frac{\partial B_z}{\partial\phi} - \frac{\partial B_\phi}{\partial z},\\
(\nabla\wedge\mathbf{B})_\phi &= \frac{\partial B_\varpi}{\partial z} - \frac{\partial B_z}{\partial\varpi},\\
(\nabla\wedge\mathbf{B})_z &= \frac{1}{\varpi}\left[\frac{\partial}{\partial\varpi}(\varpi B_\phi) - \frac{\partial B_\varpi}{\partial\phi}\right].
\end{aligned} \tag{A15}$$

D. Viscosity Expressions

D.1. VISCOUS FORCE IN VECTOR OPERATOR FORM

Equation (2.82) for the viscous force density $\mathbf{F}_\mathrm{v}$ follows from the tensor expressions (2.80a,b,c). These give

$$F_{\mathrm{v}i} = \frac{\partial}{\partial x_j}\left(\rho\nu\frac{\partial v_i}{\partial x_j}\right) + \frac{\partial}{\partial x_j}\left(\rho\nu\frac{\partial v_j}{\partial x_i}\right) - \frac{2}{3}\frac{\partial}{\partial x_i}(\rho\nu\nabla\cdot\mathbf{v}). \tag{A16}$$

The first term is

$$\frac{\partial}{\partial x_j}\left(\rho\nu\frac{\partial v_i}{\partial x_j}\right) = \frac{\partial}{\partial x_j}(\rho\nu)\frac{\partial v_i}{\partial x_j} + \rho\nu\frac{\partial^2 v_i}{\partial x_j\partial x_j} = (\nabla(\rho\nu)\cdot\nabla)\mathbf{v} + \rho\nu\nabla^2\mathbf{v}.$$

The second term is

$$\begin{aligned}\frac{\partial}{\partial x_j}\left(\rho\nu\frac{\partial v_j}{\partial x_i}\right) =& \frac{\partial}{\partial x_j}\left(\frac{\partial}{\partial x_i}(\rho\nu v_j) - v_j\frac{\partial}{\partial x_i}(\rho\nu)\right)\\ =& \frac{\partial}{\partial x_i}\left(\frac{\partial}{\partial x_j}(\rho\nu v_j)\right) - \frac{\partial v_j}{\partial x_j}\frac{\partial}{\partial x_i}(\rho\nu) - v_j\frac{\partial}{\partial x_j}\left(\frac{\partial}{\partial x_i}(\rho\nu)\right)\\ =& \nabla\left[\nabla\cdot(\rho\nu\mathbf{v})\right] - (\nabla\cdot\mathbf{v})\nabla(\rho\nu) - (\mathbf{v}\cdot\nabla)\nabla(\rho\nu).\end{aligned}$$

The last term in (A16) is a gradient, so its addition to the first two terms gives

$$\begin{aligned}\mathbf{F}_\mathrm{v} =& \rho\nu\nabla^2\mathbf{v} + \nabla\left(\nabla\cdot(\rho\nu\mathbf{v}) - \frac{2}{3}\rho\nu\nabla\cdot\mathbf{v}\right)\\ &+ (\nabla(\rho\nu)\cdot\nabla)\mathbf{v} - (\mathbf{v}\cdot\nabla)\nabla(\rho\nu) - (\nabla\cdot\mathbf{v})\nabla(\rho\nu).\end{aligned}$$

Use of (A6) then yields

$$\begin{aligned}\mathbf{F}_\mathrm{v} =& \rho\nu\nabla^2\mathbf{v} + \nabla\left(\nabla\cdot(\rho\nu\mathbf{v}) - \frac{2}{3}\rho\nu\nabla\cdot\mathbf{v}\right)\\ &+ \nabla\wedge\left[\mathbf{v}\wedge\nabla(\rho\nu)\right] - [\nabla^2(\rho\nu)]\mathbf{v}.\end{aligned} \tag{A17}$$

D.2. RATE OF STRAIN TENSOR IN CYLINDRICAL COORDINATES

$$\begin{aligned}e_{\varpi\varpi} &= \frac{1}{\varpi}\frac{\partial}{\partial\varpi}(\varpi v_\varpi), \quad e_{\varpi\phi} = \frac{1}{2}\left(\varpi\frac{\partial}{\partial\varpi}\left(\frac{v_\phi}{\varpi}\right) + \frac{1}{\varpi}\frac{\partial v_\varpi}{\partial\phi}\right),\\ e_{\varpi z} &= \frac{1}{2}\left(\frac{\partial v_\varpi}{\partial z} + \frac{\partial v_z}{\partial\varpi}\right), \quad e_{\phi\phi} = \frac{1}{\varpi}\frac{\partial v_\phi}{\partial\phi},\\ e_{\phi z} &= \frac{1}{2}\left(\frac{\partial v_z}{\partial\phi} + \frac{\partial v_\phi}{\partial z}\right), \quad e_{zz} = \frac{\partial v_z}{\partial z}.\end{aligned} \tag{A18}$$

E. Orthogonal Functions

E.1. BESSEL FUNCTIONS

Bessel's equation is

$$\frac{d^2y}{dx^2} + \frac{1}{x}\frac{dy}{dx} + \left(1 - \frac{\nu^2}{x^2}\right) y = 0. \tag{A19}$$

For ν a non-integer the general solution is

$$y = AJ_\nu(x) + BJ_{-\nu}(x), \tag{A20}$$

where A and B are constants and $J_\nu(x)$ is a Bessel function of the first kind of order ν. For $\nu = n$, an integer, $J_n(x)$ and $J_{-n}(x)$ are linearly dependent. The general solution of (A19) in this case is

$$y = AJ_n(x) + BY_n(x), \tag{A21}$$

where $Y_n(x)$ is a Bessel function of the second kind, which is singular at $x = 0$.

The functions $J_n(x)$ are given by

$$J_n(x) = \sum_{m=0}^{\infty} \frac{(-1)^m}{m!(m+n)!} \left(\frac{x}{2}\right)^{n+2m}. \tag{A22}$$

This set of functions has the orthogonality relation

$$\int_0^a J_n\left(\frac{\alpha_m}{a} r\right) J_n\left(\frac{\alpha_l}{a} r\right) r dr = -\frac{a^2}{2} J_{n+1}(\alpha_m) J_{n-1}(\alpha_l) \delta_{ml}, \tag{A23}$$

where α_m and α_l are zeros of J_n.

Bessel functions of order $n + \frac{1}{2}$, where n is an integer, are expressible as

$$J_{n+\frac{1}{2}}(x) = (-1)^n \left(\frac{2}{\pi}\right)^{\frac{1}{2}} x^{n+\frac{1}{2}} \frac{d^n}{(xdx)^n} \left(\frac{\sin x}{x}\right). \tag{A24}$$

E.2. ASSOCIATED LEGENDRE FUNCTIONS

The associated Legendre equation is

$$\frac{d^2y}{dx^2} - \frac{2x}{1-x^2}\frac{dy}{dx} + \left[\frac{l(l+1)}{1-x^2} - \frac{m^2}{(1-x^2)^2}\right] y = 0. \tag{A25}$$

Solutions, $P_l^m(x)$, finite at $x = \pm 1$ only occur for l and m as integers. The general solution is

$$y = AP_l^m(x) + BQ_l^m(x), \tag{A26}$$

where the second solution $Q_l^m(x)$ is singular at $x = \pm 1$.

Due to the form $l(l+1)$ in (A25), negative values of l lead to the same set of functions as positive values, so $l > 0$ is used. The functions P_l^{-m} and P_l^m are linearly dependent and the choice $P_l^{-m} = P_l^m$ can be made. Legendre functions, free of singularity, are given by

$$P_l^{|m|}(x) = \frac{(1-x^2)^{\frac{|m|}{2}}}{2^l l!} \left(\frac{d}{dx}\right)^{|m|+l} (x^2-1)^l, \tag{A27}$$

where $|m| \le l$. Their orthogonality relation is

$$\int_{-1}^{1} P_l^{|m|}(x) P_n^{|m|}(x) dx = \frac{2}{2l+1} \frac{(l+|m|)!}{(l-|m|)!} \delta_{ln}. \tag{A28}$$

Spherical harmonics are defined as

$$Y_l^m(\theta, \phi) = P_l^{|m|}(\cos\theta) e^{im\phi}. \tag{A29}$$

Noting that $x = \cos\theta$ and using the operator L^2 defined by (2.146), gives

$$\begin{aligned} L^2 Y_l^m &= -\left[\frac{1}{\sin\theta}\frac{d}{d\theta}\left(\sin\theta \frac{dP_l^{|m|}}{d\theta}\right) - \frac{m^2}{\sin^2\theta} P_l^{|m|}\right] e^{im\phi} \\ &= -\left[(1-x^2)\frac{d^2 P_l^{|m|}}{dx^2} - 2x\frac{dP_l^{|m|}}{dx} - \frac{m^2}{1-x^2} P_l^{|m|}\right] e^{im\phi} \\ &= l(l+1) P_l^{|m|} e^{im\phi}, \end{aligned}$$

where the last equality follows from (A25). Hence Y_l^m obey the eigenvalue equation

$$L^2 Y_l^m = l(l+1) Y_l^m. \tag{A30}$$

It follows from (A28) and (A29) that spherical harmonics have the orthogonality relation on the unit sphere of

$$\int_{4\pi} Y_l^m Y_n^{-r} d\Omega = \frac{4\pi}{2l+1} \frac{(l+|m|)!}{(l-|m|)!} \delta_{ln} \delta_{mr}, \tag{A31}$$

where the differential solid angle $d\Omega = \sin\theta d\theta d\phi$.

F. Elliptic Integrals

Elliptic integrals of the first, second and third kinds are;

$$F(\phi, k) = \int_0^\phi \frac{d\alpha}{(1 - k^2 \sin^2 \alpha)^{\frac{1}{2}}}, \tag{A32}$$

$$E(\phi, k) = \int_0^\phi (1 - k^2 \sin^2 \alpha)^{\frac{1}{2}} d\alpha, \tag{A33}$$

$$\Pi(\phi, p, k) = \int_0^\phi \frac{d\alpha}{(1 + p \sin^2 \alpha)(1 - k^2 \sin^2 \alpha)^{\frac{1}{2}}}. \tag{A34}$$

Complete elliptic integrals have $\phi = \pi/2$.

G. Gravitational Torque on a Spheroid

If the primary star is non-spherical, due to non-radial internal forces, then it will experience a torque resulting from interaction with the secondary's gravitational field. The simplest way of finding this torque is to calculate the torque a distorted primary exerts on the centre of mass of the secondary, and then use the fact that the torque on the primary is equal and opposite to this.

The case of distortion symmetric about an axis is considered. This principal axis is taken to be along the $\mathbf{e}_3$ direction. Figure A1 shows a mass element in the primary at position $\mathbf{r}'$ relative to its centre of mass O. The centre of mass of the secondary is at P. The gravitational potential at P is

$$U = -\int_{M_\mathrm{p}} \frac{G dm}{(r^2 + r'^2 - 2rr' \cos\theta')^{\frac{1}{2}}}.$$

This can be written as

$$U = -\frac{G}{r} \int_{M_\mathrm{p}} \frac{dm}{[1 - 2(r'/r)\cos\theta' + (r'/r)^2]^{\frac{1}{2}}}.$$

Since the orbital separation significantly exceeds the mean radius of the primary, it follows that $r'/r \ll 1$ in the above integrand over the range of integration. Binomial expansion of the integrand to second order in r'/r, noting that O is the primary's centre of mass, then enables the potential to be expressed as

$$U = -\frac{GM_\mathrm{p}}{r} - \frac{G}{r^3}\left(I_O - \frac{3}{2}I\right), \tag{A35}$$

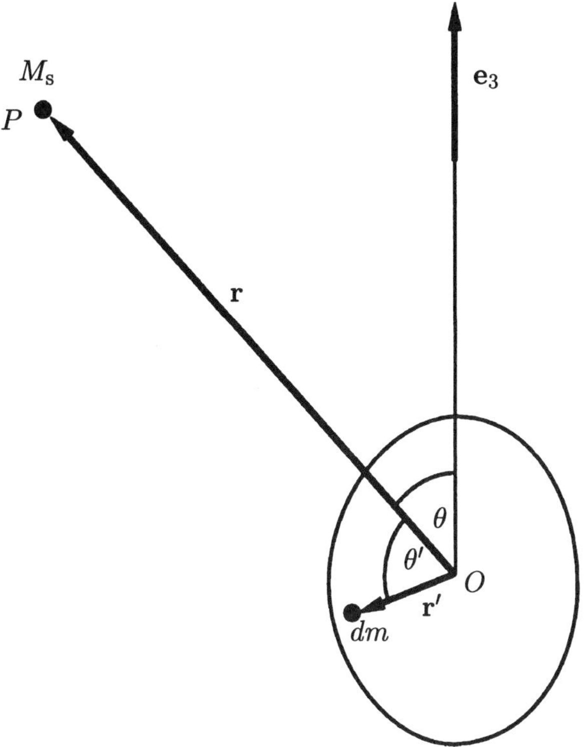

Figure A1. Coordinates for calculating the gravitational torque on the primary.

where

$$I_O = \int_{M_p} r'^2 dm, \qquad I = \int_{M_p} r'^2 \sin^2\theta' dm. \tag{A36}$$

I_O is the moment of inertia of the primary about O, while I is the moment of inertia about OP. It is simple to show that I_O is related to the principal moments of inertia by

$$I_O = \frac{1}{2}(I_1 + I_2 + I_3).$$

For a symmetric body $I_1 = I_2 = I_\perp$, the moment of inertia about any axis in the equatorial plane. Hence

$$I_O = I_\perp + \frac{1}{2} I_3. \tag{A37}$$

The moment of inertia about OP can be written as

$$I = \lambda^2 I_1 + \mu^2 I_2 + \nu^2 I_3,$$

where λ, μ and ν are the direction cosines of OP along the principal axes, given by

$$\lambda = \frac{x}{r}, \quad \mu = \frac{y}{r}, \quad \nu = \frac{z}{r},$$

taking the positive z-direction along $\mathbf{e}_3$. It follows that

$$I = I_\perp + (I_3 - I_\perp)\frac{z^2}{r^2}. \tag{A38}$$

Equations (A35), (A37) and (A38) give the potential as

$$U = -\frac{GM_\mathrm{p}}{r} - \frac{G(I_3 - I_\perp)}{2r^3} + \frac{3G(I_3 - I_\perp)}{2r^5}z^2. \tag{A39}$$

Since the gravitational force on M_s at P is $-\nabla U$, the torque exerted on the primary is

$$\mathbf{T}_\mathrm{g} = M_\mathrm{s}\mathbf{r} \wedge (\nabla U)_P.$$

Noting that $z = r\cos\theta$, where θ is the angle between OP and $\mathbf{e}_3$, gives

$$\mathbf{T}_\mathrm{g} = M_\mathrm{s}\left(\frac{\partial U}{\partial \theta}\right)_P \hat{\mathbf{r}} \wedge \hat{\boldsymbol{\theta}}. \tag{A40}$$

The unit vector $\hat{\boldsymbol{\theta}}$ is related to $\mathbf{e}_3$ by

$$\sin\theta\,\hat{\boldsymbol{\theta}} = \cos\theta\,\hat{\mathbf{r}} - \mathbf{e}_3.$$

Use of this and (A39) in (A40) yields

$$\mathbf{T}_\mathrm{g} = \frac{3GM_\mathrm{s}(I_3 - I_\perp)}{D^3}\cos\theta\,\hat{\mathbf{r}} \wedge \mathbf{e}_3, \tag{A41}$$

where D is the orbital separation.

INDEX

CARDIFF
UWCC LIBRARY